Work, Health, and Environment

DEMOCRACY AND ECOLOGY
A Guilford Series

Published in conjunction
with the Center for Political Ecology

JAMES O'CONNOR
Series Editor

IS CAPITALISM SUSTAINABLE?
POLITICAL ECONOMY AND THE POLITICS OF ECOLOGY
Martin O'Connor, Editor

GREEN PRODUCTION
TOWARD AN ENVIRONMENTAL RATIONALITY
Enrique Leff

MINDING NATURE
THE PHILOSOPHERS OF ECOLOGY
David Macauley, Editor

THE GREENING OF MARXISM
Ted Benton, Editor

WORK, HEALTH, AND ENVIRONMENT
OLD PROBLEMS, NEW SOLUTIONS
Charles Levenstein and John Wooding, Editors

Work, Health, and Environment

OLD PROBLEMS, NEW SOLUTIONS

Edited by
CHARLES LEVENSTEIN
JOHN WOODING

Foreword by
ANTHONY MAZZOCCHI

THE GUILFORD PRESS
New York London

A Division of Guilford Publications, Inc.
72 Spring Street, New York, NY 10012

Printed in the United States of America

This book is printed on acid-free paper.

Last digit is print number: 9 8 7 6 5 4 3 2 1

Library of Congress Cataloging-in-Publication Data

Work, health, and environment: old problems, new solutions / Charles Levenstein and John Wooding, editors; foreword by Anthony Mazzocchi.
p. cm.—(Democracy and ecology)
Includes bibliographical references and index.
ISBN 1-57230-233-X.—ISBN 1-57230-234-8 (pbk.)
1. Industrial hygiene—Government policy—United States.
2. Industrial hygiene—Political aspects—United States.
3. Environmental health—Government policy—United States.
4. Environmental health—Political aspects—United States.
5. Industrial safety—Government policy—United States.
6. Industrial safety—Political aspects—United States.
I. Levenstein, Charles. II. Wooding, John. III. Series.
RC967.W665 1997
616.9'803'0973—dc21 97-17255
CIP

Introduction to the Democracy and Ecology Series

This book series titled "Democracy and Ecology" is a contribution to the debates on the future of the global environment and "free market economy" and the prospects of radical green and democratic movements in the world today. While some call the post-Cold War period the "end of history," others sense that we may be living at its beginning. These scholars and activists believe that the seemingly all-powerful and reified world of global capital is creating more economic, social, political, and ecological problems than the world's ruling and political classes are able to resolve. There is a feeling that we are living through a general crisis, a turning point or divide that will create great dangers, and also opportunities for a nonexploitative, socially just, democratic ecological society. Many think that our species is learning how to regulate the relationship that we have with ourselves and the rest of nature in ways that defend ecological values and sensibilities, as well as right the exploitation and injustice that disfigure the present world order. Others are asking hard questions about what went wrong with the worlds that global capitalism and state socialism made, and about the kind of life that might be rebuilt from the wreckage of ecologically and socially bankrupt ways of working and living. The "Democracy and Ecology" series rehearses these and related questions, poses new ones, and tries to respond to them, if only tentatively and provisionally, because the stakes are so high, and since "time-honored slogans and time-worn formulae" have become part of the problem, not the solution.

James O'Connor
Series Editor

Foreword

This volume brings together some of the most critical and important articles published in the journal *New Solutions*. In so doing, it collects together material about occupational and environmental health problems of interest to a broad audience.

A quarter of a century ago protests from organized labor and from environmental and public interest groups heralded the beginning of a new wave of environmental activism. Passage of the Coal Mine Safety and Health Act in 1969, the Occupational Safety and Health Act in 1969, and the National Environmental Protection Act in 1970 promised to provide workers and the public alike with significant protection from hazardous materials and processes, both on the job and in their communities.

In those days, those most active in the movement took it for granted that the interests of workers and community members in a safe and healthful workplace were inseparable. That I should be asked, as a labor official active in environmental issues, to chair the first Earth Day ceremonies in April 1970 at New York's Union Square struck no one as unusual. The labor movement and the environmental movement were working together on a common cause.

Things have changed. The end of the postwar economic boom in 1973 resulted in the radical restructuring of the U.S. economy and the loss of hundreds of thousands of unionized manufacturing jobs. Workers became increasingly reluctant to complain about health and safety problems, even as the grassroots environmental movement—spurred on by a growing number of tragic incidents and catastrophic accidents—became ever more vocal in demanding an end to the unchecked, heedless growth of the hazardous waste stream and, as much as possible, the complete elimination of toxic materials from the production cycle. This put the trade unions and the environmental movement on the collision course that they find themselves on to this day.

Those of us in the trade union movement, who represent people employed in industries that either use or produce hazardous materials, now find ourselves caught on the horns of a difficult dilemma. On the one hand, our first concern is to protect the jobs, incomes, and working conditions of our members. That is what they elected us to do, and, if we can't do it, they are within their rights to replace us—which they do. On the other hand, people who work in hazardous industries, as most Oil, Chemical, and Atomic Workers (OCAW) members do, want safe jobs and a healthy environment, just as everyone else does. And they expect us to do everything we can to ensure that their employers provide them with a workplace free from all recognized hazards, as the Occupational Safety and Health Act (OSH Act) requires.

The problem is that these two goals often seem to conflict with each other. Ask workers at home if they support government efforts to clean up the air or the water, and they will most likely say yes. Ask them at work, after the company has told them that new OSHA or Environmental Protection Agency (EPA) regulations threaten their jobs, and they will most likely say no. They are right both times. Most people are astute analysts of their own best interests. Workers know that they are the first to die from exposure to hazardous materials, both at the workplace and in the community. They also know that their families can't live on clean air and clean water alone. And they know that if they didn't have a job, or one that paid as well as many of the jobs in the petrochemical and atomic industries do, they would live less well and die a lot sooner.

Both the trade unions and the environmental movement need to find a way to help working people escape from this cruel choice. We can't go on forever always putting off until tomorrow the necessary task of cleaning up the mess we make today. We are rapidly running out of corners into which the dirt can be swept. At the same time, we can't expect most people voluntarily to give up all the benefits of modern industry, which has enabled more of us to enjoy a higher standard of living than has ever been true before, without offering a realistic alternative—a different, if not a better, way to live.

This is not an easy task. It requires, first of all, that we each—community activist as well as trade unionist—know as much as we can about the scientific, medical, and economic issues involved. What are the hazards we face? How can they harm us? What alternatives exist? We must, together, find answers to such questions. The only acceptable way forward from this point is one that both respects the limits of mother nature's tolerance for too much more garbage and acknowledges the equal right of us all—petrochemical worker as well as college professor—to a decent standard of living.

The Oil, Chemical, and Atomic Workers International Union, under the auspices of its Alice Hamilton Library, undertook publishing *New Solutions: A Journal of Environmental and Occupational Health Policy* in order to help with this common search. We wanted this journal to be a place where trade unionists and community activists could learn about the latest scientific and medical research, and where scientists and physicians could learn more about the concerns of working people and their families. We want it to be a forum where trade unionists and environmentalists can work together to find answers to the critical questions that confront both our movements.

Necessarily, then, *New Solutions* has not simply been a platform. It has become an argument. As such, it includes articles with which many of us disagree, and to which we might even take strong exception. Certainly, in my opinion, from the first issue the journal includes such articles. But that is the nature of an argument. We are committed to the journal because we are committed to finding a way for our members to escape the cruel choice between their livelihood or their lives, now forced upon them by both industry and government policy alike. We welcomed *New Solutions* as an effort to find the way forward to a time when working people everywhere can enjoy both wealth and health, can live both well and long.

We are also committed to it because we believe that, only if and when scientists and policymakers submit their research and proposals for action to the scrutiny of the general public, can we find acceptable solutions to our common problems. We want *New Solutions* to be a place where all the different issues of environmental and occupational health policy that we now face, and that we will have to face in the coming years, can be argued out. Whatever the current tension between the interests of the workplace and those of the community, from the standpoint of workers, they are, in the long run, inseparable. *New Solutions* is intended to be an opportunity to stake out and then to share that common ground on a wide range of issues of concern to us all.

I hope, therefore, that this reader will lead you to the journal and the ongoing debate about the critical issues reflected in its pages.

Anthony Mazzocchi
Special Assistant to the President
Oil, Chemical, and Atomic Workers
International Union
Publisher, *New Solutions*

Preface

Death, disease, and injury caused by occupational exposure to dangerous working conditions are major public health problems. The issues of health and safety at work, however, remain low on the national agenda. While the problems facing the environment and concern about the pollution of our air, water, and land have rightly drawn a great deal of attention, the health problems facing workers have never really been given the attention they deserve. In an effort to change this situation *New Solutions: A Journal of Environmental and Occupational Health Policy* was begun at the instigation of Anthony Mazzocchi of the Oil, Chemical, and Atomic Workers International Union. Given this resource, we thought we might provide a useful service if we brought together some of the exemplary articles that have appeared in the journal and organized them around themes that reflect our concern about the work environment. This volume is the outcome.

The goal of the journal is to bring the discussion of occupational health and safety and its relationship to environmental issues to the forefront. By providing an outlet for analysis, comment, and perspectives on occupational and environmental health, the journal functions to bring concrete, contemporary, and relevant discussions of the work environment to a wide audience. Over the past five years the journal has been on the cutting edge of these issues. In its pages some of the best activists, professionals, specialists, and academics have written, reported on, and argued about a wide range of problems facing workers, communities, unions, activists, and environmentalists. The journal contains a rich collection of the very best discussions of the politics of occupational and environmental health.

We see this text as a useful resource for specialists and nonspecialists alike. It will, we hope, provide a unique analysis of the complex web of issues surrounding the health of workers and the environment.

CHARLES LEVENSTEIN
JOHN WOODING

Acknowledgments

We would like to extend our thanks to the contributors to *New Solutions*, some of whose work has been selected for this volume. Additionally, we would like to thank the members of the editorial and advisory boards of *New Solutions* and the peer reviewers of the chapters published in this reader for their hard work and for the dedication they have shown to the publication of a progressive policy journal. Perhaps most important of this group is Mary Lee Dunn, who has managed the journal, but who has also helped us with editorial preparation of the book.

We owe a special debt to the Oil, Chemical, and Atomic Workers International Union, to the publisher of the journal, Anthony Mazzocchi, and to the union's President, Robert Wages, who intellectual, material, and moral support have made *New Solutions* possible. Additionally, we thank them for permission to reprint the chapters included in this book.

Finally, we began this project some years ago with Dr. Fred Sperounis, Vice Chancelor of the University of Massachusetts Lowell, who had to withdraw because of his administrative responsibilities. We owe him an acknowledgement for his support and friendship over many years. Of course, any errors in this reader can only be attributable to the editors.

Contents

Section B. Case Studies: Industrial Disease

Section C. The Use of Science in Policymaking

PART I

THE POLITICAL ECONOMY OF THE WORK ENVIRONMENT

The themes we elaborate here and around which we have organized the articles from *New Solutions* emphasize that the problem of occupational and environmental health starts in the factory, the office, and the warehouse—in short, at the point of production. Work and the organization of work are where we begin our analysis of these issues. What we elaborate here is a *political economy of the work environment*. Although production is at the crux of understanding what determines the health of workers and the environment, we recognize that the relationship of political institutions and regulatory bodies, legal systems, social conflict, belief systems, and the dynamics of race, class, and gender all deeply affect how we perceive the causes of and solutions to the problems we are concerned with here. This is the social "web" to which we return again and again. We have, therefore, constructed this volume in four parts. The first addresses the general political and economic issues that define the work environment. The second looks at the role of regulation and public policy. The third takes up social and political conflicts that surround health and safety issues, and the last part provides discussion of where these issues might take us. Each part is divided into sections dealing with specific issues and is introduced by some general comments on how the theme of the book is developing.

Part I consists of three chapters. In the first, Levenstein and Tuminaro set out the "political economy of occupational disease."

In their discussion they take issue with those who would define occupational disease narrowly. As they note: "The notion that disease is a social construction is not a new one. Nevertheless, . . . there is no systematic exposition of social, political, and economic factors that significantly shape our understanding of work-related health hazards and their elimination or control" (p. 3). In essence, that is the theme of this volume—that we need a more comprehensive appreciation of the factors involved. Levenstein and Tuminaro discuss in detail the production, perception, control, and compensation of occupational disease. In so doing, they construct a political economy of the work environment that serves as an analytical framework for all aspects of occupational and environmental health problems. The set of hypotheses they put forward demands serious consideration by all those concerned about these problems. By putting to rest the misconception that disease and injury are the result simply of technical problems susceptible to technical solutions, they thrust the problems of the work environment into the political realm. We believe that a full understanding of these problems can only proceed, therefore, from the analytical assumptions that Levenstein and Tuminaro lay out.

The two linked chapters by Kuhn and Wooding suggest that we might begin our discussion of work environment problems by looking at the structure of work. The authors provide a detailed description of the changes undergone by the American economy during the past 30 years or so. As with other advanced industrial economies, the United States has undergone a significant shift from heavy manufacturing to an expanded service sector. As Kuhn and Wooding point out, it is not exactly clear what constitutes this sector. What is clear, however, is that the changing structure of work deriving from not only domestic but also international forces has had a significant impact on the lives and health of Americans.

Kuhn and Wooding paint a picture of the American economy in which fewer and fewer Americans have job security, full-time employment, and health benefits. Even more striking is the decline in union membership, as the economy shifts away from the traditional, unionized manufacturing sector. These developments, the authors argue, expose workers to a host of health problems that go beyond direct workplace exposure and may lead to greater stress in general, as well as widespread job insecurity and concern about the availability of health care. The point here is that occupational and environmental health problems should be viewed in the context of these global shifts in production.

1

The Political Economy of Occupational Disease

CHARLES LEVENSTEIN
DOMINICK J. TUMINARO

Economists addressing the issue of occupational disease typically take on the role of administrative analyst, using such well-worn tools as cost/benefit analysis or cost-effectiveness analysis and attempting to extrapolate from scientific studies and industry estimates of the cost of regulatory measures in order to give guidance to decision makers. The critique of such methodologies is well known and will not be dwelled upon here. Rather, our purpose is to indicate an alternative approach to the economics of occupational disease that relies on a broader construction of political-economic analysis and that suggests the importance of investigating the social history of occupational disease, especially paying attention to the interplay of science and political economy. (1)

The notion that disease is a social construction is not a new one. (2) Nevertheless, occupational medicine, industrial hygiene, occupational epidemiology—and economics—are taught and practiced with the fiction that "policy" considerations are separate from and peripheral to hard science. In particular, there is no systematic exposition of social, political, and economic factors that significantly shape our understanding of work-related health hazards and their elimination or control.

Discussion of the political-economic aspects of occupational disease requires attention to four broad areas of concern: (a) the production of disease; (b) the perception or recognition of disease; (c) control measures; and (d) compensation for affected workers. To a certain extent these categories are arbitrary since the overriding principles and structure of the political economy provide the context for our discussion. The central principle of profit making, the debate about the appropriate role of government in industry and the rise of the interventionist state, and the playing out of class and interest group conflict through government are recurring themes. In the United States, the weakness of labor as a political and economic force appears to be a critical influence on the resolution of occupational health issues. The role of the public health scientist and advocate is of particular interest in this context.

THE PRODUCTION OF DISEASE

Occupational disease emanates from the production of goods and services, which involves the exposure of workers to materials, machines, technologies, and work practices that are hazardous to their health and well-being.

Technology and Disease

The choice of technology in production is an engineering and political-economic decision. (3) The use of particular materials, the organization of work, the employment of particular machines, and their arrangement in relation to one another are subject to economic and social imperatives and constraints. For instance, the introduction of the automatic loom in cotton textile production was, at least in part, the result of labor–management conflict in the industry as well as interfirm competition. (4) The decision to manufacture fluorescent lights, and the resulting hazards of beryllium exposure, was part of a strategy for Sylvania to win an increased share of the market. (5) Implied in the search for efficient technologies of production, at minimum, are notions of minimizing costs and effective control of the labor process by management. Because employer decisions on choice of technology rarely consider the full cost of occupational health hazards, current efforts to remedy dangerous work situations may involve expensive retro-fitting of equipment, reorganization of production and difficult retraining of labor, or the unknowable costs of technology-forcing regulation.

The Level of Production

At times, the scale of operations or the intensity of use of equipment or labor can have a substantial impact on hazardous exposures. Increased production requirements for World War II, for instance, may have produced greater dust exposures for asbestos or cotton textile workers. (6) Increasing shift-work may lead to accelerated depreciation of equipment and the failure of dust controls. Particularly susceptible people, ordinarily barred from employment, may be utilized during periods of high demand. Child labor legislation, for instance, is relaxed during wars. (7) And industrialization may increase the sheer numbers of workers exposed to dangerous environments. (8)

(Unemployment and underemployment, of course, have their own negative impact on health; the risk of hazardous employment may appear attractive to workers faced with debilatory poverty. The net result of industrialization may be a change in the structure of the cases of morbidity and mortality among workers.)

The Impact of Industry Structure

A more subtle aspect of the generation of occupational disease involves interfirm competition and political-economic definition of industries. Vertical integration in an industry tends to be limited by the potential for stabilizing and planning production: service stations, for instance, are highly competitive small operations and are not considered part of the highly concentrated automobile industry. Similarly, agricultural production is a relatively competitive sector related to the more highly concentrated industrial users of "raw" materials. Decisions concerning production methods and technology made because of the economic or social objectives in one sector may have implications for the occupational hazards realized in another. For instance, mechanization of cotton picking resulted in a higher trash content in dust in textile mills, probably increasing the incidence of byssinosis experienced by mill workers. On the other hand, the industrialization of agriculture and the development of cash crops for emerging industries in developing countries may change the nature of hazards experienced by farm workers. The generation of such hazards may go unrecognized even in countries where industrial development is planned because of the narrow focus of development strategists. (9)

The Labor Market

The relative scarcity of labor as a whole or of particular types of labor will influence technical choices in production, the resultant hazards associated with labor in particular industries, the options available to labor, and the pressure exerted on managements to improve working conditions. The influence of trade unions may be limited by adverse levels of employment. In some industries, employers, anxious to maintain an adequate labor supply and required to pay more attention to occupational hazards during periods of high employment, may relax standards during periods of declining demand.

Further, in the face of particular labor shortages, employers may introduce technical changes with unknown hazards in order to reduce the costs and power of skilled workers. Finally, considerations about control of the labor process may influence employers to institute technical and administrative changes in production methods with consequences for the health of employees.

THE PERCEPTION OF DISEASE

On the one hand, occupational disease is generated by technologies employing labor in production. On the other hand, while ill health may appear to be an "objective" matter, political-economic considerations are important in the perception of disease. Observers of worker health occupy different, and sometimes opposed, locations in the system of production. Therefore, they bring to their understandings of occupational disease viewpoints directed or constrained by other, sometimes more important, determinants of their positions. (10) While some investigators of occupational health may have that role as their primary responsibility, they also may have conflicting institutional responsibilities. Other investigators have such a peripheral connection with the system of production and its control that they may be blind to the industrial etiology of disease.

Workers

Presumably, workers themselves are most likely to be aware of their own occupational hazards and their ill health where the relationship between exposure and disease is clearcut. Occupational health literature is replete with examples of workers alerting health professionals to hazardous situations. (11) However, the increasing complexity of industrial production and the insistence of employers on control over the production process prevent workers from having information about more subtle, though sometimes devastating, hazards. Workers rely more and more on managers and health professionals to interpret disease for them. Thus, although a worker may "know" he/she is sick, the occupational etiology of the illness may be unknown or only suspected.

The economic condition of workers may act against their "knowing" the source of their illness. Maintaining family income may be of much higher priority to a worker than the correct diagnosis of the source of an illness. Indeed, if a worker suspects a health hazard on the job and yet perceives no reasonable employment alternatives nor the possibility of changing the work environment, the worker may not want to "know" about the hazard. Further, workers may believe that some hazards and illness are simply part of the job—always have been and always will be—and therefore do not "know" they exist or acknowledge that such conditions are "abnormal." The realities of working class life require primary attention to the economic survival of the family; thus, the hazards to a breadwinner may be accepted "fatalistically."

To the extent that working people depend on management and/or health professionals for interpretation of their illnesses, the above considerations contribute to worker acceptance of assurances that the work environment is "safe." In addition, the cost of medical care is substantial enough to discourage workers from using the medical care system.

Workers will mobilize around occupational health hazards and perceived disease if the hazard is believed to be drastic and the effects serious; if clear alternatives seem to be available; if their bargaining position vis-a-vis employers seems to be strong; or if new avenues for correcting hazardous situations seem to be feasible. Workers in trade unions may or may not find these organizations to be responsive to their health concerns. Some trade unions promote attention to industrial hazards; others seem fearful that worker militancy about such issues makes for political insurgency; and some may promote interest in occupational health as part of an overall bargaining strategy, to be traded off against other demands. To the extent that trade union leadership believes that managerial control of technological choices is inviolable, bargaining strategists may direct member attention to other "winnable" issues.

Finally, it should be noted that worker expectations about their own health may be quite low. To the extent that ill health is presumed, worker attention to occupational hazards will be negligible. Indeed, some workers, particularly low-wage, minority workers, may feel that they may be able to get their jobs because of the dangers involved and may be fearful that any attention to "cleaning up" the workplace will cost them their employment.

Management

Management's perceptions of worker ill health and occupationally derived disease must be strongly conditioned by economic considerations. In certain circumstances, employers may be concerned with the toll of hazardous exposures as expressed in absenteeism and general debilitation of its workforce, particularly if production is affected. In addition, management may worry about public outcry and/or media attention about industrial hazards, in part because of its fears of government regulation or other public intervention, and in part because of concern for "image" and public relations effects. Also, workers' compensation insurance coverage can be affected by casual managerial attitudes toward worker health. Finally, management attention may be drawn to health issues because of collective bargaining pressures or, even where there is no union, because of difficulties in recruiting employees to hazardous occupations or the possibilities of employee unionization as a consequence of health-related grievances.

Nevertheless, there is great incentive to employers to "externalize" the costs of occupational disease by recognizing workers' ill health but attributing it to personal habits or to the community environment. Employers, for instance, may be supportive of community efforts to deal with lung disease, such as tuberculosis or smoking-related disease, while denying the existence or seriousness of occupational exposures to respiratory hazards.

Employers may differ, however, in their attitudes toward industrial hazards depending on interfirm competition. A technologically "progressive" firm may see advantage for itself in identifying the hazardous exposures embodied in old technologies: gaining a respectable public image, justifying its own investment policy in public health terms, and promoting government regulation of exposures at the expense of its more backward competitors. Employer "paternalism"—serious attention to worker health hazards—may have such economic justification.

On the other hand, workers' compensation payments may appear to be a "dead weight loss" to employers. A firm that supports control of exposures to prevent disease may, nevertheless, dispute individual workers' claims for compensation, thus accepting the statistical existence of disease but challenging specific cases. (12)

Other social factors may affect management's perception of disease. Symptoms may be interpreted as being the result of the poor health status of "inferior" racial stock or of lower class social behavior. (13) The "Monday morning feeling" characteristic of byssinosis, for instance, has been attributed to workers' drinking habits on weekends. On the other hand, Alice Hamilton's ability to draw management's attention to occupational disease depended, at least in part, on her social class connection. (14)

Health Professionals

A critical notion in understanding the perceptions of health professionals is that of "occupational role." (15) On one hand, physicians, nurses, and industrial hygienists are trained in scientific, medical, and public health disciplines; on the other, they earn their incomes in a variety of locations relative to the system of production. Frequently, the first indications of a hazardous industrial exposure arise in contact between the worker and his/her own physician, or the community-based medical care delivery system. Outside of the company-dominated, single-industry town, however, primary health care personnel have no control and little access to the workplace. Perhaps for this reason, physicians and nurses receive so little training in occupational health and generally give short shrift to occupational considerations in their practice. (16) When primary care personnel do attend to industrial hazards, they may be constrained in their advice to workers by the limited opportunities for employment which many workers face. Further, physicians specializing in workers' compensation cases deal mainly with accidents and, in any event, are not directly concerned with the control or elimination of exposures, but with the compensability of the injury. They have greater access to production—presumably their findings would filter through to the management through experience rating of insurance premiums or through direct controls asserted by insurance companies. But such physicians, since they are in the employ of economically interested actors (that is, insurance companies), may discount ill-defined or difficult-to-diagnose industrial disease. Similarly, physicians and nurses employed by companies have conflicting demands placed on them by their professional orientation and their economic obligations to employers. The health of the corporation can require that attention be paid to hazards that result in injury to the economic condition of the firm. Therefore, the variety of pressures affecting management similarly influence the company "doc" and the industrial nurse. (17) The emphasis in such practice is frequently on preemployment physicals, "loss control," reducing workers' compensation expenses, and in ensuring a level of health commensurate with economically functioning employees. Recently, "health promotion in the workplace," focusing on personal behavior rather than occupational hazards, has gained the attention of companies and health professionals concerned about rising medical care costs. (18)

Industrial hygienists are generally more closely related to production in the firm, although just as company physicians and nurses, they are usually located in staff, rather than line, departments in the corporation. This fundamental distinction is significant, since staff departments are essentially advisory to the management of production, although OSHA and other regulatory agencies may have elevated the importance of health and safety-related professionals in the corporate hierarchy. As distinct from medical professionals, the training of industrial hygienists is more closely related to engineering disciplines, including both the written and unwritten economics of cost-efficient production.

Health and safety personnel have at least three different orientations to the perception of industrial disease in addition to the economic constraints of corporate employment. The engineering ethic heavily stresses efficiency in production. The scientific ethic stresses a conservative approach to information about health effects, leaning toward the requirement of conclusive proof of disease related to particular occupational exposures. The clinical medicine or public health ethic focuses on the condition of the worker, stressing preventive measures where there is an indication that injury may be occurring. Thus, economic concerns influence the perception of worker illness through the interaction of occupational roles and professional orientation, depending on the particular profession involved and the location of the particular professional in the scheme of production.

The State and the Legal System

Understanding how the perception of occupational illness is mediated by the state and the legal system requires focus on three concepts or characteristics of modern government in capitalist society: hegemony, class conflict, and bureaucracy. The structural idea of "hegemony," or capitalist dominance, suggests that through the state and its concomitant legal order, the dominant class successfully presents itself as the guardian and guarantor of the interest and sentiments of the whole society, including subordinate classes. (19) A somewhat crudely instrumental theory would reduce the role of the state and the rule of law to that of merely serving the interests of the dominant capitalist class, of intervening in social-economic matters in order to mediate and contain class antagonisms in a manner which preserves the legitimacy and stability of the system. Social problems, such as occupational disease, which principally affect the working class, would accordingly be downplayed and defined as marginal to the fundamentally satisfactory existent class relations. Recent history, for example, reveals a marked tendency by business to contest occupational disease claims filed under state workers' compensation laws. While the citing of scientific uncertainty concerning many issues about occupational disease does provide a putative legal basis for such controversions, the actual impact of this tactic is to shift the economic burden of occupational disease from managers, who exercise control over the work environment, to workers, who already bear the physical and social costs of the disease. But if the state and law can be seen as instruments of class power, they also must be understood as central arenas of class conflict.

Even if the state and its legal system function to mediate and reinforce existing class relations while serving ideologically to legitimate and mystify dominant class power, there are limits and constraints. It is inherent in the nature of the hegemonic state and in the forms and traditions of law that they cannot be reserved for the exclusive use of the dominant class. In the history of occupational health, the black lung movement provides an example of the exercise of counterhegemonic power in its successful attempt to impose its own perception of this occupational disease in a broader medical construct. (20) The coal workers struggled for a definition of their illness, fundamentally a political conception, which reflected their collective experience as victims of a preventable disease resulting from political-economic decisions taken by the coal industry. The specific political-economic weight of their organized actions, both legal (lobbying) and extralegal (wildcat strikes), forced the enactment and administration of law responsive to their interests.

Significant class conflict and its reverberation in the state and the legal system are currently evident in respect to compensation and prevention of asbestos-related disease. Although claims are contested and compensation denied, claimants and their lawyers elicit from the legal system a recognition of disease and some degree of remedy through resort to the social welfare laws and through third-party tort liability suits against manufacturers. Even as statutes of limitations barred many claims in certain states, lawyers for claimants persuaded courts to adopt "discovery rules" while legislators introduced amendments to afford some remedy to workers. (21)

Similarly, the perceived failure of the legal regime to prevent occupational disease by the imposition of liability under compensation laws and tort law (so-called market deterrence) led to regulation by the state in the form of the Occupational Safety and Health Act. The establishment of the National Institute for Occupational Safety and Health (NIOSH) under the act constitutes a potentially significant institutional structure which may affect the perception of occupational hazards and disease. It can be seen as one more legal outcome of class conflict and of partially successful struggles by subordinated classes to extract concessions and impose inhibitions on the exercise of power by industry.

Of course, perceptions of occupational disease which have developed as legal outcomes of class conflict are not immutable by virtue of being embodied in law. During periods when the underlying political-economic climate is concerned with fiscal austerity, cost-effectiveness, and deregulation, erosion of such gains can and does occur. Thus, the extent that the state in modern capitalist democracy constitutes a central location for class conflict, victories, losses, and political compromise is reflected in the official and unofficial stance of regulatory agencies and the legal system in general.

It should be taken into account, however, that the state and its legal system are not directly administered for the most part by capitalists, but by a middle class stratum of professionals and bureaucrats. This raises the issue of the impact of bureaucracy and of bureaucrats, with their particular and distinct political, economic, and professional interests. This may seem to be a further nuance of the foregoing theme of class conflict, and to a certain extent it does embody the variety of considerations included in the above discussion of "professional" perceptions of disease. It is here, however, that we may most usefully explore the significance of a "public health" or "professional" ethic as a distinct political force partially independent of orthodox notions of class conflict.

Lawyers and other legal professionals, including judges and legally trained government bureaucrats, have played a partially autonomous, significant role in the state and the legal system as a central arena of conflict between the fundamental classes. Through adjudication they mediate and legitimate the contending perceptions of industrial disease. Such professionals who administer "the Law" come to believe in the long tradition in which they have been steeped, in the particular forms and character of law, its own independent history and internal logic of evolution. They have a profound personal and professional stake in maintaining the apparently neutral stance of law as a body of objective rules applied as logical criteria with reference to standards of equity and universality. Thus, the implementation and shaping of occupational health and safety laws and perceptions may take turns that are reflective of neither management nor labor, but may rest rather on the specific political weight or processes of the bureaucracy. The significance of this bureaucratic legal, quasi-scientific stance in an occupational health subsystem of capitalism should not be underestimated.

THE CONTROL OF DISEASE

Political-economic influences on measures to control disease are of no small significance since the introduction of controls raises questions about the appropriate allocation of scarce public and private resources. Further, the choice of control measures may reflect underlying notions of private property rights, the proper role of the state, and the appropriate social control of technology. These are highly politicized choices in the current climate of debate over regulation and deregulation of industry.

Industrial Hygiene

Industrial hygiene approaches involve the elimination or reduction of hazardous exposures to workers through three fundamental measures: engineering controls, personal protection, and administrative measures. Although the use of these "tools" appears to be a technical matter, there is no question that economic considerations are important. Engineering controls may involve substantial capital expenditures and may be linked to the overall investment policy of the firm. Personal protection frequently appears to be less expensive in the short run but may involve substantial maintenance costs over the longer haul. Administrative measures, such as job rotation and worker education, sometimes appear to be relatively inexpensive ways to deal with minimal hazards. On the other hand, financial costs may not be the only consideration for firms: the degree of control over the choices of technology and of labor processes may be a conditioning influence for both labor and management. (22) Administrative measures involving worker participation—for example, labeling of materials and worker surveillance schemes—have been attacked by management organizations on cost grounds, but the real opposition seems to be founded on control concerns.

A basic economic issue in the choice of industrial hygiene controls involves the appropriate way to figure the cost of future expenditures. Future costs must be discounted by an interest rate indicating the return on alternative use of funds. Therefore, current expenditures may loom large compared to ones which will be made in the future. This, of course, leads management to be more wary of engineering controls with their attendant capital requirements. The discount rate chosen by the firm may differ quite substantially from a social discount rate. In any event, any choice may involve guesses about quite distant future economic conditions. (23)

Clinical Approaches and Epidemiology

Clinical approaches to the control of occupational disease are most strongly influenced by the sharp separation of primary health care and occupational medicine. The lack of access to the workplace results in a focus on removing the worker from the exposure, rather than eliminating or reducing the exposure. When the disease is not specific to an occupational hazard, the clinician frequently will ignore occupational histories and focus on personal habits (smoking for example) and therapy for the disease instead. Particularly where employment opportunities are limited, workers may receive medical care for occupational disease with the issue of removal from exposure never arising. Social class, racial, and ethnic considerations may also be relevant to the situation, particularly when physicians observe undesirable worker behavior such as excessive consumption of alcohol, smoking, or "malingering." (24)

Epidemiology's roots in medicine raise related problems. Because most epidemiological studies are conducted by investigators from outside the firm and industry, the characterization of exposure may be inadequate. Such statistical studies can only be suggestive of hazardous situations. Since control of exposure data still rests in the hands of management (when that data exist at all)—as is the control of the exposure itself—the effectiveness and value of epidemiological studies rest upon the cooperation of management. Economic concerns thus have an indirect way of influencing studies of occupational disease.

The Importance of Technical and Economic Considerations

Perhaps the key underlying thread of political economy constraining industrial hygiene and clinical-epidemiological approaches to occupational disease is the nature of the hypothesis that each of these disciplines sets out to test. The existence of disease and warranted controls is examined in a context that is taken as given. Yet the particular economic and technological settings are determined by particular political and economic arrangements and are not inevitable nor sacrosanct. (25) The leeway afforded to industrial hygiene appears to be a bit wider than that to health professions because of the former's more intimate connection with the system of production. Still, the outside health professional may be freer to express concern or alarm about occupational disease in that a peripheral position imposes fewer constraints.

The Political Economy of Regulation

The previous discussion of the role of the state applies even more aptly to control issues than to disease recognition. Limits to regulation are imposed (at least in the United States) by the imperatives of the capitalist firm, though it is not completely clear how far those limits can be pushed. Economists are especially sensitive to the devastating consequences of regulatory efforts for the capitalist firm. All the same, most firms seem to adjust reasonably well to new regulations, and some even manage to increase profits with new, less hazardous plants and equipment. (26) Nevertheless, regulators must take into account the impact of new standards on industry, which means that the fundamental rule of profit-making will be protected.

At the same time, the rule-making process has become a new arena for conflict and compromise between economic and political actors. Beyond that, the struggle over enforcement of the rules provides a further battleground for labor and management. It is plain that well-organized political forces marshall their experts for these battles. Those who bewail the adversarial nature of health and safety regulation are probably not doing as well as they had hoped in the controversy.

The Political Economy of Research

Occupational health research has been affected and shaped by political-economic factors in a variety of ways:

a. Funding sources may influence the choice of problems to be investigated and the direction of research. Research priorities may be strongly influenced by political and economic actors.
b. The definition of the problem to be investigated may reflect economic interests.
c. The scientific disciplines to be involved in the research may reflect extraneous political considerations, depending on the relative political power or current popularity of particular disciplines.
d. The range of problem solutions considered may be subject to political and/or economic constraints.

e. Notions of efficiency in research methodology—reflecting economic constraints on researchers—frequently influence the conduct of research.
f. Political and economic considerations may influence the publication of research findings. The occupational location of researchers—whether in private industry, government agencies, research institutes, or universities—can inhibit the flow of scientific information.

COMPENSATION FOR OCCUPATIONAL DISEASE

Attempts by workers to gain protection from the economic effects of occupational disease affect the definition of disease and subsequent research on occupational health hazards as well as research on and the implementation of control technologies. (27) Scientists may require evidence of an incontrovertible relationship between an exposure and health effect in order to put their stamp of approval on the occupational etiology of a worker's illness. Much more rigorous definitions of disease develop for occupational illnesses because of the issues of financial compensation than for the "ordinary diseases of life," which involve no employer liability. Medical researchers are unhappy about "subjective" reports of symptoms in the case of occupationally related illness and search for "objective" technologies by which to make distinctions between disease caused by a particular exposure as opposed to illness merely associated with particular working situations. The significance of smoking becomes exaggerated because of the concerns of employers about financial liability.

Principles of Workers' Compensation

At least in part, the function of workers' compensation insurance programs is to expedite financial relief to workers who fall ill from occupational injury. Rather than permitting (or requiring) that injured employees take their employers to court, workers' compensation is a no-fault insurance program, providing some level of income maintenance and medical benefits. On the other hand, workers' compensation enables employers to avoid potentially disastrous damage suits and to insure against economic liability for occupational hazards. By and large, the system is geared toward handling accidental injuries rather than occupational disease, although some states' original legislation provided for benefits to workers with employment-related illness. Ordinary diseases of life, however, are not compensated. Thus, although workers' compensation is supposed to be no-fault insurance, in the case of disease, employees must demonstrate the occupational etiology of the illness. This can be extremely difficult when the illness is not recognizable as specific to a particular exposure or where the illness takes years to develop, as is frequently the case. The burden of proof is the employee's in instances where employers or insurance carriers challenge the occupational nature of the illness. Financial considerations, then, are extremely important in the definition of diseases potentially compensable under this system. The political strength of labor and management is brought to bear in controversial situations, because a disease may be listed as specifically compensable by state legislatures or workers' compensation commissions. Similarly, political weight can be brought to bear in the selection of commission members.

On the other hand, if workers bring no claims for compensation, the occupational etiology of an illness may go undetermined. Disease may go untreated or treated by medical practitioners as an ordinary disease of life.

The Uses of Medical Uncertainty

Medical uncertainty may exist about the etiology of occupational disease for a number of reasons:

a. Safe conditions may be presumed, as in the case of the cotton textile industry in the United States, pre-1960s. The illness of workers may go undefined.
b. Workers may not file compensation claims, in which case there may be no particular incentive for scientific research.
c. If there is no clear impact on workers' productivity or if an ample supply of replacement labor is available, employers may find it economically safer to avoid, prevent, or squelch scientific research.
d. The research problem may be difficult—that is, may require resources and attention beyond the financial capabilities of the interested parties.
e. If substantial economic interests are involved, a plethora of research may be generated, resulting in the obscuring of the issues in question.
f. Finally, neither epidemiology nor medicine is an exact science.

Eventually, expert medical witnesses enter into a politically and economically loaded workers' compensation setting and are asked to testify as to the work-relatedness of the claimant's illness. Different constructions of information about the disease and about the nature of the worker's exposure frequently are possible. Political and economic bias may interact with the interpretation of medical information.

Compensation and Prevention: Incentives

The workers' compensation system is presumed to encourage preventive measures for the elimination or reduction of occupational hazards through experience ratings of premiums. Yet, the insurance aspects of workers' compensation enable employers to plan their expenditures, preventing cataclysmic financial burdens, and thereby *discourage* safety measures. In addition, the successful political efforts of employers to minimize workers' compensation payments, while preventing suits, further reduce safety incentives.

The emphasis management and labor place on compensation issues also greatly detracts from efforts to develop useful prevention information. Workers may press for a broad definition of disease in order to ensure that sick workers get income maintenance—but such a broad definition may only serve to mask the specific causes of disease and deflect research efforts from preventive measures that *may* require targeting of specific harmful agents. (Economics, again, may be a consideration in this discussion of prevention, since cost-effective control techniques may require knowledge of the specific agent.) On the other hand, employer resistance to the costs of compensation may result in research that obscures specific agents or situations in order to establish the disease as "ordinary." In any event, the debate about compensation and its potentially large costs often keeps attention from being directed to preventive measures. In some instances, the debate shifts scientific attention to the influence of smoking, again pulling resources away from controlling the occupational environment. Finally, even though a political resolution of a compensation controversy might be successful, the pressure for further investigation of the disease could easily wane.

We have presented a *structure of hypotheses* concerning the social determinants of occupational disease. They are the key questions one must ask in analyzing the social history of occupational health, in studies of particular diseases, in cross-disease comparisons, in related institutional studies, and for cross-national comparisons.

NOTES

1. This chapter has benefited particularly from suggestions by David Wegman, Barbara Rosenkrantz, Peter Barth, Stephanie Woolhandler, and Ian Greaves.

2. George Rosen, *The History of Miners' Diseases*, Schuman's, New York, 1943.

3. David F. Noble, *American by Design*, Oxford University Press, New York, 1979.

4. Charles Levenstein et al., "Labor and Byssinosis, 1941-1969," in David Rosner and Gerald Markowitz, *Dying for Work*, Indiana U. Press, Bloomington, 1987.

5. Craig Zwerling, "Salem Sarcoid: The Origins of Beryllium Disease," in Rosner and Markowitz, ibid.

6. Rosner and Markowitz, ibid.

7. Jeremy P. Felt, *Hostages of Fortune, Child Labor Reform in New York State*, Syracuse University Press, Syracuse, NY, 1965.

8. Thomas M. Legge, "Industrial Diseases Under the Medieval Trade Guilds," *Journal of Industrial Hygiene*, February 1920, Vol. I, No. 10, pp. 476-477.

9. See B. S. Levy and C. Levenstein (eds.), *Environment and Health in Eastern Europe*, MSH, Boston, 1990.

10. Ronald Bayer, *The Health and Safety of Workers*, Oxford University Press, New York, 1988.

11. Barry S. Levy and David H. Wegman, *Occupational Health* (2nd ed.), Little Brown, Boston, 1988.

12. Peter S. Barth and H. Allan Hunt, *Workers' Compensation and Work-Related Illnesses and Diseases*, MIT, Cambridge, 1980.

13. *American Textile Reporter*, July 10, 1969.

14. Alice Hamilton, *Exploring the Dangerous Trades*, Little Brown, Boston, 1943.

15. Helga Nowotny, "Controversies in Science: Remarks on the different modes of production of knowledge and their use," *Zeitschrift fur Soziologie*, Jg. 4, Heft 1, Jan. 1975, pp. 34-35.

16. Levy and Wegman, op. cit.

17. See Bayer, op. cit.

18. Paul Kotin and Lois A. Gaul, "Smoking in the Workplace: A Hazard Ignored," (Editorial) *AJPH*, June 1980.

19. A. Gramsci, *Prison Notebook*, International Publishers, New York, 1971.

20. Barbara E. Smith, "Black Lung: The Social Production of Disease," *International J. of Health Services*, Vol. II, No. 3, 1981.

21. See Barth and Hunt, op. cit.

22. Harry Braverman, *Labor and Monopoly Capital*, Monthly Review Press, New York.

23. Leslie I. Boden, "Cost-benefit Analysis: Caveat Emptor," *AJPH*, 1979; 69: 1211.

24. Dan Berman, *Death on the Job*, Monthly Review Press, New York.

25. Vicente Navarro, *Medicine Under Capitalism*, Prodist, New York, 1976.

26. J. Ives, *The Export of Hazard*, Routledge and Regan Paul, London, 1985.

27. See Barth and Hunt, op. cit.

2

The Changing Structure of Work in the United States

PART 1—THE IMPACT ON INCOME AND BENEFITS

SARAH KUHN
JOHN WOODING

It has become a commonplace to say that the United States is a "service" economy. The evidence is compelling: service occupations and industries are by far the largest—and still growing—share of economic activity. In spite of their dominant position within the economy, we know less about service occupations and industries than we know about the comparatively tiny manufacturing sector. In particular, we are only beginning to form ideas about the effect this economic transformation is having on jobs, employment conditions, and on health and welfare.

The increase in low-paid service-sector employment has resulted in less access to good job benefits for most workers. Fewer employees have decent health care insurance provided by employers, paid vacation, and opportunities for developing a career path within a company. Most Americans are now working longer hours for less money (in real terms) than their parents. The increase in stress-related injuries and illness is but one manifestation of the toll this takes on most workers.

Therefore, a full understanding of occupational and environmental disease, the causes of new epidemics (for example, repetitive motion injuries), the increase in stress-related diseases, and the decline in the availability of good health care for the majority of the population, requires that these problems be understood within the context of the changing structure of work in the United States and the impact of an increasingly global economy.

PERIOD OF TRANSITION

It is not an exaggeration to say that the United States is in transition to a new era in the production of goods and services. Extraordinary changes have taken place in the last two decades: dramatic increases in international competition, the transition from a creditor to a debtor nation, deep challenges to the sense that the United States has of its place in the world, an extraordinary increase in the rate of capital mobility, and the internationalization of consumer and capital markets. In addition, the breakdown of the "labor accords" of the postwar era and the conservative industrial relations system that accompanied them, the increase in economic inequality within the United States and the erosion of the "social safety net," the growing experimentation by segments of the business community in the face of unprecedented competitive pressure, have all contributed to the increased economic insecurity of the American worker. It may be too early to see how these forces will play themselves out, yet some features are already becoming clear.

Increased competition, especially from European and Japanese corporations, and from multinationals in Newly Industrialized Countries (NICs), has forced workers in the manufacturing sector into a deeper identification of their own interests with those of their employers. Weakened unions, often forced into concessions, have increasingly seemed to adopt—understandably—the position that "what's good for General Motors is good for America." Although some unions have had the vision to recast the debate, generally workers face a "take-it-or-leave-it" bargain with management.

Corporations are responding to increased competition in a variety of ways, but among these is increased capital mobility (1) and the increasing use of a contingent or "just-in-time" employment strategy. This has the effect of externalizing risk to the individual worker and the community. At the same time, the "social safety net" has eroded as a result of the policies of the Reagan and Bush administrations, leaving many people without an adequate level of social services or the help necessary to make the adjustment as communities and individuals bear the risks of the marketplace.

INCREASING INEQUALITY

One consequence of this is increasing inequality of the income distribution. Blacks, Hispanics, and many women are particularly disadvantaged, and have higher rates of poverty and (with the exception of women) unemployment than do white men. One in five U.S. children lives below the poverty line. While unemployment is not as high as in Europe, underemployment is a pervasive problem, with the growth of low-wage jobs masking what might otherwise be higher levels of unemployment. (2, 3) White males, who once dominated the labor market, are now in the minority, although they continue to hold positions of privilege.

The erosion of the middle class, the increasingly unequal distribution of income, the rising incidence of poverty (particularly among children), increasing temporary and part-time work, declining availability of health benefits, weakened social, environmental, and occupational safety and health regulation: these are all undesirable and potentially destabilizing developments that directly affect the public health and welfare.

This chapter discusses in detail some of these trends and identifies a number of interrelated phenomena that have a direct bearing on the well-being of American workers and their communities. We believe that to understand public health problems in general requires an appreciation of what is going on in the American and global economy. In what follows, therefore, we provide an overview of the changing structure of work in the American economy, and discuss the impact it has had on incomes, benefits, and the nature of work organization. In a subsequent chapter (Chapter 3), we will suggest that these developments have added to the health problems facing American workers—especially those unemployed, underemployed, facing job insecurity, or working temporarily or part time. Our purpose is to map these developments and suggest how they might be linked to the many issues now facing those concerned with public health.

WHAT HAS HAPPENED TO AMERICAN INDUSTRY?

Over the last century, the size of the manufacturing sector as a proportion of total employment has changed only slightly, from about 25 percent in 1880 to a peak of 35 percent in 1952, and tending toward its 1880s level again during the 1980s. The most dramatic change over that period has been the rise of the service sector, which grew from about a quarter of employment in 1880 to nearly 70 percent in 1992, offsetting the steep decline in agricultural employment.[1] Between 1940 and 1980, employment in retail trade more than tripled, as did finance, insurance, and real estate, while employment in miscellaneous services, such as personal and business services, grew nearly five-fold. As can be seen in Table 2.1, these trends have continued and today some 85 million Americans work in "service-producing" industries. Certain service occupations have shown dramatic

TABLE 2.1. Employment Level by Industry (in Thousands)

Industry	1984	1985	1986	1987	1988	1989	1990	1991	1992
Total employment	94,408	97,386	99,344	101,958	105,210	107,896	109,419	108,256	108,519
Private sector	78,384	80,992	82,651	84,948	87,824	90,117	91,115	89,854	89,866
Goods-producing	24,718	24,842	24,533	24,674	25,125	25,254	24,905	23,745	23,142
Mining	966	927	777	717	713	692	709	689	631
Construction	4,380	4,668	4,810	4,958	5,098	5,171	5,120	4,650	4,471
Manufacturing	19,372	19,248	18,947	18,999	19,314	19,391	19,076	18,406	18,040
Service-producing	69,690	72,544	74,811	77,284	80,086	82,642	84,514	84,511	85,377
Transportation, public utilities	5,156	5,233	5,247	5,362	5,514	5,625	5,793	5,762	5,709
Wholesale trade	5,568	5,727	5,761	5,848	6,030	6,187	6,173	6,081	5,045
Retail trade	16,512	17,315	17,880	18,422	19,023	19,475	19,601	19,284	19,346
Finance, insurance, real estate	5,684	5,948	6,273	6,533	6,630	6,668	6,709	6,646	6,571
Services	20,746	21,927	22,957	24,110	25,504	26,907	27,934	28,336	29,053
Government	16,024	16,394	16,693	17,010	17,386	17,779	18,304	18,402	18,653
Federal	2,807	2,875	2,899	2,943	2,971	2,988	3,085	2,966	2,969
State	3,734	3,832	3,893	3,967	4,076	4,182	4,305	4,355	4,403
Local	9,482	9,687	9,901	10,100	10,339	10,609	10,914	11,081	11,281

Source: Monthly Labor Review, V. 116, N. 7, July 1993, p. 86.

expansion between 1972 and 1986: executive, administrative, and managerial workers; technicians and related support workers increased by roughly 75 percent, and professional and sales occupations also grew substantially (Table 2.2). This trend has continued into the 1990s (Table 2.3). In short, the statistical evidence suggests that the U.S. economy has, indeed, become a service economy.

But what does this mean? There is a certain amount of confusion in defining the term "service."[2] The ambiguity of the goods/services distinction points to a deeper economic fact: the extraordinary interdependence of manufacturing and services. As the U.S. Office of Technology Assessment notes, "any view of the future that sees the services taking the place of manufacturing has pushed the distinctions between them too far." (4, p. 3) Manufacturing industries are heavily, and increasingly, dependent on services as suppliers of inputs—business services, transportation, communications, and so forth—and as customers. Quinn (5) estimates that as much as 85 percent of the communications and related information technology sold in the United States in 1985 was purchased by information industries, predominantly in the service sector. Services industries are also heavily dependent on manufacturing for equipment and supplies, and for their patronage as consumers of services. (6) This interdependence is represented in Figure 2.1. It is this inextricable linking of manufacturing and services that suggests that the economy, although transforming, cannot be called "postindustrial."

More important for the argument being developed here is the impact that these structural transformations have had on the lives and working conditions of American workers. These transformations occur in a variety of forms. We consider them at length below.

The Impact on Income Distribution

The fact that virtually all net new job creation in the United States is now in the service sector is creating anxiety among observers who see the service sector primarily as a source of low-wage jobs. Because average hourly real wages for all workers have not grown since 1973 (2) and real average weekly earnings have fallen (see Figure 2.2), these fears appear well grounded.

If we compare the industries with the highest growth between 1979 and 1989 and those with the lowest, we find that wages in the declining industries are disproportionately above-average (see Table 2.3). The industries that generated the most jobs during this period (retail trade and services) had median weekly earnings of $276 and $357 respectively. In contrast, manufacturing, which lost nearly a million and a half jobs in this period, had median weekly earnings of $415. The growth industry with

TABLE 2.2. Employment by Broad Occupational Group for 1986 and Projected for 2000, and Change in Employment for Selected Periods

Major occupational group	1986 (in thousands)	2000 (in thousands)	Percent change in employment 1972–1979	1979–1986	1972–1986	1986–2000
Total employment	111,619	133,026	20.3	10.9	33.4	19.2
Executive, administrative, and managerial workers	10,583	13,616	34.9	28.7	73.7	28.7
Professional workers	13,538	17,192	29.8	21.4	57.5	27.0
Technicians, related support workers	3,726	5,151	39.9	24.7	74.5	38.2
Sales workers	12,606	16,334	24.3	24.4	54.6	29.6
Administrative support workers, including clerical	19,851	22,109	23.5	9.5	35.2	11.4
Private household workers	981	955	-23.0	-11.5	-31.9	-2.6
Services workers, except private household workers	16,555	21,962	25.7	16.0	45.9	32.7
Precision production, craft, and repair workers	13,923	15,590	21.7	6.5	29.6	12.0
Operators, fabricators, and laborers	16,300	16,724	8.7	-9.2	-1.3	2.6
Farming, forestry, and fishing workers	3,556	3,393	-5.1	-5.6	-10.4	-4.6

Note: Estimates of 1986 employment, the base year for the 2000 projections, were derived primarily from data collected in the Occupational Employment Statistics (OES) surveys. The 1972–1986 rates of change and subperiods were derived from the Current Population Survey data, because comparable data were not available for 1972 and 1979 from the OES surveys.

Source: Kutscher, Ronald E. "Growth of Service Employment in the United States." In *Technology in Services: Policies for Growth, Trade, and Employment*. Edited by Bruce R. Guile and James Brian Quinn. Washington, DC: National Academy Press, 1988: p. 52.

TABLE 2.3. Employment Growth by Sector, 1979–1989 (in Thousands)

	Employment			Industry share of job	Median weekly
Industry sector	1979	1989	Job growth	growth	earnings, 1989
Goods-producing	26,461	25,634	-827	-4.4%	
Mining	958	722	-236	-1.3	$565
Construction	4,463	5,300	837	4.5	431
Manufacturing	21,040	19,612	-1,428	-7.6	415
Durable goods	12,760	11,536	-1,224	-6.5	445
Nondurable goods	8,280	8,076	-204	-1.1	373
Service-producing	63,363	82,947	19,584	104.4%	
Transportation, communications, utilities	5,136	5,705	569	3.0	$502
Wholesale	5,204	6,234	1,030	5.5	412
Retail	14,989	19,575	4,586	24.4	276
Finance, insurance, real estate	4,975	6,814	1,839	9.8	406
Services	17,112	26,892	9,780	52.1	357
Government	15,947	17,728	1,781	9.5	472
Total	89,824	108,581	18,757	100.0	

Source: Lawrence Misae, David Frankel. *The State of Working America 1990–91*. Washington, DC: Economic Policy Institute.

the highest median weekly wage (finance, insurance, and real estate) had weekly wages well below mining, construction, and manufacturing.

A look at the occupations projected to generate the largest number of jobs up to the year 2000 tells a similar story. Eight of the 10 occupations have earnings below the 1987 median, and half of the occupations could not generate enough earnings to support a family of four above the 1987 poverty threshold (Figure 2.3).

These figures give an impression that is incomplete, however, for the service sector includes high-paying as well as low-paying industries and occupations. The most striking characteristic of wage distribution in the service sector is not the proliferation of low-wage jobs, but that wages are distributed more unequally in the private service sector than in manufacturing.

There is general agreement in the literature that income inequality has increased in the United States over the last decade and a half or more, with an associated decline in the proportion of middle-income or "middle class" earners (see Figures 2.4 and 2.5).[3] Because this phenomenon is so important, both from a social and a public policy point of view, it has become clear that to measure an industry by the average, or even the median, wage paid is to miss an essential point: the average may be stable or even rising while *at the same time low-wage jobs are proliferating*. This has led to a

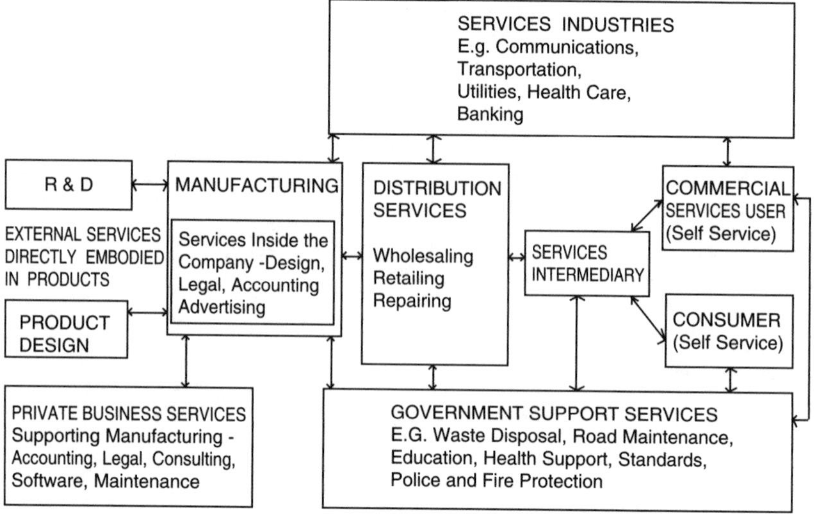

FIGURE 2.1. Some mutual interactions among manufacturing and service activities. Doube-headed arrows indicate that each party benefits from the presence of the other in the trade. *Source:* James Brian Quinn, "Service Technology and Manufacturing: Cornerstones of the U.S. Economy." In *Managing Innovation: Cases from the Services Industries*. Edited by Bruce R. Guile and James Brian Quinn. Washington, DC: National Academy Press, 1988: p. 20.

situation of increasing income inequality that weighs particularly heavily on women and minorities.

The Inequalities of Race and Gender

A major area of concern is that the inequalities arising from race and gender are getting worse. A recent study analyzing data on wages and salaries of individuals drawn from the decennial census of 1960, 1970, and 1980 found that inequality among men, especially nonblack men, increased, while inequality among women fell (7). One limitation of the study is that the category "nonblack" includes both whites and Hispanics who generally have very different labor market outcomes. The increased inequality among nonblack men for the period may indicate the growth of a low-wage Hispanic population.

In 1960 the earnings of black women were 60 percent those of nonblack women, but by 1980 the two groups had converged in mean earnings. This, too, is likely due in part to the inclusion of Hispanic women in the nonblack category; it also reflects the success of black women in moving out of the lowest-paying domestic service jobs. Within-group inequality among black women also fell during the period, becoming the same as that of nonblack women. Overall, inequality among the four race/gender groups fell, while within-group inequality grew (mostly due to increased inequality among nonblack men, which more than offset increased equality among black women).

The same study also found that "changes in the sectoral composition of employment accounted for almost two-thirds of the overall increase in inequality, and that inequality was further increased by rising irregularly *within* industrial sectors" (p. 11). Neither the marked increase in the labor force participation of women nor shifts in regional patterns were significant to the changing pattern of inequality, but the authors note two other factors that they were unable to investigate: the fact that the sectors with the most highly skilled and educated work forces were those with the greatest increase in inequality, pointing to occupational change as a factor (although since these industries also tended to be the fastest growing, we may simply be witnessing a scale effect); and the fact that Tilly, Bluestone, and Harrison (8) found that one-third of the increase in inequality between 1978 and 1984 was due to inequality in hours worked, not in wages.[4]

These findings suggest that investigation of within-industry changes in employment practices and industrial structure will be fruitful for deepening our understanding of growing earnings inequality. Clearly, however, racism and sexism combine with the shifting pattern of manufacturing and

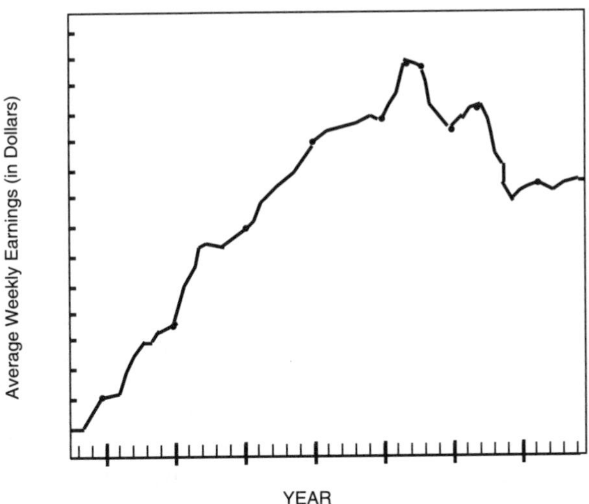

FIGURE 2.2. Real average weekly earnings, 1947–1986 (in 1986 dollars). *Source:* Bennett Harrison and Barry Bluestone. *The Great U-Turn: Corporate Restructuring and the Polarizing of America*. New York, NY: Basic Books, 1988: p. 6.

TABLE 2.4 Job Openings by Major Occupational Group, Projected 1990–2005 (in Thousands)

Occupational group	Job openings due to replacement needs	Job openings due to employment increases	Total job openings
All occupations	38,837	26,891	65,728
Managerial	3,085	3,414	6,499
Professional specialty	4,281	5,107	9,388
Technicians	1,200	1,551	2,751
Marketing and sales	5,379	3,401	8,780
Administrative support	6,413	3,309	9,722
Service	7,403	5,830	13,233
Precision production	4,764	2,068	6,832
Operators	5,449	1,734	7,183
Agriculture-related	863	477	1,340

Source: U.S. Dept. of Labor, Bureau of Labor Statistics, *Occupational Outlook Quarterly*, Fall 1991, p. 29

employment to put an increasingly inequitable burden on minorities and women. In addition, the growth of new immigration from Mexico and Southeast Asia creates a pool of unorganized and easily exploited labor.

Contingent, Part-Time, and Temporary Work

In the U.S. economy, the standard "job" has been typified by full-time, full-year employment, with a long-term commitment between employer and employee. One of the most striking recent trends in the U.S. workplace is the breakdown of traditional job ladders. Under the old model of employment, in which long-term commitments were made implicitly between employer and employee, there was a potential for moving up through the ranks, learning on the job and through employer-sponsored training. While many never were able to take advantage of this system, for some it was the key to a lifetime of career advancement. These old avenues, which made possible the rise from newsboy to newspaper tycoon typified in the "Horatio Alger" stories, are increasingly constricted or disrupted. In a convincingly argued book, Thierry Noyelle (9) makes the case that internal labor markets are breaking down, and that firms in the "sunrise" industries exhibit quite different forms of work organization. He attributes these changes to a variety of circumstances, including:

a. Computer technology as applied in workplaces, which has the effect of universalizing and homogenizing skill, making it easier for workers to move among companies and for employers to hire from outside.

b. Equal Employment Opportunity enforcement, which has challenged old methods of hiring and promotion because they tended to work to the disadvantage of those who were not white and male.
c. Geographic mobility of back-office operations, which can use new technology to operate in locations remote from customer service and headquarters sites. This severs some of the ties that might otherwise have been the basis for promotion ladders.
d. "Vertical disintegration" (a term borrowed from Christopherson and Storper, 1985) of firms, leading to greater use of the market rather than internal transactions, and a use of labor characterized by "assembling modules of expertise." (9, p. 101)

The consequence of these developments is that more and more workers hold "contingent" positions. Increasing numbers of workers are employed part-time rather than full-time, and an increasing proportion of those are part-time involuntarily rather than by choice. The number of those working two jobs, working as temporary rather than permanent workers, and acting as self-employed has also grown. (10) All these workers are arguably "contingent" in that their economic fate is more vulnerable to the

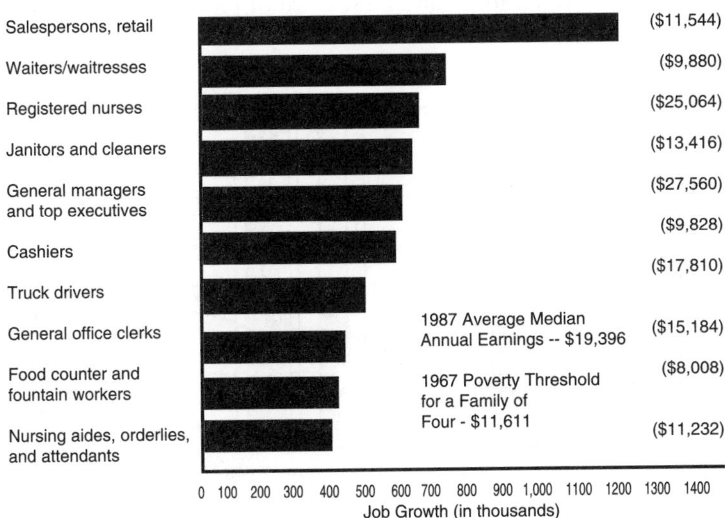

FIGURE 2.3. Occupations with largest job growth (projected), 1986–2000. *Source:* John J. Sweeney and Karen Nussbaum. *Solutions for the New Work Force: Policies for a New Social Contract*. Cabin John, MD: Seven Locks Press, 1989.

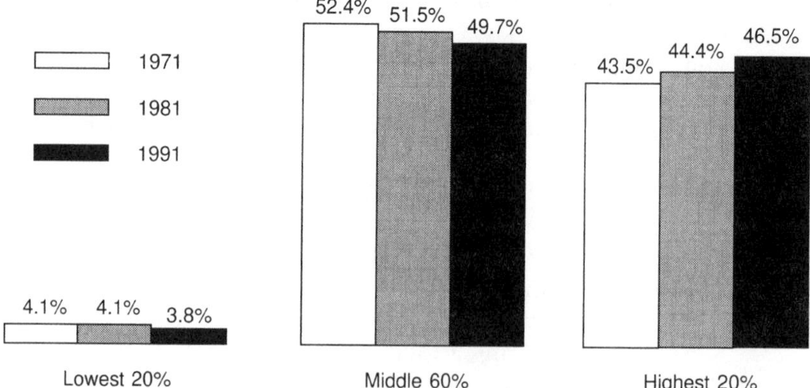

FIGURE 2.4. Share of aggregate household income, by quintile: 1971, 1981, and 1991. *Source:* U.S. Bureau of the Census, current *Population Reports, Series* p. 60, No. 180, "Money Income of Households, Families and Persons in the United States: 1991."

vagaries of the marketplace and employers' practices than is the fate of the long-term, full-time year-round worker.

The literature on contingent work frequently makes the distinction between contingent, or "secondary," workers and "core" workers. Core workers, according to Belous (11), have strong affiliation with an employer, have a long-term attachment and a measure of job stability, and have an

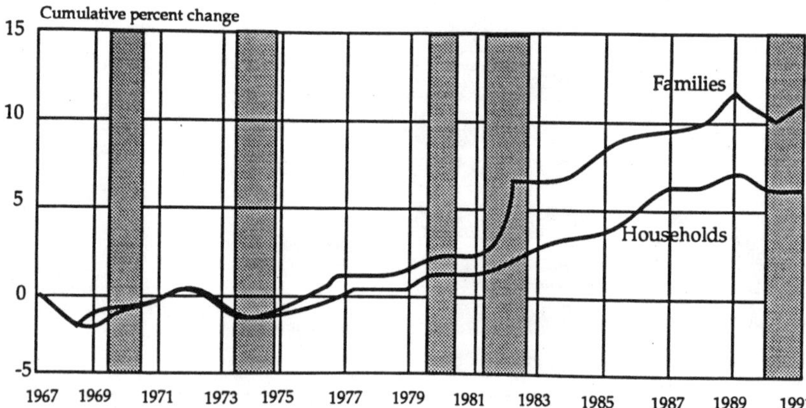

FIGURE 2.5. Percent change in household and family inequality as measured by the Gini Index: 1967–1991. (3) Shaded areas, recessionary periods; the data points represent the midpoints of the respective years. *Source:* U.S. Bureau of the Census, current *Population Reports, Series* p. 60, No. 180, "Money Income of Households, Families and Persons in the United States: 1991."

implicit contract with the employer that their commitment and cooperation will yield security and advancement. Contingent workers, by contrast, have a weak affiliation with any particular employer, have only short-term attachment and no job security. A single employer may employ both types of workers.

The shift toward contingent employment is one manifestation of deep and pervasive changes taking place in corporate structure and strategy. As international competition has heated up and the pace of change accelerated, companies have responded in a variety of ways. As Christopherson notes:

> The risks associated with uncertain input costs and production for uncertain markets have been dealt with by: outsourcing or subcontracting particular production activities, just-in-time inventory systems, network subsidiary firms, *and flexible or contingent labor strategies*. All of these responses attempt to reduce the risks associated with uncertain product and service demand and fluctuating input costs by transferring production transactions to the market where inputs, including labor, can be purchased on an as-needed basis. (10, emphasis in the original)

In particular, groups of workers previously sheltered from the vicissitudes of the marketplace have been affected as those who had the greatest expectation of a lifetime relationship with their employer are laid off.

More Contingent Work

There is no official government definition or measure of contingent work, and there are significant problems in trying to estimate the size of the contingent work force. Most analysts include part-time workers, those employed in the temporary help industry, and those who are self-employed. This method misses many—such as those temporary workers employed directly rather than through an agency—and double-counts others (40 percent of temporary agency employees work part time, according to Plewes [11]). Belous (12) estimates the size of the contingent work force at 29–35 million people in 1987, up from 25–29 million in 1980. This means that more than *a quarter of the U.S. labor force are contingent workers*, and the proportion is growing.

Another major factor affecting the working conditions and general benefits accruing to workers is the increase in part-time workers, the largest segment of the contingent labor force. In 1985, 18.6 million people, or just over 17 percent of the labor force, were employed part-time. (13) These figures undercount the number holding part-time jobs, however. First, the number of people who worked part-time for *a portion* of the year was about twice as large. (14) Second, those who work 35 or more hours a week by combining two part-time jobs or one part-time and one full-time are not counted. (11) By 1990, 18 percent or 19.6 million workers (almost one in five of all workers) were working part-time. (15)

Part-time work is classified as either *voluntary* (those who work part-time out of preference) or *involuntary* (those who work part-time because they cannot find full-time work). Over the last decade, most of the growth in part-time work has been involuntary part-time.[5] The extent of part-time work that is not fully voluntary may be understated, furthermore, since those who take part-time work because of child care or transportation problems or because of illness are considered to be working part-time voluntarily.

Along with the distinction between voluntary and involuntary part-time work, Tilly (16) divides part-time jobs into "retention" and "secondary" work. "Retention" jobs are relatively high in skill, training, and responsibility; low in turnover; have higher pay and better benefits; and are located on promotion ladders (although part-timers may not be promoted while working a reduced schedule). "Secondary" part-time jobs are the opposite of retention jobs—less well paid, more marginal, and so forth. Retention part-time work is usually exceptional, negotiated by an individual with his or her employer, often in occupations not characterized by part-time work. As evidence of the high quality of retention part-time jobs, part-time women in professional and managerial occupations earn *more* on an hourly basis than their full-time counterparts. (18) Employers may agree to create retention part-time jobs as a way of keeping a valued employee (for example, a woman with young children who would otherwise leave the company). Retention part-time employees, then, are essentially part of the "core" work force, even though they are working a part-time schedule. By contrast, secondary part-time jobs are usually found in groups, as whole categories of employment are given part-time hours.

Part-Time Work Pays Less

Secondary part-time jobs predominate, with the result that part-time jobs have a mean hourly wage that is about 40 percent below the hourly wage of their full-time counterparts. About half of this discrepancy can be attributed to the different observed characteristics of part-time workers (sex, race, education, experience) and by the fact that part-time jobs are concentrated in low-wage industries. Part-time workers are also likely to have less tenure. (16)

Clearly part-time employees are less likely than their full-time counterparts to have good fringe benefits, but how much less likely is a matter of disagreement in the literature.[6] What is clear is that part-time workers are much less likely to have good benefits of the kind typical of unionized employees in the manufacturing sector.

Part-time work is especially prevalent among young workers, older workers, and white women. Younger and older workers may choose part-time schedules to accommodate schooling (in the case of the young) or semi-retirement (in the case of the old). White women may be combining child care with market work, and have a family income that allows them to cut back on earned income. Although whites are more likely than blacks to work part-time schedules, blacks are more likely to work part-time involuntarily. (11)

Related to the growth of contingent and part-time work is, of course, the increase in temporary employment. Estimating the size of the temporary work force is a difficult proposition, however, since data are not collected regarding the permanence of the employment contract. Most analysts use the size of the labor force employed through temporary help service agencies as a proxy for the temporary labor force, but Abraham has estimated that these figures capture only about half of employees working on temporary contracts. (10)

In 1987, employment in the temporary help industry was nearly one million, up from 340,000 in 1978. By 1990, just over one and a quarter million workers were employed as temporaries. Since 1982 temporary employment has grown nearly three times as fast as overall employment. (15) During the course of a year, however, it is estimated that more than three times as many people work as temporaries. The largest group of temporary workers—43 percent—are in clerical positions, but the second largest category of temporary employees are those who do industrial work. (11) Industrial temporary employment has been growing at a faster rate than office employment, increasing from one-third of the industry in 1972 to 45 percent in 1982. (17)

The Temporary Force

The temporary industry's labor force is substantially female and young. Employment in temporary work, when voluntary, is advantageous to those who need a great deal of flexibility in work hours, or those who are entering or reentering the job market. However, one out of five temporary industry workers is black (11), nearly twice their representation in the labor force as a whole, and this seems more likely to be a result of blacks' relative disadvantage in the labor market, rather than employee choice.

One last category of work is that of the self-employed. About 10 percent of the U.S. labor force is self-employed, with about 7 percent operating unincorporated businesses while the remainder are incorporated.

Professionals in many industries act as independent contractors: business consultants, graphic designers, computer programmers, and so forth.

For many, high pay and control over their work compensate for the insecurity of working on a project-by-project basis. At the other end of the spectrum are "independent contractors" who perform such tasks as homework in the garment and electronics industries, and some kinds of home-based clerical work. Five thousand to 10,000 workers in the United States now work as home-based clericals, and this number may rise as some of the companies known to have home-based clerical programs expand their use of this form of labor. According to Christopherson and Noyelle, "the attractions of home-based work for the firm are substantial. They include increased productivity, elimination of benefit costs, reduction of turnover, and reduced costs related to off-hours computer utilization as well as decreased office space needs." (10, p. 17)

As with part-time and temporary workers, there are differences between voluntary and involuntary independent contractors. Some may prefer their independent status, while others would prefer a regular relationship with their employer.

Among the self-employed, women are more likely than men to own so-called casual businesses—defined by the fact that they have low earnings. Over one-third of "casual" business owners work full-time at their businesses, resulting in a very large discrepancy in the annualized earnings of self-employed men and women.

It is clear from this brief review that there are some significant changes in the pattern of employment in the United States. More people are working part-time and in temporary positions, and on a "contingent" basis. The traditional pattern of fairly stable long-term employment in an industry or firm is eroding. The impact on workers' lives of this development: the loss of certainty of employment, reduced benefits, lower incomes, and increased pessimism about continual improvements in standards of living are significant and play out through a range of social and political phenomena. These changes, as we will discuss in the next chapter, also raise questions about skill, stress, and safety at work.

There is, however, a further effect of changing patterns of economic structure and work organization that has a direct consequence for workers: the availability and extensiveness of job benefits.

THE EFFECT ON JOB BENEFITS

Slightly more than one-quarter of total employee compensation came to employees in the form of benefits in 1987 (compensation is defined as the sum of wages or salaries plus benefits). The U.S. Department of Labor's Bureau of Labor Statistics recognizes six categories of benefits: paid leave, supplemental pay, insurance benefits, retirement and savings benefits, legally required benefits, and other benefits. In 1987, roughly three-quarters of compensation costs were in wages and salaries, while just over a quarter were distributed in the form of benefits. Figure 2.6 shows the proportion of employer compensation costs by category. Workers in higher-paid industries received a larger fraction of total compensation in the form of benefits than did workers in lower-paid industries. (18)

The rate of increase in employer costs for employee benefits has fallen substantially since 1980. The bulk of this decrease in growth rate is explained by the fact that the rate of wage and salary growth also fell during this period, and two-thirds of total benefit costs are in categories whose costs are tightly linked to wage costs. In the early 1980s, the cost of benefits was rising more quickly than that of wages, and by 1985 employer cost reduction measures brought the rate of benefits cost growth into line with that of wages. (19)

Pay for holidays, vacations, and sick time are all forms of paid leave and may be considered a job benefit. From 1980 to 1987, the average number of days of paid leave remained unchanged. In 1986, the average number of days of paid vacation for private employees in medium- and large-sized firms was 8.8 after one year of service, 15.8 after 10 years, and 20.6 after 20 years. The number of sick leave days allowed was roughly double the number of vacation days. Employees averaged 10 paid holidays per year.

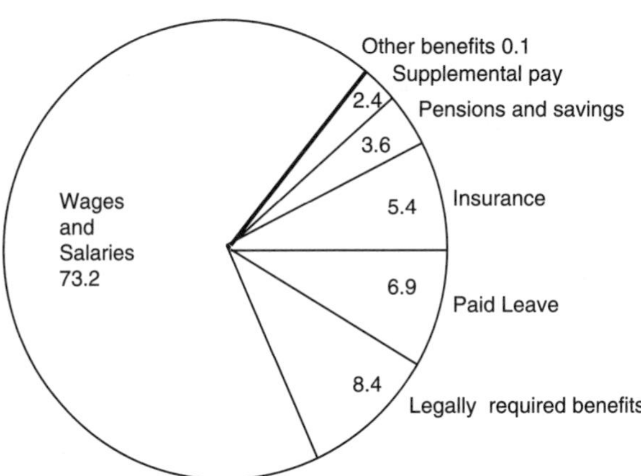

FIGURE 2.6. Relative importance of components of employer costs for compensation in private industry, March 1987. *Source:* Felicia Nathan, "Analyzing Employers' Costs for Wages, Salaries, and Benefits." *Monthly Labor Review*, October 1987: p. 5.

Some employees are also covered by "personal leave" provisions. (19) Today these numbers remain relatively unchanged, placing the United States well behind most other industrialized societies in the amount of annual vacation time given to workers.

Flextime Systems

"Flextime," while not a form of paid leave, is also a benefit that allows flexibility in work hours. Flextime systems allow employees themselves or their departments to set arrival and departure times at work, within certain limits. These plans are generally popular with employees because they allow commuters to avoid rush hour, parents to plan better child care coverage, and the "morning person" to work at the time most effective for them. Forty-four percent of all firms offered flextime in 1987. Small firms (10–49 employees) were more likely than large firms (250 employees or more) to provide flexible hours.

Some jobs provide supplemental pay (overtime) for work done outside the standard work week. Blue-collar employees are far more likely than white-collar or service workers to receive premium pay for weekend or holiday work, or to be paid shift differentials for evening and night work. Lump-sum payments (or stock options and "at-risk pay" for some white-collar employees) are becoming increasingly popular with employers as they seek to reduce fixed costs by shifting to variable costs where possible. Such plans are still not very widespread, as supplemental pay still averages less than 3 percent of total employer costs for compensation. (20, 18)

Along with these benefits, employers may provide life, health, sickness, and accident insurance for their employees, paid for in whole or in part by the company. In recent years health insurance has been the most talked-about benefit in this group, as health care costs, and thus health care insurance premiums, continue their steep climb. The trend toward larger employee contributions to health care coverage is pervasive. In 1980, 72 percent of employees had individual coverage paid for entirely by their employers; the figure fell to just over 50 percent in 1986. During the same period, full payment for family coverage by employers fell from 51 to 35 percent. (19)

Employers have used other methods as well for reducing their costs for employee medical insurance. Many employers have established self-funded plans rather than using commercial plans. Employers also have fine-tuned their reimbursement structure to discourage the use of certain services, and have increased their reliance on Health Maintenance Organizations, which charge a flat fee. (19, 20) There is no indication in the literature of who bears the costs of these changes—to what extent the employee pays, with higher contributions and less complete medical coverage, or whether the health care system itself is also receiving less revenue. Sweeney and Nussbaum (21) report a drop in the proportion of workers covered by employer health plans. At the same time, the labor force has grown considerably, leaving a much larger number of employees uncovered. Most uninsured workers (65 percent) are in service industries, Sweeney and Nussbaum report, and more than two-thirds of uninsured workers are in small businesses with fewer than 100 employees. In addition, fewer than one-quarter of part-time and temporary workers have health coverage through their employer.

Changing Demographics

Many changes have taken place during the last decade with respect to benefits. In addition to those discussed above, which are often motivated at least in part by employers' desires to cut or control costs, some changes are coming about as the result of the changing demographic structure of the work force and employees' interest in changing benefits. One innovation, which affects not the types of benefits offered but rather the way they are administered, is the move by some companies to offer "flexible" benefits. Such plans allow employees to choose a set number of benefits from a larger "menu" of options offered by their employer. This means that in a work force with an increasing number of two-earner families, couples can eliminate redundant coverage and opt instead for the benefits most suited to their circumstances—child care assistance, for example.

Child care is one of the most talked-about benefit issues today. In 1987, more than half of all children under age six had mothers in the work force. (22) The United States is almost alone in the industrialized world in having no national child care program. Families pay for child care themselves, leave the labor force, prevail on relatives or friends for help, or leave older children unattended after school (in this last category there were about a million children in the mid-1980s).

In a 1987 survey of employment establishments, just over 60 percent reported having work-schedule or leave policies for at least some of their employees that would aid in child care. Eleven percent of all establishments actually provided child care benefits or services, while well over a third have no provisions of any sort. Maternity leave is available from some employers, but employees with or without maternity leave may still use other sorts of leave (sick leave, vacation leave, personal leave), if available, to assist with pregnancy and childbearing. Elder care support, like child care assistance, is of growing interest to the working population. In many cases it is handled, by those few companies who offer elder care assistance, in a way similar to child care. Health insurance for retirees has been offered in the past, but with the aging of the population and the increase in health care costs it is arousing more interest. (20) Add to this an increased interest in corporate prevention programs such as "wellness" programs, and one begins to form a picture of a changing and evolving benefits scene in the United States.

CONCLUSIONS

This brief overview has presented some basic information about the changing structure of work in the United States and its impact on the incomes and benefits available to workers. As others have noted, these changes are caused by a number of interrelated phenomena: changes in the the world economy affecting the competitive position of the United States, the pattern of industrial production in a mature capitalist economy, and changing power relations within the U.S. political economy. We detail this information here to provide a characterization of work in the United States in the 1990s that is important, we believe, to our further understanding of the health and welfare of Americans. In the next chapter, we will show in what ways these changes affect the health and safety of workers and the environment and why we think it is important to pay attention to these developments.

NOTES

[1]Figures from Michael Urquhart, "The Employment Shift to Services: Where Did It Come From?" *Monthly Labor Review*, April 1984: p. 16.

[2]In an attempt to clarify the term "services," Gershuny (22), for example, distinguishes three meanings in common use:

a. Service occupations: "people who are not themselves directly involved in the physical manipulation or transformation of materials." That is, service is a type of activity (for example, the absence of manufacturing).

b. Service industries are those which produce intangible products, making anyone employed in these industries a service worker, regardless of the type of activity they perform.

c. A "particular sort of final consumption," in which products are consumed at about the same time that they are produced.

[3]Figure 2.5 shows the Gini Index of inequality. The Gini index ranges from 0 indicating perfect equality to 1 indicating perfect inequality. Between 1970 and 1990 the Gini index for household income grew from .394 to .428, signifying a significant increase in income inequality.

[4]Loveman and Tilly (23), surveying the literature on job quality and on income inequality, find the following points of agreement among those who want to explain increased inequality:

a. "The changing gender composition of employment is not important in explaining the growth of low-earning jobs or increased earnings inequality."

b. "The shifting age distribution has played a small role in explaining the growth of low-earnings jobs."

c. "Shifts in industrial and occupational mix explain little of the increase in earnings inequality, nor do shifts in occupational mix explain the rise in full-time, year-round jobs with low earnings."

d. "The distribution of earnings within industries and occupations has become more unequal. Wage differentials between industries and between occupations have also widened."

e. "Shifts in industrial structure have contributed to the increase in low-earnings employment but evidence is mixed regarding the extent of its contribution." (23, p. 63)

[5]Tilly (14) reports that 2.4 percent of the 2.9 percent increase in part-time jobs between 1969 and 1987 was involuntary part-time work.

[6]Tilly (16) says that part-time workers are from 11 percent to 33 percent less likely to have job-related fringe benefits, while Belous (12) asserts that fewer than half as many part-time workers have health insurance as do full-time workers. Axel (11) cites a Census Bureau study that found that only one of every 15 part-timers was covered by group health insurance.

REFERENCES

1. Bluestone, Barry, and Bennett Harrison. *The Deindustrialization of America: Plant Closing, Community Abandonment, and the Dismantling of Basic Industry*. New York, NY: Basic Books, Inc., 1982.
2. Appelbaum, Eileen, and Peter Albin. "Shifts in Employment, Occupational Structure, and Educational Attainment: 1973, 1979, and 1987." In *Skills, Wages and Productivity in Services: The U.S. National Report to the Center for Educational Research and Innovation*. Edited by Thierry J. Noyelle. Organization for Economic Co-operation and Development, October 1989.
3. Kuhn, Sarah, and Barry Bluestone. "Economic Restructuring and the Female Labor Market: The Impact of Industrial Change on Women." In *Women, Households, and the Economy*. Edited by Lourdes Beneria and Catharine R. Stimpson. New Brunswick, NJ: Rutgers University Press, 1987.
4. U.S. Congress, Office of Technology Assessment. *International Competition in Services*. Washington, DC: U.S. Government Printing Office, July 1987.
5. Quinn, James Brian. "Technology in Services: Past Myths and Future Challenges." In *Technology in Services: Policies for Growth, Trade, and Employment*. Edited by Bruce R. Guile and James Brian Quinn. Washington, DC: National Academy Press, 1988[b].
6. Duchin, Faye. "Role of Services in the U.S. Economy." In *Technology in Services: Policies for Growth, Trade, and Employment*. Edited by Bruce R. Guile and James Brian Quinn. Washington, DC: National Academy Press, 1988.
7. Grubb, W. Norton, and Robert H. Wilson. "Sources of Increasing Inequality in Wages and Salaries, 1960–80." *Monthly Labor Review*, April 1989: pp. 3–13.
8. Tilly, Chris, Barry Bluestone, and Bennett Harrison. "The Reasons for Increasing Wage and Salary Inequality, 1978–1984." University of Massachusetts

at Boston: The John W. McCormack Institute of Public Affairs, February 1987.

9. Noyelle, Thierry J. *Beyond Industrial Dualism: Market and Job Segmentation in the New Economy*. Boulder, CO: Westview Press, 1987.
10. Christopherson, Susan. "Production Organization and Worktime: The Emergence of a Contingent Labor Market." In *Flexible Workstyles: A Look at Contingent Labor*. Women's Bureau, U.S. Department of Labor, 1988.
11. Plewes, Thomas J. "Understanding the Data on Part-Time and Temporary Employment." In *Flexible Workstyles: A Look at Contingent Labor*. Women's Bureau, U.S. Department of Labor, 1988.
12. Belous, Richard S. "How Human Resource Systems Adjust to the Shift Toward Contingent Workers." *Monthly Labor Review*, March 1988: pp. 7–12.
13. Axel, Helen. "Part-Time Employment: Crosscurrents of Change." In *Flexible Workstyles: A Look at Contingent Labor*. Women's Bureau, U.S. Department of Labor, 1988.
14. Tilly, Chris. "Short Hours, Short Shrift: Understanding Part-Time Employment." For the Economic Policy Institute, July 1988.
15. Callaghan, Polly, and Heidi Hartmann. *Contingent Work: A Chart Book on Part-Time and Temporary Employment*, Washington, DC: Economic Policy Institute, 1992.
16. Tilly, Chris. "Two Faces of Part Time Work: Good and Bad Part-Time Jobs in U.S. Service Industries." In *Part Time Work: Opportunity or Dead End?* Edited by Kitty Lund and Barbara Warme. Praeger, 1989.
17. Christopherson, Susan, and Thierry Noyelle. "The Contingent Worker: New Employment and Benefit Options for the Year 2000." Quarterly report, Conservation of Human Resources, Columbia University, October 1988.
18. Nathan, Felicia. "Analyzing Employers' Costs for Wages, Salaries, and Benefits." *Monthly Labor Review*, October 1987: pp. 3–11.
19. Braden, Bradley R. "Increases for Employer Costs for Employee Benefits Dampen Dramatically." *Monthly Labor Review*, July 1988: pp. 3–7.
20. Norwood, Janet. "Measuring the Cost and Incidence of Employee Benefits." *Monthly Labor Review*, August 1988: pp. 3–8.
21. Sweeney, John J., and Karen Nussbaum. *Solutions For the New Work Force: Policies for a New Social Contract*. Cabin John, MD: Seven Locks Press, 1989.
22. Gershuny, Jonathan I. "The Future of Service Employment." In *The Emerging Service Economy*. Edited by Orio Giarini. Oxford, England: Pergamon Press, 1987.
23. Hayghe, Howard V. "Employers and Child Care: What Role Do They Play?" *Monthly Labor Review*, September 1988: pp. 38–44.
24. Loveman, Gary W., and Chris Tilly. "Good Jobs or Bad Jobs: What Does the Evidence Say?" *New England Economic Review*. January/February 1988: pp. 46–65.
25. Nine to Five (9 to 5), National Association of Working Women. *The 9 to 5 National Survey on Women and Stress*. Cleveland, OH: National Association of Working Women, 1984.

3

The Changing Structure of Work in the United States

PART 2—THE IMPLICATIONS FOR HEALTH AND WELFARE

SARAH KUHN
JOHN WOODING

In Chapter 2, we laid out some of the significant features of the rapidly changing structure of work in the United States. We argued that these phenomena derive from the maturation of the American economy, the application of new technologies, intense foreign competition, the rapid expansion of global trade, and extensive international interdependence. These features and their impact are intensified by a new international division of labor and the extensive and rapid movement of capital.[1]

Clearly, the changing character of the U.S. economy has altered the nature and experience of work for many Americans. It also has enormous social and political implications. The decrease in heavy manufacturing, the expansion of the service sector, and the concomitant changes in benefits and unionization have all contributed to changes in the overall welfare of working people. But the structural changes we described have had a profound impact on the complexion of employer–employee relations that goes well beyond the restructuring of work and the availability and extent of benefits such as health coverage and vacation time.

What we will argue here is that these changes have resulted in some clearly identifiable problems that shape the overall welfare of Americans and have important health consequences in and outside of the workplace.

In broad terms, we suggest that technological and economic changes place an increasing burden on workers, including but not limited to new chemical and ergonomic hazards. Further, working now takes up more and more time for those in full-time work, resulting in a significant reduction of leisure time and a negative impact on the quality of life for many working Americans.

We relate these problems not only to the issue of the restructuring of work as a response to global economic changes of the 1970s and 1980s but also to the decreasing ability of many workers to resist these changes. This results, in particular, from the decline in trade union power in the workplace and the dominance of an ideology that promotes the rights of corporations and limits the acceptability and legitimacy of state interventions to protect worker and environmental health. In the final part of this article, we propose a number of areas in which solutions to these problems might be pursued.

CHANGING WORK, CHANGING HAZARDS

The change in work and work organization, reflecting international economic changes, accompanies significant transformations of technology and production. While the impact of technology on work has been discussed extensively elsewhere, the overall effects may be briefly noted here.[2]

The literature of the 1970s was divided between those who saw technology as upgrading jobs and skill requirements on the one hand, and those who believed that technology "deskilled" the majority of jobs. The literature of the 1980s and 1990s, however, delivers a consistent message: work organization, skill requirements, autonomy, and other features of the automated workplace are not predetermined. Instead, they are the outcome of a host of factors such as organization size and culture, the design of hardware and software, organization politics, industry, and so forth. (1–3)

Effects on jobs include "job enrichment" (in which task variety, skill, and often autonomy are increased), "deskilling" (the removal of these features, creating narrower and more repetitive work), and "job enlargement" (an increased workload). (5) Furthermore, the effects on jobs are not static, since the organization and content of work may change significantly before, during, and after the implementation phase of a new technology. Early inefficiencies may give way to increased efficiency, varied and nonroutine work may become tedious as "bugs" are removed from the system, and so forth. (3, 4)

Even though the effect of technology is indeterminate, technology is not neutral. The impact technology and automation will have in the workplace will depend on the social organization of work, the amount of control workers have over their working conditions, the availability of alternative employment, and the extent of competitive economic pressures. Increasing use of technology in the rapidly expanding service sector—particularly in clerical and office work—raises a number of significant issues for the health and autonomy of workers.

We have tended to think of the office, for example, as a safe place, by contrast to the dirty, noisy, and physically hazardous industrial shop floor. This perception is generally accurate, and is borne out by industrial statistics on the incidence of illness and accidents, which show that industries like finance, insurance, and real estate, as well as consumer services, have significantly below-average incidence of illness and injuries. Service sector work is far from free of health hazards, however, although they tend to be less visible and dramatic than those in manufacturing settings. A number of problems, such as eyestrain, musculo-skeletal problems, reproductive hazards, other exposures (including chemicals and other pollutants in the air), noise, poor light, overcrowding, and stress-related illness, have been associated with office work and many other service occupations (for example, janitors and garbage collectors).

Many jobs now generate a host of ergonomic problems as technology and automation force prolonged periods of sitting and standing in front of computers, check-out scanners, automated systems, and the like. The frequent and repetitive motions required by these jobs strain the musculoskeletal system and have created a wide range of repetitive motion injuries.

STRESS

Perhaps the most significant human health consequence of many of the changes we have discussed is the increase in stress felt by most workers across a range of occupations. Stress-related illness is prevalent in the automated workplace, the offices of private and public corporations, small and large retail outlets, and a host of service industries. The state of chronic heightened arousal characteristic of those under stress can lead to lowered immune function, ulcers, cardiovascular problems, anxiety, and depression. Stress on the job is exacerbated as organizations frequently respond to increased competitive pressures by passing this pressure along to employees. The U.S. Office of Technology Assessment predicts that stress-related illness may be the greatest public health problem faced by office workers in the future.[3]

Autonomy on the job bears an important relationship to stress and health outcomes, since the ability of a worker to control her or his work environment can make the difference between experiencing added workload as a challenge or as a source of negative stress. Control over the pace and methods of work is repeatedly cited as key to worker health, satisfaction, and well-being. (1, 3, 5) To offer just a single example: women who reported having a heavy workload and limited job control were at three times greater risk for coronary heart disease than women who had heavy workloads combined with control over their work. (3) Lack of work autonomy is particularly prevalent in lower-level service occupations and is characteristic of much part-time, temporary, and contingent work.

WORKPLACE MONITORING

In addition to the stress problems resulting from work speed-up, longer hours, and increased responsibility, technology used to monitor employee performance adds to the burdens placed on workers. Monitoring of employee performance acts as a constraint on worker autonomy. In the United States, the controversy over employer monitoring of workers is framed as a conflict between the right of employers to manage the enterprise, to act to reduce costs, and to avoid potential liability on the one hand, and employees' right to privacy and individual freedom on the other. (3) Computerized information technology has increased the speed and scope of potential employee monitoring, expanding the detail and comprehensiveness of information available to management about employee performance (2), and created new tensions and conflicts between employer and employee.[4]

Some evidence exists to suggest that the use of electronic monitoring contributes to negative stress and deteriorating quality of work life. Nearly half of those monitored by computer report their work "very stressful" compared to less than one-third of those not so monitored, and excessive supervision or monitoring was found to be most highly correlated with stress-related medical problems. (5) The U.S. Office of Technology Assessment asserts, however, that there is still very little research that allows us to differentiate the effects of electronic monitoring from those of job and equipment design, machine pacing, and so forth. (6)

The impact of such monitoring, in combination with competitive pressures that force management to seek myriad methods to increase worker productivity, not only increases the stress felt by the individual worker but also exacerbates the tensions deriving from alienation at work and job insecurity. (20)

LESS TIME, MORE WORK

As we suggested in the preceding chapter, American workers are now working longer and harder than at any time in the postwar period. Juliet

Schor calculates that the average employed person in the United States today works 163 hours more a year than she or he would have 20 years ago. These extra hours are worked in all industries and by many kinds of workers. (7) This expansion of work not only adds to overall stress in the workplace but clearly reduces the amount of time available for relaxation, family, recreation, and leisure pursuits. Longer work hours for those Americans in full-time employment obviously reduce leisure time, as the demands of family life, work, and self-improvement have burgeoned with two-earner families and a decrease in real incomes since the 1970s. Automation and technological sophistication seem to have resulted not, as much of the sociological literature of the 1950s and 1960s predicted, in increased leisure time for workers, but rather quite the reverse.

WORK, POWER, AND TRADE UNIONS

The restructuring of work has also brought about a restructuring of power in the workplace. The increase in nonunion service-sector employment, heightened economic competition, and the changed demographic characteristics of the work force have significantly reduced the ability of workers to resist demands for speed-up, and monitoring, and to fight against the introduction of technologies that create new occupational hazards. In particular, the decline in union power limits the ability of workers to resist pressures for wage cutbacks and to gain long-term employment security.

Organized labor in the United States is now weaker numerically and politically than at any time in its postwar history. This decline began in the 1970s and increased precipitously by the late 1970s and early 1980s. It is reflected across the whole range of union activity: loss of negotiating strength at the bargaining table, the decrease in union membership, the decline in strike activity, and the vast increase in the number of "concessionary" collective bargaining agreements between unions and industry.

A flavor for the magnitude of this change can be gained by just looking at a few figures. In 1982, 45 percent of all workers covered by collective bargaining agreements had their wages frozen or reduced. In 1983, 37 percent of workers suffered a similar fate; 22 percent in 1984. These concessions dropped off as the economy moved out of recession but they left workers in major industries (particularly textiles, rubber, lumber, metal fabrication, transportation, and mining—in other words, in most of the traditional manufacturing sector), with a significant decrease in their standard of living. Industrial actions for better wages or conditions—strikes—have also declined: the number of work stoppages involving more than one thousand workers reached a peak in 1974 (424 stoppages) and dropped steadily until 1984. (8)

The decline in union membership is also significant. In 1969, unions represented about one quarter of the total labor force. By 1993, union membership was 16.6 million in both public and private sectors (reflecting an increase from 16.4 million in 1992—the first increase in 14 years). In 1993, approximately 11.2 percent of workers (7 million workers) were organized in the private sector; a further 6.6 million were organized in the public sector, the only area experiencing an increase in union membership.[5]

Absolute numbers are even more dramatic in the 10 years between 1979 and 1989, some unions lost enormously. For example, the United Steelworkers (USWA) lost 408,000 members; the Oil, Chemical and Atomic Workers (OCAW) 75,000; and the International Ladies Garment Workers (ILGWU) lost 161,000 members. While some unions (predominantly in the service and public sectors, notably Service Employees International [SEIU] and the American Federation of State, County and Municipal Employees [AFSCME]), made gains, the overall membership picture for American unions remains bleak.

Along with these problems, organized labor has suffered from the antilabor pronouncements of both the Reagan and Bush administrations that have built up a public antagonism to labor unions. Cuts in the "social wage" (unemployment insurance, welfare programs, food stamps, and so forth) have also hit workers in America (the social wage functions to reduce the impact of the reserve army of unemployed labor on employed workers). Without such programs, the competition for jobs becomes more fierce and the bargaining strength of employed workers is further reduced.

The decline of unionization has many causes and would require a long discussion, but one contributing factor has been the changing industrial mix in the U.S. economy. Industries with high levels of unionization are the infrastructural industries (transportation, communication, and public utilities), government, and the goods-producing sector. Service-sector industries, the growth areas of the economy, have low rates of unionization. Furthermore, part-time workers in all industries are rarely organized (just over 7 percent in 1985), while temporary workers and independent contractors are also rarely represented by unions. (9) The opportunity for organizing exists, and some unions and locals are pursuing it actively. If unions can make the adjustments necessary to appeal to service employees, as some have already done, there seems reason to believe that more workers can gain union protection: 40 percent of service workers told a Harris poll that they would welcome a union, as opposed to 25 percent in manufacturing. (10)

Unions and other forms of protection for workers—such as protective labor legislation and a strong social welfare infrastructure—are important because much of the change in work, as well as in the nature of the employment contract in the last decade or so, seems to have taken place at the behest of management rather than as a result of worker preference. (9, 11)

NEW PROBLEMS, NEW SOLUTIONS

In the changing American workplace, much is made of leaner, meaner production. In many workplaces, management has achieved remarkable productive gains and improved the bottom line. That this was possible in the 1980s had much to do with new technologies, increased automation, and capital flight to countries where labor is cheap and government regulation weak. But it also reflects a changing set of beliefs about how management's (and the state's) responsibilities to workers are defined. Stressing the virtues of trickle-down and emphasizing the need of capital to be free from governmental, that is to say, social, constraints (environmental and health and safety regulation in particular, taxation in general) created a new ethic of personal responsibility and reaffirmed the seeming virtues of the market. What little sense of obligation there had been of employer to employee (to provide relatively secure employment over the long term, to provide steadily increasing income, to offer some rudimentary set of benefits) has broken down as the structural pressures for change are played out in the form of ideological assertions of the rights of capital over those of workers and citizens. Attacks on government regulation and organized labor, often spearheaded by business, have exacerbated job instability and insecurity. There is no longer a commitment by the government to full employment.

In spite of this, there have been some positive developments in the structure of work in the last several decades. Workers in the new, service-based jobs are at less risk from life-threatening industrial injuries, and are generally safer at work than those working in heavy industry. Although the glass ceiling and job segregation are still pervasive, there is more economic opportunity for women and some members of minority groups. There is some evidence that jobs are becoming more skilled, and perhaps, therefore, more interesting. To reinforce these positive changes and reduce the negative ones, what might be done to ameliorate the problems we have outlined above? Our review suggests that what is required is the development of a basic political program essential to public health.

Full Employment or a Superfund for Workers?

At one time, this country (like many in western Europe) made a commitment to maintaining a level of full employment for its citizens. That commitment was predicated on government's willingness to use public investment, monetary, and fiscal policies to ensure that most Americans had a job sufficient to provide a minimal standard of living. Since unemployment weakens workers' economic and political voice and, therefore, the ability to protect public health, the political demand for full employment remains a powerful and important strategy.

Since, however, a rapidly changing economy will necessitate redundancies and layoffs, and some types of work and production are dangerous to the health of workers and the environment, it may be preferable to provide programs for workers temporarily facing unemployment. The demand for a Superfund for Workers has merit. It should provide income, training, and security for workers from a wide range of industries and professions.

A National Health Program

Some system of guaranteed publicly funded health care is not only necessary as a replacement for the employer-based, high-cost private system currently in crisis but is indispensable if workers are to regain any kind of security about the availability and quality of health care. Further, a system of guaranteed health care would prevent "job-lock" where workers remain tied to an employer because of health insurance. It also would provide coverage for the increasing number of Americans in part-time and contingent work who have no health benefits.

Labor Law Reform

Organized labor should seek to expand its efforts to build union membership in the new industries and among younger workers, minorities, and women. Unions also should seek support from members of communities affected by plant closings, environmental threats, and the consequences of investment decisions. But the development of traditional unions under the economic and political conditions we have described is and will be extremely difficult. The broader public must be convinced to strengthen labor laws, including OSHA. Effective public health efforts require a strong labor movement.

Reform of Corporate Governance

It is clear that public control over investment decisions must be broadened. The increasing mobility of all forms of capital and the enormous economic and social consequences of capital flight need to be addressed. Workers and communities should not be held hostage to investment decisions. Without some control over investment, security of employment, environmental health, and community identity are constantly threatened. Some form of return to the original meaning of a "corporate charter," reinforcing the idea that corporations are social creations and exist at the pleasure of the public, is necessary, and it must outline the obligations that companies have to workers and communities.[6]

Community-Based Political Organizing

New groups of workers (immigrants and minorities, part-timers, temporary workers, and professionals working independently) are not linked by the commonalty of their work; rather they may be linked by the commonalty of their community. This suggests that organizing efforts for change and struggles to resist the impact on general health and welfare of the restructuring of the U.S. economy might more profitably be pursued in the context of community and neighborhood rather than within the confines of the rapidly eroding "traditional" workplace. Unions need to develop new community-based forms to be responsive to these changed circumstances. The enormously successful Right-to-Know coalitions of local labor organizations, environmental, and public interest groups suggest the shape of a new popular political form of organizing. The rise of the environmental justice movement further suggests that such new forms hold promise for meaningful change.

Reduce Income Inequality

The dramatic income inequality resulting from the changes we describe must be lessened. Many of the physical, mental, and emotional ills we discuss above can arguably be called "diseases of poverty" or "diseases of powerlessness," and can be attributed to increased income inequality.[7] While well-paid professionals are not immune to stress and ergonomic hazards, to name two examples, their ability to avoid these problems is far greater, both because of professionals' greater power in the workplace and because their high income and class status provide them with more options. Furthermore, many other social problems, such as the maldistribution of health care, can arguably be traced to income inequality.

A steeply progressive income tax is one possible measure to reduce income inequality while also helping to fund improved health care for all and other vital social services; this measure does not, however, address the underlying causes of increased inequality.

The American economic landscape has changed considerably. Some of these changes are beneficial. Many of the new industries and jobs are less dangerous than those in the heavy manufacturing sector. There are now more opportunities for women and some minorities to obtain decent work and occupational mobility. Yet accompanying these changes is a decline in security, living standards, and access to meaningful work for many. What we have described in this chapter reflects permanent and structural changes in the economic system. Protecting the health of Americans requires, therefore, a political response—that labor and the public health movement turn to political action.

ACKNOWLEDGMENTS

The authors would like to thank the following for many helpful comments and suggestions in the writing of this article: Barbara Baran, Charles Levenstein, Charles Richardson, and three anonymous reviewers for *New Solutions*.

NOTES

[1]Sarah Kuhn and John Wooding, "The Changing Structure of Work in the United States: Part 1—The Impact on Income and Benefits," Chapter 2, this volume.

[2]See for example Bright (12), Woodward (13), Bell (14), Braverman (15), Gallie (16), and Hirschhorn (18).

[3]A major survey on workplace stress, conducted by 9 to 5, the National Association of Working Women, found that two-thirds of respondents who worked with automated equipment reported that it made their job more interesting and enjoyable than before (while 9 percent said it made their jobs more boring and monotonous). A slight majority (54 percent) said that their work was easier and less stressful than before automation (18). The findings of the U.S. Office of Technology Assessment, consistent with this, are that automation can either increase or decrease workload, depending on a host of factors (3).

[4]The U.S. Office of Technology Assessment (6) identifies three types of electronic monitoring of office activities:

- Computer-based monitoring: the use of a computer to automatically record data about the work activity, such as number of keystrokes, time per transaction, and nature of transactions.
- Service observation: supervisors listen electronically to employee interaction with customers. New technology allows this to be done completely silently, without the knowledge of the worker or customer.
- Telephone call accounting: records origin, destination, and length of telephone calls. This is generally used for reducing telephone costs, sometimes by limiting or prohibiting personal calls.

[5]U.S. Department of Labor, Bureau of Labor Statistics, 1993.

[6]See the ideas suggested in Richard Grossman and Frank T. Adams (19).

[7]For a discussion of the ethical dimension of the health care system and the link between unfairness and income inequality, see Ronald Dworkin, "Will Clinton's Plan Be Fair?" *New York Review of Books*, Jan. 13, 1994 (Vol. 41, Nos. 1 and 2), p. 20.

REFERENCES

1. Karasek, Robert, and T. Theorell. *Healthy Work*. New York: Basic Books, 1990.
2. Hartmann, Heidi I., Robert E. Kraut, and Louise A. Tilly. *Computer Chips and*

Paper Clips: Technology and Women's Employment. Washington, DC: National Academy Press, 1986.

3. U.S. Congress, Office of Technology Assessment. *Automation of America's Offices*. Washington, DC: U.S. Government Printing Office, December 1985.
4. Greenbaum, Joan, Sydney Pullman, and Sharon Szymanski. *Effects of Office Automation on the Public Sector Workforce: Case Study*. U.S. Congress, Office of Technology Assessment, contractor report, April 1985.
5. Nine to Five (9 to 5), National Association of Working Women. *Hidden Victims: Clerical Workers, Automation, and the Changing Economy*. Cleveland, OH: National Association of Working Women, 1985.
6. U.S. Congress, Office of Technology Assessment. *The Electronic Supervisor: New Technology, New Tensions*. Washington, DC: U.S. Government Printing Office, September 1987.
7. Schor, Juliet, *The Overworked American: The Unexpected Decline of Leisure*, New York: Basic Books, 1991.
8. Wren, Robert, "The Decline of American Labor," *Socialist Review*, Vol. 15, Nos. 4 & 5, 1985, pp. 89-101.
9. Appelbaum, Eileen, and Judith Gregory. "Union Responses to Contingent Work: Are Win-Win Outcomes Possible?" In *Flexible Workstyles: A Look at Contingent Labor*. Women's Bureau, U.S. Department of Labor, 1988.
10. Green, James, and Chris Tilly. "Service Unionism: Directions for Organizing." *Labor Law Journal*, August 1987: pp. 486-495.
11. Christopherson, Susan, and Thierry Noyelle. "The Contingent Worker: New Employment and Benefit Options for the Year 2000." Quarterly Report, Conservation of Human Resources, Columbia University, October 1988.
12. Bright, James R. "Does Automation Raise Skill Requirements?" *Harvard Business Review* 36:4, July/August 1958.
13. Woodward, Joan. *Industrial Organization: Theory and Practice*. London, 1965.
14. Bell, Daniel. *The Coming of Postindustrial Society: A Venture in Social Forecasting*. New York, NY: Basic Books, Inc., 1973.
15. Braverman, Harry, *Labor and Monopoly Capital*, New York: Monthly Review Press, 1974.
16. Gallie, Duncan. In *Search of the New Working Class: Automation and Social Integration Within the Capitalist Enterprise*. Cambridge, England: Cambridge University Press, 1978.
17. Hirschhorn, Larry. *Beyond Mechanization: Work and Technology in a Postindustrial Age*. Cambridge, MA: The MIT Press, 1984.
18. Nine to Five (9 to 5), National Association of Working Women. *The 9 to 5 National Survey on Women and Stress*. Cleveland, OH: National Association of Working Women, 1984.
19. Grossman, Richard, and Frank T. Adams,"Taking Care of Business: Citizenship and the Charter of Incorporation," *New Solutions*, Vol. 3, No. 3, pp 7-18.
20. Attewell, Paul. "Big Brother and the Sweatshop: Computer Surveillance in the Automated Office." Sociological Theory 5, Spring 1987: pp. 87-99.

PART II

REGULATION AND PUBLIC POLICY

One of the major ways in which countries control threats to the health of workers and the environment is to impose government regulations. Such regulations typically require business and industry to comply with specific standards for the conditions of work and production. These government standards are ubiquitous—nearly all industrialized countries impose some statutory and regulatory requirements about the conditions of work—but how the regulatory policies are arrived at, what is regulated, and how the regulations are enforced vary widely from country to country.

Standard setting and enforcement in the regulation of the work environment is a controversial and highly politicized process in the United States. How such regulatory policy comes about in a particular country, however, will depend upon the interaction of a variety of forces and factors: historical traditions; assumptions about the role of government; the relative strength of workers' movements and management-employer organizations; political representation for workers' (social democratic or labor parties); and the legal system in place for defending citizens' rights. But the efficacy of such regulations will also depend on specific economic, social, and political conditions that may change over time.

In the United States, the provisions of the Occupational Safety and Health Act of 1970 (OSH Act) created something of a break with past forms of regulation. In giving wide regulatory power to the federal government through the offices of the Occupational Health and Safety Administration (OSHA), it established significant legal protection for most American workers. A bitter lobbying and legislative struggle culminated when President Richard M. Nixon signed the act into law in December 1970. The act was hailed as a triumph for workers, with AFL-CIO President George Meaney pronouncing it "a long step . . . towards a safe and healthy workplace."

Indeed, it did—for the first time—create a federal law mandating that each employer "furnish to each of his employees employment and a place of employment which are free from recognized hazards that are causing or are likely to cause death or serious physical harm to his employees." It also created three agencies, one to set and enforce mandatory health and safety standards (OSHA), one to research occupational hazards (the National Institute for Occupational Safety and Health—NIOSH), and one to review contested enforcement actions (the Occupational Safety and Health Review Commission—OSHRC).

OSHA began operations in 1971. It is part of the Department of Labor and is headed by a presidentially appointed Assistant Secretary for Occupational Safety and Health. OSHA sets health and safety standards, inspects workplaces to ensure compliance, and designs abatement plans and sets fines for employers found to be violating standards. OSHA also provides for worker and employer education and training through grant programs and through publications and advice. In addition, OSHA also partially funds the operation of state agencies operating their own state plans in conformity with the requirements of the 1970 act (states may have their own plans, provided they meet federal OSHA standards). The second agency created by the OSH Act, the National Institute for Occupational Safety and Health, is part of the U.S. Public Health Service, itself part of the Department of Health and Human Services (HHS). NIOSH is headed by a Director appointed by the Secretary of HHS, a cabinet-level appointee. It does research on health and safety hazards and develops criteria documents, ostensibly for use by OSHA in setting standards or evaluating workplace hazards.

In addition, the Occupational Safety and Health Review Commission serves to adjudicate disputes over OSHA enforcement actions. The OSHRC has three members appointed by the President, with the advice and consent of the Senate, who serve for staggered terms of 6 years. It reviews and resolves disputes concerning OSHA citations and penalties.

The legal framework of the 1970 act emphasizes that the law is to be implemented through inspection, enforcement, and standard-setting activities. These inspection and enforcement activities (and the general legal framework) are supplemented or supplanted in some cases by individual state plans. The OSH Act established a means by which states may regulate occupational safety and health subject to federal oversight and approval. Those plans must be at least as effective as federal ones, which means that, in some cases, rare ones, as it turns out, states may actually have *more stringent* requirements than OSHA.

In addition to OSHA and the instrumentalities of state plans, occupational safety and health regulation is provided by a number of other state and federal agencies. For example, all of the following are in some way involved with such regulation; broadly: the Mine Safety and Health Administration (health and safety of coal miners); the Federal Railroad Administration (railroad employees); the Department of Transportation (which includes the Bureau of Motor Carrier Safety, the Federal Aviation Administration, and the U.S. Coast Guard, involved, respectively, with employees working on motor vehicles involved in interstate commerce, airline flight and ground crews, and seamen on Coast Guard-inspected and -certified vessels); the Department of Energy (employees in government-owned, contractor-operated facilities); the Nuclear Regulatory Commission (worker exposed to radiation hazards from materials licensed by the NRC); and the Environmental Protection Agency (workers involved with the mixing and application of pesticides, and farm field workers).

All these agencies operate under a variety of federal regulations and laws under which, effectively, a large part of the American workforce is now covered. They are supplemented further by a variety of state agencies that deal with public, environmental, and occupational health and safety.

This system, in place since 1971, has been under constant attack by both labor and industry—but for very different reasons. The passage and implementation of OSHAct has created unprecedented political conflicts about the role of government in the regulation of private enterprise. Clearly, the activities of OSHA were (and are) highly circumscribed by such factors as: (a) traditional American resistance to government interference (especially at the federal level) in local business and commerce; (b) OSHA's vulnerability to political influence; (c) a relatively weak and fragmented labor movement; and (d) the overall balance of political forces in postwar America.

Part II of this book deals with natural questions relating to government regulation and assesses the efficacy of our current regulatory and public policy attitude toward the work environment. We have subdivided Part II into three sections: (A) Issues in Regulation; (B) Case Studies: Industrial Disease; and (C) The Use of Science in Policymaking.

Section A, "Issues in Regulation," consists of three chapters dealing with specific elements of regulatory policy for the work environment.

Noble's assessment of OSHA after 20 years' experience (Chapter 4) suggests that there may be more substantial obstacles to successful government protection of workers from occupationally induced injury and illness than are initially apparent. Noble posits that the structure of American industry and the particular features of that economy must be closely analyzed to properly understand the success and failures of OSHA's strategies. The intrinsic character of a primarily *profit-based* free enterprise system might be said to militate against proper and adequate provisions for a safe working environment. Moreover, the fact that the key investment decisions are left in the hands of private individuals (namely, owners) means that workers have a powerful incentive to accept employer control. Our economy is heterogeneous: a wide variety of products and services are made and developed by a large number of relatively small firms—although wealth and power gravitate toward large multinational corporations, some of which are foreign-based. This structure of American industry has implications for the regulatory strategies employed to control working conditions. Noble argues that both the strategies that OSHA has pursued and its own institutional structure are poorly suited to fully addressing and resolving the key issues.

In an analysis of the regulation of working conditions in American mines (Chapter 5), Weeks argues that the government's strategies for controlling occupational disease in the mining industry provide an important case study. Here, as elsewhere, mining is among the most dangerous occupations: various safety problems and particularly the disease pneumoconiosis ("black lung") have both killed and disabled thousands of miners. In the United States the Federal Coal Mine Health and Safety Act of 1969 provides a strong basis for the protection of miners. Weeks argues, however, that poor enforcement and a succession of "regulatory reforms" (only arguably so) have, in fact, *weakened* the government's protection of mineworkers from hazardous dust, have reduced medical surveillance techniques, and have diluted compensation programs for black lung. Weeks argues that the original plan developed in the 1969 act was an effective public health commitment to a disease-prevention strategy. That strategy, however, has been seriously weakened by piecemeal reforms and the multiplicity of agencies involved in controlling black lung. Weeks also notes that, as in all issues concerning occupational disease, the commitment to control exposure competes with the needs of production.

Despite the obvious importance of providing meaningful regulatory protection of the work environment, the United States (like other industrialized countries) provides for a system of compensation to workers injured on the job. These workers' compensation regulations have long been a characteristic of the strategy to deal with the problems arising from an unsafe and unhealthy working environment. Beckwith argues that our current system does not provide a clear financial incentive for employers to invest in occupational injury and illness prevention and, additionally, does not provide adequate compensation for those who have been injured. Beckwith (in Chapter 6) shows that the system currently in place not only provides few incentives to make the workplace safer but also transfers most of the cost of injuries either to employees or to employers and institutionalizes them in such a way that injured workers end up collecting only 3 to 5 percent of the funds—making them *doubly* victims of dangerous working conditions. Beckwith also points out that the system does very little to compensate victims of occupational disease.

Sections B and C provide case studies of industrial disease and the use of science in policymaking, respectively. Section B consists of five chapters (7–11) covering issues ranging from multiple chemical sensitivity to occupational stress and sexual harassment. These chapters provide the basis for discussion of particular standards, as in the critiques of benzene and lead standards. For example, Silbergeld, Landrigan, Froines, and Pfeffer, in their critique of the lead standard, illustrate the need for constant scrutiny of permissible exposure levels in light of scientific progress.

The politics of chemical control are further developed in Cullen's chapter on multiple chemical sensitivity, and in Swartz and Clapp's discussion of cancer theories. Landisbergis et al. take the discussion of work environment regulation into the difficult area of occupational stress and work organization. Finally, Kasinsky shows how the endemic problem of sexual harassment should be considered an occupational hazard. Each of these chapters provides an opportunity to discuss the complex ways in which the social, political, and economic factors blend and interact with matters of science.

Finally, in order to properly evaluate the emerging field of risk assessment, we present seven perspectives in Section C (Chapters 12–18) representing critical evaluations of this quasiscience. Silbergeld asks that we look for progressive alternatives, while Ginsburg is highly critical of the politics of risk assessment. Smith, Kelsey, and Christiani review the use of risk assessment in occupational health and give it mixed grades. Wartenberg and Chess assert that thorough systematic review of relevant scientific information can empower communities concerned about environmental risk. O'Brien raises fundamental ethical questions about the technique. And finally, Tarlau discusses industrial hygiene as a tool for work environment improvement and raises important questions about mainstream industrial hygiene practices.

SECTION A

ISSUES IN REGULATION

4

OSHA at 20

REGULATORY STRATEGY AND INSTITUTIONAL STRUCTURE IN THE WORK ENVIRONMENT

CHARLES NOBLE

What government does about occupational health and safety is a rather reliable measure of its class affiliations—whether to the 120 million workers who drive the American economy, or to the substantially smaller number who own and control it. Unfortunately, by this measure, nearly 20 years of federal regulation of the work environment do not inspire much confidence in the political commitments of the American state, or in its capacity to impose public standards of conduct on the corporate economy. The Occupational Safety and Health Administration's (OSHA's) history of glacial standard setting and feeble enforcement has simply left too many workers exposed to too many on-the-job hazards.

For most of OSHA's history, labor and public health activists who have looked to the agency for support in their efforts to improve working conditions have searched for ways to make good on the 90th Congress's statutory commitments to safe and healthy workplaces. They have defended the goals and methods of the Occupational Safety and Health Act of 1970 (OSH Act) that created OSHA, and lobbied Congress on behalf of the agency, or, more accurately, what they believed the agency could, under better management, become. They have asked for more for and from OSHA—more money from Congress, more attention to standard setting and more vigorous enforcement from the agency itself. But until recently, labor and public health activists have not advocated amending the OSH Act. Throughout the 1970s, efforts to revise the law were efforts to restrict or gut it; and these were fought vigorously by the agency's supporters. Ronald

Reagan's administrative deregulation of OSHA in the early 1980s encouraged worker safety and health activists in this tactic, and they met the agency's refusal to set or enforce standards with petitions, lawsuits, and lobbying to force OSHA to take its statutory obligations seriously.

Now, many people sympathetic to the Act's purposes are calling for substantial revision of its provisions. Proposals are pending, for example, that would change the Act's standard-setting process, its penalty structure, and its enforcement scheme. With the OSH Act's 20th anniversary nearly upon us, interest in wholesale reform is likely to quicken.

On balance, this is, I think, a good thing. Whether or not it makes sense to try to amend the OSH Act in the current political climate—a subject to which I will return—OSHA's long-standing problems do suggest that we need to think hard not only about specific agency policies, but also about the general strategy of workplace reform adopted by Congress in 1970. That strategy, variously called "command and control" or "administrative" regulation, relies on federal and state agencies to inspect firms and enforce health and safety standards set by the government. In this chapter, I would like to examine both the record of agency policies and the logic and desirability of this general strategy.

To anticipate my conclusions, I will argue that the political economy in which OSHA intervenes—including, most importantly, the private institutions that organize work, and the public institutions through which any effort to regulate the work environment must pass—creates powerful obstacles to improving working conditions. These must be taken seriously by reformers. I also will argue that the regulatory strategies followed by the Act and the agency do not fully address these obstacles. Finally, I will suggest that, to be politically effective, the movement for workplace regulation must build alliances with community and environmental movements seeking to protect the land, air, and water. The sections that follow take up and elaborate these points. In a final section, I consider recent proposals to revive the agency and to reform the OSH Act, and suggest where reformers might most usefully concentrate their efforts.

THE POLITICAL ECONOMY

Too often, government regulation of business is viewed as a simple, bipolar relationship between the regulatory agency and the regulated interest. But, in fact, any effort to impose public standards of conduct on private firms involves a multitude of public and private institutions, ranging from the several branches of government, to the multinational corporations that dominate the economy. Moreover, relations among these institutions do as much to determine regulatory policy as day-to-day decision making within the agency itself. Thus, in order to evaluate OSHA and the OSH Act, we must first establish the basic features of the regulatory environment in which both intervene—including the structure and organization of the private economy, the state, and the political process. We can then consider how that environment shapes what OSHA does, and the prospects for workplace reform.

The Economy

The obvious place to begin is with the private economy. Three points are essential here. First, because most enterprises are owned and managed as private capitalist firms, work and the labor process are organized to maximize profits. To this end, employers attempt to maximize their control over the workplace—to obtain the greatest possible work effort from their employees. They also attempt to minimize the costs of production. In combination, under most conditions, these two employer strategies will increase the hazards that workers face and discourage efforts by the firm to undertake the sorts of investments, and promote the kinds of work practices, that make work environments safe and healthy. (1)

Second, in a private capitalist economy, efforts by workers to regulate property, redistribute wealth, or exercise direct control over the workplace are likely to frighten investors and encourage managers to shift production to more congenial (that is, more profitable) settings. Thus, because most employees depend on private investment for jobs and incomes, they have powerful incentives to accept employer control over work and investment and concentrate their efforts on the pursuit or, as in the current climate, defense of immediate economic gains, most importantly, wages, hours, medical insurance, and the like. (2) As a consequence, working conditions, and other "quality of worklife" issues, do not often top the agendas of union negotiators, even when they capture the interest of rank-and-file workers.

Third, the United States economy is quite diverse and heterogeneous. Not only do American firms undertake a wide variety of productive activities, a good deal of that economic activity is carried out by relatively small firms. Although wealth and power are concentrated in large, multinational corporations, smaller firms play an often unappreciated role. Nearly one-quarter of all U.S. workers are employed in enterprises that hire fewer than 20 employees, more than half in establishments that employ fewer than 100 (see Table 4.1). More important, nearly nine of 10 private firms employ fewer than 20 employees (see Table 4.2). Because of this diversity and heterogeneity of firms, regulators find it very difficult to fashion detailed standards that apply equally well across the economy, or to forge economy-wide agreements that will satisfy the concerns of all businesses.

TABLE 4.1. Distribution of Employees by Establishment Size, 1986

	All industries	Manufacturing	Construction
Less than 100	55.9%	27.8%	74.8%
Less than 20	26.7%	7.4%	40.3%

Source: U.S. Department of Commerce. Bureau of the Census. *Statistical Abstract of the United States, 1989*. Table No. 859. p. 523.

Moreover, the abundance of cost-conscious small businesses assures a large and easily mobilized constituency against workplace regulation.

The Political System

Obstacles to workplace reform arising from the organization of the private economy are exacerbated by two basic features of the American state. The first has to do with how the government intervenes in the economy. In most important respects, the American state is classically "liberal," that is, government respects and enforces private property rights. When it does intervene in the economy, government does so at arm's length from the production process. Despite the various ways in which public policies regulate and support the corporate sector, public officials do not enjoy control over the primary means of organizing production, most importantly the rate and composition of investment, technological innovation and its diffusion, and the organization of the labor process. As a result, public officials remain dependent, as workers do, on decisions of private firms to invest in the national economy. Predictably, this "structural dependence" of the state on private capital tends to make political leaders wary of capital strikes, and cautious about opposing the demands of corporate capitalists or imposing large costs on them. (3, 4) In addition, public policies are likely to be reactive rather than proactive, as public officials are left searching for ways to clean up after the corporate sector.

The state also is highly "federated," or internally fragmented, and this powerfully shapes what government can do to protect workers. State power is splintered by federalism, which divides government institutions vertically,

TABLE 4.2. Distribution of Establishments by Employee Size, 1986

	All industries	Manufacturing	Construction
Less than 100	88.5%	89.6%	99.0%
Less than 20	87.5%	64.8%	90.0%

Source: U.S. Department of Commerce. Bureau of the Census. *Statistical Abstract of the United States, 1989*. Table No. 859. p. 523.

and by the separation of powers, which redivides them horizontally. This fragmentation causes an extreme (in comparative perspective) decentralization of state power. It also produces a "checks-and-balances" system that, while celebrated in civics courses and political science texts as a bulwark against government tyranny, discourages coordinated, expeditious policymaking. In addition, it provides easy access to conservative, well-organized interests intent on blocking reformist policy initiatives. Finally, this fragmentation of state power makes it difficult to use government to facilitate bargaining among private interests on new policy arrangements, and to enforce whatever arrangements might result from that bargaining. As a result, public policy is biased toward the maintenance of the status quo—in this case, private control over work and the undersupply of healthy and safe workplaces.

The distinctive organization of the American political process further discourages workplace reform. Here, to an unusual degree, interest representation is organized along "pluralist" rather than "corporatist" lines. In corporatist systems, labor and capital are represented by economy-wide, national associations that enjoy state-sanctioned monopoly representation of workers and employers. Representatives of these associations meet with public officials in highly centralized institutional settings where they can bargain face-to-face over issues of common concern. Government then enforces the agreements that result. Clearly, these sorts of associations and institutions do not exist in the United States. Rather, (some) workers and (most) employers are organized into interest groups that bargain in and around the state, not as exclusive representatives of entire sectors, but as more narrowly based associations of like-minded economic actors who agree on a particular organizational form and strategy for advancing their common economic interests. (5, 6)

The absence of corporatist forms of interest representation shapes public policies toward work in two ways. First, it makes direct negotiations among capital, labor, and the state about the work environment difficult. Second, because centralized, encompassing associations do not exist to discipline their own members, negotiators cannot assume that the deals that are reached can be easily enforced.

Efforts to serve workers' interests through the political process are further complicated by the exceptional weakness of the American labor movement. (7) Few American workers are organized. Indeed, as Table 4.3 indicates, the United States exhibits the lowest level of union density (that is, the percentage of workers unionized) of 17 Western capitalist democracies. Moreover, those workers who are organized belong to an unusually decentralized labor movement in which coordination, and therefore, effective collective action, is difficult and, in some cases, illegal. (8)

The American labor movement also is exceptionally weak politically. Alone among workers' movements in the Western capitalist democracies, it does not advance its interests through an independent labor-dominated

TABLE 4.3. Levels of Union Density in the West, 1984–1985

Sweden	95%
Finland	85%
Belgium	77%
Denmark	70%
Norway	61%
Austria	61%
Australia	57%
United Kingdom	52%
Ireland	51%
New Zealand	46%
Italy	45%
West Germany	42%
Netherlands	37%
Canada	37%
Switzerland	35%
France	28%
U.S.	18%

Note: Union density in Belgium and new Zealand is for 1979.
Source: Richard B. Freeman. "Labour Market Institutions and Economic Performance." *Economic Policy*, April 1988, p. 66.

political party, whether Labor, Socialist, Social Democratic, or Communist. Instead, the labor movement works within a highly diffuse Democratic Party where it functions in coalition with a variety of other actors, including corporate interests that can and do veto labor's efforts to dedicate the party to the pursuit of worker demands. (9) In addition, because workers, as voters, are steadily withdrawing from the electoral system, organized labor's clout within the political system continues to wane. (10, 11) This, in turn, weakens whatever incentives political actors might otherwise have to pursue worker interests like occupational safety and health.

In effect, the American political economy rewards employers who undersupply workplace health and safety, and simultaneously discourages public officials from taking effective action to counteract this tendency. These highly federated, liberal-pluralist institutional arrangements do not preclude reform, as the wave of health, safety and environmental laws passed in the 1960s and 1970s makes clear. In fact, bursts of innovative policymaking punctuate American political history as economic, social, and political crises and the movements they give rise to have forced government to respond to new issues and new interests.

There are, however, few guarantees that this sort of "big-bang" approach to policymaking will generate appropriate solutions to the problems targeted. (12) Given the political process described above, it is more reasonable to expect the opposite: new government institutions, forced onto a recalcitrant business community in a moment of crisis, will be partial in their approach, limited in their impact, and effectively resisted once the crisis passes and the movement for reform subsides.

REGULATORY STRATEGY

The tendencies and obstacles described above are systemic in the sense that they result from the structural organization of the economy and the constitutional design of the state. We should, I think, be clear about what this means for reformers. Since the basic institutions of the state and economy are not likely to change dramatically in the near future, effective occupational safety and health regulation is a challenge under the best of circumstances.

But, as health and safety activists already know, there is room to maneuver, even within a liberal-pluralist state and capitalist economy. To maximize the possibilities of change, however, it is important to pay careful attention to how different institutional arrangements shape what government can do about hazards in the workplace, and how different reform strategies affect our ability to regulate the work environment.

Two points are particularly important here. The first is drawn from the comparative record of industrial relations in the capitalist democracies of Western Europe and North America. This record makes clear that occupational safety and health regulation works best when it is part of a broader, more radical strategy of reform than is normally considered feasible or appropriate in the United States. This strategy should aim to build alternative institutions that encourage workers to act collectively in support of their interests, and make sure that those interests are effectively represented in regulatory decision making. In this way, regulators can draw on the interest and expertise of rank-and-file employees and the political support of organized workers (1).

At a minimum, this means establishing worker health and safety committees that empower employees to take part in enforcement at the point of production. It also includes building public decision-making mechanisms that limit the ability of regulated industries to block, delay, or dilute standard setting and enforcement. And, in the long run, it entails building economic institutions that free public officials from their structural dependence on private firms by promoting social control of investment and technological change.

The second point concerns the political bases of support for workplace reform: given the small percentage of American workers who belong to unions, and the organizational weakness of those unions, occupational safety and health reformers must look beyond the labor movement for support. Efforts to require independent worker health and safety committees, or to insulate the standard-setting process from corporate influence, will need the support of the broadest possible political coalitions. Potential allies exist, including, most prominently, the environmental movement, with its own reasons to care about corporate power generally and safety and health in particular.

Unfortunately, neither the OSH Act nor OSHA fully embrace the implications for reform of the political economy in which they intervene. Too little is done to promote collective action by workers at the workplace, to link workplace regulation to a wider environmental agenda, or to free the state from its crippling dependence on the employment and investment decisions of private producers.

The OSH Act

The problem with the OSH Act can be stated simply: it imposes an underdeveloped regulatory process onto an otherwise unchanged economic environment. The results are predictable: OSHA is asked to do much with little, and in the face of intense business opposition.

The OSH Act actually melds two different approaches to workplace safety and health. The first is a rather conventional form of administrative regulation. The Department of Labor is given discretionary power to set and enforce health and safety standards on regulated firms (with the proviso that health standards are to protect workers "to the extent feasible"). These standards are set through a complex, cumbersome, hybrid rulemaking process that combines informal rule making and formal hearings. Standards are to be based on "substantial evidence" (established facts) contained in the rule-making record and are reviewable by the Courts of Appeal. Subject to review by an independent Occupational Safety and Health Review Commission (OSHRC), the Department of Labor is empowered to enforce these standards by inspecting workplaces and imposing civil and criminal penalties on employers who fail to comply. The Act further imposes a "general duty" on employers to provide healthy and safe workplaces (though the precise meaning of this provision is disputed). States can, with approval from OSHA, operate their own plans, if those plans provide employees with the same sorts of rights and protection afforded to them by the Act. Finally, the Act places the National Institute for Occupational Safety and Health (NIOSH) in the Department of Health, Education and Welfare (now Health and Human Services) to provide OSHA with the science it needs to regulate.

The Act also contains a second approach to regulation built around the idea of citizen rights to participate in rule making and enforcement. Rights of this sort were novel at the time of the Act's passage and reflect the influence of the public interest movement, which believed that corporate power over the government could be limited by empowering citizens to participate more directly in administrative decision making. In the case of the OSH Act, workers have rights to participate in standard setting, workplace inspections, and the monitoring of hazards; to have access to information about hazards and agency findings; to appeal agency rulings to the courts; and to be protected from employer discrimination for exercising these rights.

There is much to recommend here. The Act's language about workers' health and safety rights is strong and substantive. On paper, the OSH Act is one of the crowning reform achievements of the 1960s. The creation of a single federal agency both to set and enforce workplace standards simplifies administration. Placement of the agency within an existing executive department makes clear who is ultimately responsible for regulatory policy. The Act's attempt to institutionalize the public interest movement's belief in the virtues of citizen participation also makes sense.

Other statutory provisions, however, unnecessarily complicate life for OSHA, or create too many avenues for regulated interests to block agency action. The "substantial evidence" test for standards, for example, is stricter than the "arbitrary and capricious" standard usually applied to informal rule making and thereby invites overly broad judicial review. OSHRC, in turn, gives employers an additional opportunity to challenge agency enforcement decisions. Historically, differences between OSHA and OSHRC have created confusion and uncertainty about the scope of the agency's powers. State-level administration is difficult to monitor and, consequently, protects weak state programs from federal preemption. (13) The placement of NIOSH in the Department of Health and Human Services makes coordination between these two agencies difficult. The provisions for citizen participation in enforcement, in turn, do not do enough to assure that participation will be effective, or that the agency will benefit from the knowledge and interest of rank-and-file workers. The Act leaves too much to the initiative of individual workers or unions, and creates too few in-plant mechanisms to facilitate worker activity around health and safety. The Act does not require the establishment of worker-controlled, in-plant health and safety committees; or give those committees access to company data; or guarantee compensation for workers who participate in inspections; or protect workers' right to refuse hazardous work; or give workers a role in the selection of company health and safety professionals. All of these things should be mandatory if workers are expected to take part in the regulatory process.

Finally, the OSH Act suffers from what isn't said. Put simply, it takes the existing organization of the economy for granted: private control of investment and technology is assumed. This is not, of course, surprising. The political strategies of both organized labor and the public interest movement converged at this point in 1970. Both movements were firmly wedded to liberal capitalist values. The former sought, wherever possible, to work with rather than challenge "enlightened" private employers. Where changes were thought necessary, the labor movement preferred to use the collective-bargaining process to win more for workers within the existing political economy, not fundamentally change it. Government was expected to stimulate the private economy and provide public employment, public goods, and social benefits, not exercise control over production. (14) For their part, public interest reformers were more interested in (and successful at) reforming the state rather than the economy. When they focused on economic affairs, they sought to promote competition rather than socialize large-scale production so that it might serve public ends. Moreover, employers, as a class, would have fiercely and, given the balance of political forces at the time, probably successfully defended these "managerial prerogatives." In any case, the OSH Act does nothing to address the state's dependence on privately organized, market-driven economic growth.

OSHA

For most of its history, OSHA has done little to overcome these obstacles. Under the Nixon and Ford administrations, a hostile White House and, for the most part, indifferent administrators left the agency adrift, reluctant to set or enforce standards, let alone explore the more radical possibilities inherent in the Act.

The agency did find new life during the Carter administration. Then, OSHA embarked on an ambitious campaign to set new health standards and to defend them against demands by White House economic advisors that they be subject to cost–benefit analyses. Measures also were taken to involve workers in enforcement, including the walkaround pay rule. Agency programs also encouraged coalitional efforts in support of worker health and safety: New Directions provided critically needed public funding for the COSH movement, while agency head Eula Bingham worked with other federal health, safety, and environmental regulators in the Interagency Regulatory Liaison Group to develop a coordinated approach to economy-wide risk reduction.

These efforts met with some success. OSHA's more aggressive posture produced new, strict standards for lead, cotton dust, and benzene, for example, as well as path-breaking proposals for a generic carcinogens standard, and an economy-wide hazard communication standard. Enforcement also intensified during these years, as did compliance, though the latter is hard to measure with much certainty. (1, 15)

But the institutional and political obstacles to effective regulation outlined above took their toll. Business opposition to a revitalized OSHA quickly registered in the White House, which, in response, sent the Council on Wage and Price Stability, the Regulatory Analysis Review Group, and OMB to bring the agency around (16). In the benzene case, the Supreme Court imposed new cost–benefit criteria on rule making that further limited OSHA's powers. And while organized labor tried to defend the agency from further blows, it could not save it from the mounting corporate assault on social regulation.

The debacle of the Reagan years is, of course, well known. Managed by administrators hostile to worker protection, OSHA reversed course entirely. Budgets were slashed and enforcement staff cut. Innovative rules and programs were cancelled. The cancer policy was abandoned and the agency returned to case-by-case rule making for toxics. The Department of Labor sided with employers who sought Supreme Court relief from compliance with expensive standards. OMB imposed elaborate and often impossible hurdles for new standards, while old standards were revised to reflect the new business orientation. Worker complaints to the agency, facilitated by Bingham, were discouraged. The walkaround pay rule was revoked. Employers were encouraged to develop "voluntary" programs that might substitute for agency enforcement. And, finally, "safe" workplaces were exempted entirely from on-site inspections.

In sum, surveying the last two decades, it is, I think, fair to say that apart from its activities during the late 1970s, OSHA's policies have done little to promote workplace reform, or maximize the potential of the OSH Act. Case-by-case standard setting has been painfully slow, and unable to provide workers with up-to-date protection from hazardous substances. Major rule makings have taken several years to complete; some cases, such as the rule covering access to employee medical and exposure records, have dragged on far longer. Moreover, OMB's efforts to minimize OSHA's impact on employer costs and "paperwork" have institutionalized the representation of employer economic interests in the heart of the rule-making process.

Enforcement has been equally compromised. The White House's indifference to occupational safety and health, and Congress's failure to promote the agency's mission, have left OSHA with too few inspectors to do its job. Agency policies, in turn, have resulted in fines that are generally too small to matter to most employers. Efforts to promote voluntary compliance by employers have accomplished little, while efforts to involve workers in enforcement have been powerfully and, for the most part, successfully resisted. The policy of exempting "safe" employers from full-scale inspections based on inspection of their records produced a major scandal when it was learned that many companies had ignored and misinterpreted record-keeping requirements, or, more ominously, falsified their records. The "blockbuster" fines levied in response, while good public relations, are not likely to save the agency's enforcement effort. (17) Finally, OSHA has done such a poor job of monitoring state plans that we have no way of knowing for sure whether the states are actually implementing the Act.

RETHINKING REFORM

Clearly, substantial changes in the way the United States regulates work must be made if OSHA is to be more than a symbol of the state's fading commitment to occupational health and safety. But what exactly needs to be done? How much must be changed before OSHA can truly protect American workers? Will changes in agency policies suffice, or must the Act itself be amended? The analysis above suggests that we need reforms that, at a minimum, devolve as much enforcement as possible to workers, limit business influence over standard setting, and encourage alliances among workers and other movements interested in health, safety, and the environment, and the issue of corporate power itself. The simplest way to proceed, then, is to consider the degree to which various proposals to reform the agency and the Act meet these goals.

The relevant reform proposals can be divided into three groups. The first includes changes in OSHA policy that can be made within the provisions of the OSH Act; the second, changes that require alterations in the Act; the third, proposals that target not the agency and the Act, but the reform strategies of the labor and environmental movements.

OSHA Policies

Although the OSH Act does not tackle the major institutional barriers to reform outlined in the first section, it does provide OSHA with significant powers. An alert and aggressive administration committed to occupational safety and health could use these to promote workplace reform.

There are clearly things that OSHA could do, within the confines of the OSH Act, to expedite standard setting. It could set and maintain a tight regulatory agenda that specifies "regulatory plans" indicating how and when the government intends to regulate chemicals on its priority lists. It could expand its use of generic (for example industry-wide, multichemical, and work-practice) standards rather than rely on case-by-case standard setting. (18) Established health standards could be kept current by regularly updating the list tables of permissible exposure levels. The Department of Labor, in turn, could resist efforts by the Office of Management and Budget to block, dilute, and delay standards that business opposes.

None of these measures will eliminate business opposition. To the contrary, generic standards easily could exacerbate the problem by forcing entire industries onto the defensive. Advocates of regulatory negotiation (commonly called "regneg") claim this method of rule making can overcome business opposition by depoliticizing standard setting; it appeals for this reason to both neoliberal and neoconservative critics of OSHA and the labor movement who believe that both have mistakenly adopted an adversarial approach to regulation (19, 20). But negotiated rule making will not serve workers well unless they also enjoy the background political and institutional conditions that guarantee that labor's demands will be taken seriously at the bargaining table. Corporatist critics of American pluralism forget that corporatist negotiations benefit workers only where the labor movement is strong and the state well developed. As I argued above, neither condition holds in the United States.

OSHA also has means, within the terms of the Act, to force employers to think harder about the impact of work on workers' well-being. These include medical removal protection for employees at special risk, aggressive enforcement of the general duty clause, and full disclosure of workplace hazards to all affected employees. OSHA also could encourage cooperation among the various health and safety movements by working closely with consumer and environmental regulatory agencies, fully embracing the right-to-know movement, and incorporating a broad spectrum of health, safety, and environmental groups into worker training and education programs.

With regard to enforcement, OSHA could target its limited resources more carefully and do more to involve workers in the inspection process. At a minimum, OSHA could restore the walkaround pay rule and require employers to compensate employees who participate in agency inspections. OSHA compliance officers could work more closely with worker and union representatives during inspections. Employees could be made more aware of, and better trained to deal with, the hazards they face. (Supplementary "right-to-know" legislation, like the High Risk Occupational Disease Notification Bill now pending, could be adopted without amending the Act.) Worker rights to respond to unsafe working conditions, including the right to refuse hazardous work, could be enhanced. Employers could be encouraged (though probably not required) to establish labor-management health and safety committees with real powers, including the powers to inspect work sites regularly, to monitor health and safety programs and accident investigations, to hire and fire company health and safety personnel, to supervise the health and safety training of employees, and to shut down imminently dangerous equipment or processes (21). Finally, the agency could provide generous subsidies to encourage the further development of the COSH movement.

As with standard setting, however, there are limits to how far these reforms can be taken within the current legislative framework. Targeted enforcement is a rational response to limited resources. But it is triage, and will never solve the underlying problem: the government is unlikely ever to have enough inspectors to protect 120 million employees at work in more than four million establishments. Getting more and better information to workers will undoubtedly make them more aware of occupational hazards, but it won't guarantee that they will be able to do anything about it. Rights to act on that information, and organizational resources to support these rights, will still be needed if workers are to use what they know to win better working conditions. The OSH Act does not guarantee these rights, and the labor movement is unlikely to secure them through collective bargaining. Too few workers are organized. Those who are preoccupied with economic issues and their unions must devote increasingly scarce resources to defending the very principle of union organization in a bitterly anti-union environment.

The OSH Act

We are left, then, to consider changes in the OSH Act itself. These too can be sorted into three categories. The first has to do with what can be called "organizational" changes, that is, changes in the way the federal government organizes its institutional capacity to intervene in the workplace. These include, most prominently, eliminating OSHRC; creating an "independent" Risk Analysis Board to identify carcinogens and establish permissible exposure levels; creating new, industry-specific OSHAs; relocating NIOSH in the Department of Labor; and limiting the OSH Act's ability to preempt state-level occupational safety and health activities.

Each of these proposals could be discussed at length, but I think it is more useful to apply a general principle here that is drawn from the earlier discussion: unless there is a compelling case against it, we should prefer organizational reforms that concentrate rather than fragment government's regulatory power. Fragmentation will simply divide labor's attention, encourage interest-group divisions among workers, and provide greater access to opponents of strict regulation. On this rule, it makes sense to abolish OSHRC (its functions can be carried out by administrative law judges within the Department of Labor) and relocate NIOSH. But the further division of OSHA, including separating out its risk-analysis functions, or weakening the principle of federal preemption, should be resisted.

Health and safety activists may find the last point objectionable. After all, opponents of effective workplace regulation have attempted to use the principle of federal preemption in court against the state-level right-to-know movement and against efforts by state and local prosecutors to file criminal charges against employers who negligently kill workers whom they employ. Clearly, on these issues, the states have been in the vanguard, and are likely to remain so as long as the White House is controlled by the Right. In this context, recent court victories by the labor and public health movements against preemption of state and local right-to-know laws by OSHA's often weaker hazard communication standard are an important tactical victory. But it is important to remember the general argument against the fragmentation of state power: a divided state makes it difficult for workers to use their collective power to promote social change. It also is important to keep in mind that, historically, state and local governments have generally dragged behind federal regulators. Fortunately, recent litigation over state criminal prosecution of employers suggests a solution: legislation should establish the rights and protection in the OSH Act as a floor below which state and local officials may not descend, rather than a ceiling on their protective activities.

An additional set of proposed changes in the OSH Act has to do with standard-setting procedures. All are desirable, though none is a panacea. OSHA should, for example, be freed from the need to pass OMB's economic impact and paperwork burden tests. In a different society, these tests might make some sense. In this political economy, they surreptitiously and undemocratically insert industry interests into the heart of the regulatory process. OSHA also should be relieved of the burden of satisfying the "substantial evidence" test for its standards. Given the scientific uncertainty that surrounds health and safety, agency policy decisions are not reducible to "factual" questions, and the courts should not hold them to this standard. As the Administrative Conference of the United States has recommended, agency judgments should be subject to a standard of "arbitrariness," and the factual premises on which they are based should be held to a standard of "substantial support in the administrative record as a whole" (22).

Action-forcing, statutory commitments to hazard reduction also are desirable. The public interest reformers generally sought such limits on agency discretion, believing that detailed statutory mandates, including legislatively determined timetables, would prevent regulators from using administrative discretion to protect corporate interests. The Clean Air and Clean Water Acts contain them; the OSH Act does not (except when the agency uses a Standards Advisory Committee to develop rules). A variety of mechanisms is available, but the least intrusive would require the agency to set legally binding timetables or deadlines for each rule-making proceeding.

A final set of proposed changes in the OSH Act has to do with the enforcement process. These range from straightforward (and overdue) proposals to raise the Act's penalty structure to the more radical idea of a government-guaranteed, worker "Right-to-Act." The latter is, I think, essential to any effort to regulate working conditions. According to this idea, government would devolve enforcement powers to workers themselves through statutorily mandated, company-financed health and safety committees. These committees would enjoy extensive rights and powers, including the authority to inspect workplaces, to organize in-plant health and safety activities, to respond to imminent-danger situations, and even to cite employers for violations. Moreover, these committees would not depend on the existence of unions. All workplaces of any substantial size would be required to have them. In effect, OSHA would deputize the working class it purports to protect. They would enforce the Act. The advantages of this sort of scheme are, I think, obvious. By involving workers in enforcement, it would heighten their interest in, and contribution to, workplace health and safety. By deputizing employees, it would lighten the agency's administrative burden. And by providing a permanent, subsidized forum for cooperation among workers around shared interests, it would facilitate working-class collective action on a wide range of issues.

Does this mean, then, that it is time for health and safety activists to push for amending the OSH Act? Clearly, as the discussion above shows, the Act could be strengthened and 20 years of experience have left us with lots of good ideas about how this might be done. But good ideas do not regularly win the day in Congress. And the coalition of business-oriented Republicans and conservative Democrats that now runs Congress is not likely to support radical, worker-oriented reforms. To be sure, Congressional liberals have been able to use their committee chairmanships to block repeated conservative efforts to weaken the Act. But in the present political climate, it is unlikely that they can do much to strengthen it. Rather, the chances are good that conservatives will take advantage of any effort to amend the Act to introduce amendments that meet employers' long-standing objections to the federal regulation of working conditions. And the chances are good that they will succeed. Legislative reform should not be forgotten; indeed, the issues raised in this piece suggest that it is a precondition for truly effective regulation. But in this situation, it seems more rational to concentrate on strengthening OSHA, particularly its enforcement activities, and reorienting the way health and safety activists approach the entire problem of workplace reform.

Political Strategies

To this end, it is important to consider the political strategies of the movements in support of workplace safety and health. These too will have to change if government regulation of the workplace is to be effective. Put simply, the movement for workplace safety and health will need to become simultaneously more left and more Green.

By left, I mean simply that health and safety activists must recognize how important society's general capacity to shape democratically capital accumulation is to any effort to impose behavioral standards on business enterprises. As I argued in the first section of this chapter, privately organized production discourages workplace safety and health in two fundamental ways. First, market-based competition among profit-maximizing firms leads to the undersupply of safe and healthy workplaces. Second, the state's structural dependence on private capital accumulation leads public officials to take the interests of business more seriously than the interests of labor. Not because public officials are bad people, or misguided, or confused; but because both they and the citizens they claim to represent are so dependent on the private economy. Any effort to challenge the profits or managerial prerogatives of private employers—as effective workplace safety and health regulation does—is bound to come up against this structural limit. Social control of capital accumulation will not guarantee occupational safety and health, as the record of existing state socialist societies indicates. But it is, I think, a necessary precondition.

This sort of state role is unlikely, in turn, without a realigned political process—one in which one or another of the two political parties is committed to the representation of those working-class interests now ignored by the Republicans and Democrats. Independent political representation for workers is especially critical in the case of occupational safety and health, where employees face off against employers.

No matter how well organized, however, workers organized along traditional class lines will not be able to accomplish the sorts of changes imagined here. In the late twentieth century, the working class is too small to carry this burden; it must work with progressives in the middle class if it is to be politically effective. In any case, the working class should not carry this burden alone. After all, occupational safety and health is one facet of a larger set of interrelated environmental problems that arise at the intersection of work, economy, technology, and nature.

The movement for workplace reform, then, must also be Green. Concretely, this means close attention to the sorts of issues and organizational forms that will help mobilize the environmental movement in support of occupational safety and health, and mobilize workers in support of environmentalism. The long-standing conflicts between these two movements are familiar and need not be rehearsed here. Suffice it to say that changes will have to occur on both sides if these are to be overcome. At the rank-and-file level, many things can be done, including directly involving labor-oriented groups in environmental programs such as asbestos removal, and bringing environmental groups sensitive to labor's concerns into a broadly conceived workplace enforcement program. There also are issues, such as new source performance rules for toxic substances, that appeal to both movements and provide a natural bridge. Of course, many labor and environmental activists are already working together, particularly in local struggles. Ultimately, however, both movements will have to work more closely at the national level, and pay closer attention to the sorts of institutional considerations outlined at the outset of this chapter, if either hopes to succeed.

ACKNOWLEDGMENTS

I would like to thank Charles Levenstein, Joan Parker, Linda Rudolph, and John Wooding for comments on an earlier draft of this chapter.

NOTES

1. Noble, C. *Liberalism at Work: The Rise and Fall of OSHA*. Philadelphia: Temple University Press, 1986.

2. Przeworski, A. *Capitalism and Social Democracy*. Cambridge: Cambridge University Press, 1985.

3. Block, F. "The Ruling Class Does Not Rule: Notes on the Marxist Theory of the State." *Socialist Revolution* 33 (1977): 6–28.

4. Lindblom, C. *Politics and Markets: The World's Political-Economic Systems*. New York: Basic Books, 1977.

5. Schmitter, P.C. "Still the Century of Corporatism?" *Review of Politics*. 36(1974): 85–131.

6. Salisbury, R. "Why There Is No Corporatism in the United States." In Schmitter, P.C. and G. Lehmbruch, eds. *Trends Toward Corporatist Intermediation*. Beverly Hills, CA: Sage Publications, 1979.

7. Rogers, J. "Divide and Conquer: Further 'Reflections on the Distinctive Character of American Labor Laws.'" *Wisconsin Law Review*. 1990.

8. Wallerstein, M. "Union Growth from the Unions' Perspective." Paper delivered at the 1987 Annual Meeting of the American Political Science Association. Chicago. September 1987.

9. Ferguson, T. and J. Rogers. *Right Turn: The Decline of the Democrats and the Future of American Politics*. New York: Hill & Wang, 1986.

10. Burnham, W.D. "The United States: The Politics of Heterogeneity." In Rose, R. ed. *Electoral Behavior: A Comparative Handbook*. New York: Free Press, 1974.

11. Burnham, W.D. *The Current Crisis in American Politics*. New York: Oxford, 1982.

12. Leman, C. "Patterns of Policy Development: Social Security in the United States and Canada." *Public Policy*. 25 (1977): 261-91.

13. U.S. General Accounting Office. *OSHA's Monitoring and Evaluation of State Programs*. Washington, DC: GPO, 1988.

14. Greenstone, J.D. *Labor in American Politics*. New York: Alfred A. Knopf, 1969.

15. U.S. Office of Technology Assessment. *Preventing Illness and Injury in the Workplace*. Washington DC: GPO, 1985.

16. Noble, C. "Economic Theory in Practice: White House Review of OSHA Health Standards." In Fischer, F. and J. Forester, eds. *Confronting Values in Policy Analysis: The Politics of Criteria*. Beverly Hills, CA: Sage Publications, 1987.

17. National Safe Workplace Institute. *Unintended Consequences: The Failure of OSHA's Megafine Strategy*. Chicago, 1989.

18. McGarity, T.O. and S.A. Shapiro. *OSHA Rulemaking: Regulatory Alternatives and Legislative Reform*. Report to the Administrative Conference of the United States. Washington, DC, September 16, 1987.

19. Bacow, L. *Bargaining for Job Safety and Health*. Cambridge: MIT Press, 1980.

20. Mendeloff, J. *The Dilemma of Toxic Substances Regulation: How Overregulation Causes Underregulation at OSHA*. Cambridge: MIT Press, 1988.

21. Ruttenberg, R. *The Role of Labor-Management Committees in Safeguarding Worker Safety and Health*. U.S. Department of Labor. Bureau of Labor-Management Relations and Cooperative Programs. Washington, DC, 1989.

22. Administrative Conference of the United States. Recommendation 87-10. Regulation by the Occupational Safety and Health Administration. Washington, DC, 1987.

5

Undermining the Protections for Coal Miners

JAMES L. WEEKS

The coal mining industry provides us with numerous precedents in occupational health and safety and in labor policies and practices. It was the mining industry in which the federal government first intervened to investigate occupational safety hazards with the creation of the Bureau of Mines in 1910. More than half a century later, with the passage of the Federal Coal Mine Health and Safety Act of 1969, it was the mining industry in which the federal government first intervened (a) to compensate victims of pneumoconiosis and (b) to prevent occupational injuries, fatalities, and disease.

The purpose of this chapter is to describe the strategy for controlling occupational disease among miners as an application of classical public health strategies for control of disease and injury in the occupational environment. This strategy was initiated with the 1969 Coal Mine Health and Safety Act but is now being threatened by piecemeal reform of health and safety regulations affecting miners. It contains elements of primary, secondary, and tertiary methods of disease control.

It is important to describe control of occupational disease in the mining industry as the application of public health strategies for two reasons. First, it illustrates the effective application of disease control models in the occupational setting. Public health is the profession most clearly committed to disease prevention and has the most clearly developed strategies to accomplish prevention. When these strategies are implemented, they are effective. The strategy described below is multifaceted and involves several separate government agencies. One consequence of this is that changes in one policy may seem minor when seen in isolation. But when numerous and seemingly isolated minor changes are made, the cumulative effects can seriously weaken the overall disease control effort. This will become apparent in this analysis. Second, clarity of the commitment is important for control of occupational disease because there are strong competing interests at work. The workplace is where wealth, income, social status, and many social relations are created. These primary functions of work create barriers to defining and achieving public health objectives of disease control. On the job, top priority, rhetorical commitments notwithstanding, is given to production; everything else is second. Moreover, the workplace, even for government employees, is a protected private place, thus imposing a barrier to the practice of "public" health. Public health is an antidote to these barriers. Eula Bingham, at the end of her tenure as Assistant Secretary of Labor for Occupational Safety and Health, succinctly challenged the public health community when she said, "The field of occupational health and safety has been taken over by lawyers and economists. Public health professionals have dropped the ball."

The coal mining industry is a convenient and useful industry in which to describe efforts for occupational disease control. With a workforce of about 74,000 miners in underground mines in 1989 (down from 143,000 in 1980), the number of workers exposed to a significant hazard is sufficiently large, and there has been relatively aggressive government intervention to control disease and to keep records. The characteristics of this industry plus the historical experience give us, in effect, something analagous to a large-scale clinical trial. (1) Moreover, most workers are represented by one union, the United Mine Workers of America, making it easier for workers to participate in disease control efforts.

A PUBLIC HEALTH APPROACH TO OCCUPATIONAL DISEASE CONTROL

Control of occupational lung diseases among coal miners can be described using conventional public health terminology of primary, secondary, and tertiary prevention. Primary prevention is concerned with preventing the initiation of disease by controlling exposure to its causes. Secondary prevention is concerned with preventing disease complications early in its natural history by early diagnosis and intervention. Tertiary prevention is concerned with preventing and compensating permanent disability. In the case of occupational lung diseases commonly called black lung, primary prevention corresponds to dust control, secondary prevention to medical surveillance leading to job transfer, and tertiary prevention to the black lung compensation program. Like other progressive programs promoted (not always successfully) by the federal government during the 1960s and '70s—

such as a greater commitment to educate our children, to protect minorities, women, the handicapped, and the elderly from discrimination, to assist the poor with housing, food, and health care, and to protect the environment—the public health plan to control black lung has been threatened.

Different agencies of the federal government have undertaken different tasks. Dust control was managed first by the Mining Enforcement Safety Administration in the Bureau of Mines, then later by the Mine Safety and Health Administration (MSHA) in the Department of Labor (DOL). Responsibility for medical surveillance (the chest X-ray surveillance program) was taken by the National Institute for Occupational Safety and Health (NIOSH) in the Department of Health and Human Services. The black lung compensation program was first administered by the Social Security Administration and then was assigned to the Employment Standards Administration in DOL (physically and organizationally separated from MSHA). Black lung claims are adjudicated by the independent Benefits Review Board.

The cumulative impact of these efforts has been effective—a very important fact. Dust concentrations and the occurrence of coal workers' pneumoconiosis (CWP) have both been reduced. In 1969, the average concentration of respirable dust among continuous miner operators was about 6 mg/m^3 (2); by 1989, this average had been reduced to 1.0 mg/m^3. Based on chest X-ray surveillance data, the prevalence of category 2 or higher CWP among miners with 30 or more years' experience in underground mining was about 10 percent in 1970 and about 4 percent in 1981. For category 1 CWP, however, the corresponding prevalence rates were 12 percent in 1970 and 22 percent in 1980. These data are subject to significant limitations. (3) Moreover, more than $20 billion in compensation payments have been transferred to otherwise destitute coal mining families and their communities. In 1988, there were a total of 242,000 current beneficiaries of this program, including 57,000 ex-miners, 127,000 miners' widows, and 57,000 dependents. This is down from a peak of nearly 490,000 beneficiaries in 1974. (4)

The United Mine Workers of America provides important additional support for disease control. It was the mining industry's hazardous conditions which in part gave birth to the union in 1890. And it was union members' agitation that provided necessary political support for the Coal Mine Health and Safety Act of 1969. (5)

INSIDIOUS PROCESS WEAKENS THE PLAN

But now, each aspect of the plan to control occupational disease among miners is being weakened by an insidious process that is made easier because the federal agencies are separate. The momentum within each of the agencies has led each of them into independent paths. There is little consciousness that they form a complementary or coherent public health approach to disease control. Heads of these agencies do not meet as one body, there is no exchange of personnel from one office to another, and they do not develop policies to complement each other. Thus divided, the plan to control black lung is easier to dismantle.

To appreciate this issue, I describe below each of the elements of the plan to control black lung and proposed and implemented policy changes in each area. In this description, there is a double message. On the one hand, the details of public policy demonstrate a thorough and systematic approach to disease control. On the other, changes and proposed changes in policy reveal an equally thorough attempt to dismantle this disease control plan and to go back on commitments to prevent disease.

PRIMARY PREVENTION

Primary prevention of occupational lung diseases among miners is accomplished by controlling exposure to dust. This is acheived by three regulatory devices: performance standards, specification standards, and a permit system. The performance standard is a permissible exposure limit (PEL) to respirable coal mine dust of 2.0 mg/m^3 averaged over a work shift. A mine operator (the term used throughout the mining industry to refer to the company that operates the mine and employs miners) is required to meet this PEL at all times. It is a statutory limit set by the U.S. Congress. Therefore, changing it requires an act of Congress.

Performance of mine operators in meeting the PEL is extensively monitored. Each year, approximately 120,000 separate dust measurements are made at surface and underground mines in the United States. About 100,000 are made by mine operators, the remainder by MSHA. NIOSH and the Bureau of Mines also make a small number of measurements for research purposes. Mine operators are required to take five samples every two months on the dustiest job at each mining section where coal is being cut. If the average of these five exceeds 2.0 mg/m^3, the operator is issued a citation for noncompliance and is required to make changes and submit five additional samples until the dust concentration is reduced. MSHA is required to sample each mining section at least once each year to ensure that the correct job is being sampled and to measure exposure to respirable quartz dust.

Rare Instance of Hazard Surveillance

The benefits of this monitoring system are many. With it, it is possible to identify those mines that have problems controlling dust exposure and to take action before disease can become established. It is one of the few instances of ongoing hazard surveillance in the United States. Performance of the industry as a whole can be monitored and enforcement policies evaluated. The data also can be used for epidemiologic research.

The system of employer monitoring has some significant weaknesses, however. Since exposure above the PEL brings an automatic citation, mine operators have obvious disincentives to report exposure accurately. Manifestations of this problem are many and include at least two instances of fraud, much anecdotal information and mistrust of the program by miners, and evidence of systematic bias. (6)

The performance standard, however, is insufficient. Like some other work sites (for example, construction, logging, agriculture), a mine is a constantly changing environment. Current dust sampling methods are relatively cumbersome. Each time a sampling pump is used, its batteries must be charged, the flow rate must be calibrated, a filter must be weighed before and after sampling, and the sampling unit must be placed properly and frequently monitored during the work shift. It takes approximately two weeks from the date a sample is taken to when the operator receives the results. During that time, many things change: a mining crew advances farther into the coal seam, bits on the cutting head are changed, ventilation controls have to be repositioned, and water sprays must be readjusted. If overexposure is documented, the causes may be difficult to identify and, therefore, difficult to control. Because of these problems, specification standards also exist.

Specification standards are regulations that require an employer to take specific steps that are intended to achieve a certain result. Underground mine ventilation regulations require mine operators to maintain a certain amount of air flow (in cubic feet per minute, or CFM) at each mining section, to regulate that flow so that it sweeps across the working face (that is, the part of the coal seam from which coal is being cut), and to split incoming clean air into separate intake mine entries so that each mining section receives its own supply.

MSHA (and state mining departments) also regulates by requiring operators to obtain a permit to mine. In order to obtain a permit, an operator must submit a mine plan, including a ventilation plan, to MSHA and have it approved.

MSHA Proposes Changes in Regulations

In early 1988, MSHA proposed changes in existing underground coal mine ventilation regulations. (7) This is part of the Labor Department's review and reform of many mining regulations. In order to appreciate both the scope of specification standards and the plans to change them, it is necessary to describe and explain them in some detail.

Among the changes are proposals (a) to relax a requirement for mean entry air velocity; (b) to permit line brattices to be more than 10 feet from the working face; (c) to permit more than one set of mining equipment to be on a single split of air; and (d) to permit air coursed over a conveyor belt to be used to ventilate a working face. Technical terms are explained below.

a. Mean entry velocity (MEV) is the average velocity of fresh air entering a working section, where coal is being cut. Fresh air is essential for keeping dust away from the worker operating the continuous mining machine (the continuous miner operator), diluting it and diverting it into the return airway. Current regulations require the MEV to be at least 60 feet per minute (fpm)—at the threshold of perceptible motion. The proposal would permit an MEV lower than 60 fpm provided the dust concentraton is kept below 2.0 mg/m^3. Measuring the MEV is relatively easy and results are available immediately, unlike results of measuring exposure to respirable dust.

b. A line brattice is a roof-to-floor curtain at the side of a working place whose purpose is to direct air flow so that fresh air is provided to the continuous miner operator, dust is kept at the face, dust-laden air is diverted into the return airway that goes out of the mine. The closer the brattice is to the face, the better it performs its job. Conventional practice, supported by research by the Bureau of Mines, is to keep the brattice within 10 feet of the face. (8) Like measuring the MEV, monitoring compliance is relatively easy; all one needs is a 10-foot pole. Installing and maintaining the brattice potentially exposes miners to working under an unsupported roof, but there are procedures that enable this work to be done under supported roof.

The proposed change would permit the brattice to be more than 10 feet from the face, provided that dust concentrations are kept below 2.0 mg/m^3. Maintaining the brattice is labor intensive; each time the continuous miner cuts coal from the face, the brattice has to be moved. The purpose of the proposed change is to permit operators some flexibility (by allowing less frequent advances of the brattice) and to keep miners under supported roof.

The central problem with this proposed standard is that it replaces an easily monitored specification standard (that is, to keep the brattice within 10 feet of the face) with one that depends on a relatively complicated test of performance (keeping the dust concentration below 2.0 mg/m^3). In so doing, not only does it potentially expose miners to more dust, but it also makes it more difficult for miners on the job to monitor their own workplace.

Equipment on Separate Splits of Air

c. As regulations now read, each set of mining equipment—for example, a continuous miner, shuttle cars, and roof bolter—is required to be on a separate split of air. As fresh air is drawn into the mine, it is split many times to ventilate the several working places of a mine. Hence, the noun "split." If two or more sets of mining equipment were working on the same split, the miners downwind of one would receive the dust generated by the other. For this reason, existing regulations require each set to be on a separate split of air. In the interest of giving mine operators greater flexibility in having backup equipment available, MSHA proposes to eliminate this requirement provided that the mine operator does not operate both sets of equipment simultaneously.

This is a mere paper protection; enforcing it is a problem. Mine operators must be concerned about costs and productivity. It does not pay to allow expensive equipment to be idle. Therefore, the incentive to operate both sets of equipment simultaneously is strong and, moreover, such a practice is easy to conceal from a mine inspector. As soon as an inspector reaches a mine, the appearance of compliance is as easy as flipping a switch. Although mine inspectors are required to inspect underground mines four times each year, significantly more inspection time than OSHA inspectors, most of the time, they are absent.

d. The use of a mine entry that also contains a conveyor belt to bring air to a working face is more complicated. (9) Conveyor belts are and have been a source of fires from belt breakdown. If the mine entry with the belt (the belt entry) is used to bring air to the working face, smoke from a fire would contaminate the face and imperil workers. For this and other reasons, the Mine Act requires that belt entries must be ventilated separately from intake entries and may not be used to ventilate a working face.

Belt entries are also a source of dust. Dust can be generated when one belt dumps onto another or, if air velocity is sufficient, when air re-entrains dust, just as the wind picks up dust from a dirt road. Therefore, for dust-control purposes, it is not good practice to use dust-laden air to ventilate a working place. Alternative sources of cleaner air exist.

In recent years, many mine operators, using a provision in the Mine Act, have petitioned to use belt entries as a source of air to ventilate the working place. They claim that in some mines roof control problems are so severe that it is too dangerous to open another entry to ventilate a face. Opening another entry also is costly and reduces the amount of recoverable coal. Operators claim that they can control the risk of fires and dust associated with using belt air to ventilate the face. MSHA has been persuaded by these arguments and has granted most petitions and now seeks to change the regulations accordingly.

Combined Effect: Greater Exposure

Each of these changes—to reduce MEV, to permit the brattice to be farther away from the face, to permit more than one set of equipment on a single split of air, and to use belt air to ventilate the face—appears narrowly technical. But the combined effect is to relax ventilation controls and, in turn, to expose miners to more dust. These changes also embrace subtle but important changes in philosophy. Greater emphasis is placed on performance standards and less on specification standards. Furthermore, the Mine Act states, in the section describing ventilation requirements, "[T]he Secretary [of Labor] shall prescribe the minimum velocity and quantity of air reaching each working face of each coal mine in order to render harmless and carry away methane and other explosive gases and to reduce the level of respirable dust to the *lowest attainable level.* (10, emphasis added)

The proposed changes in ventilation regulations permit operating up to the PEL of 2.0 mg/m^3 and implicitly reject the commitment to operate at the "lowest attainable level."

These proposed changes were so troublesome to the United Mine Workers (not only because of their effects on dust but also because of their effect on methane and fire control) that considerable resources were devoted to providing comments for public hearings and about 10,000 working miners participated in public hearings.

To summarize, the plan to control dust described above proceeded not only from setting and enforcing a permissible exposure limit (PEL) of 2.0 mg/m^3, but also from standards specifying in detail how each mine should be ventilated. Performance standards, specification standards, and a permit system complemented each other. In the constantly changing and hazardous environment of an underground coal mine, complementary regulatory tools are appropriate and effective. But MSHA's proposed ventilation regulations threaten to undo this plan of primary disease prevention.

AN EXAMPLE OF SECONDARY PREVENTION

The medical surveillance program is an example of secondary prevention. Under the 1969 and 1977 Mine Acts, every mine operator is required to provide underground miners with medical examinations. MSHA and NIOSH are required to develop regulations to implement this statutory requirement. The intent was simple: miners with early stages of disease caused by exposure to coal mine dust should be identified and their exposure reduced in order to halt disease progression. Miners with simple coal workers' pneumoconiosis (CWP) are at significantly greater risk of developing complicated CWP. (11)

Chest X-ray films are to be made according to protocol at a facility certified by NIOSH, and they are to be interpreted by readers also certified by NIOSH. Miners may have films made at any certified facility, but most often they go to a facility chosen by the mine operator. Miners with positive films are then given the opportunity to request transfer to a job with a lower concentration of respirable dust (1.0 mg/m^3 or less). Since CWP is irreversible, miners may exercise or waive this right at any time.

In principle, work at a job with reduced exposure to dust can be achieved by lowering dust concentration to less than 1.0 mg/m^3 at the miner's existing job. In practice, however, it almost always means that the miner is transferred to another job. Thus, the program is commonly referred to as the transfer program.

Though the statutory language providing for medical examinations is broad, NIOSH and MSHA require operators only to offer a chest X-ray examination, thus adopting a narrow and restrictive definition of disease that limits the ability of the medical surveillance program to control disease. The Mine Act states:

> [A]ny miner who, in the judgment of the Secretary of Health and Human Services based upon such reading [of a chest X-ray film] *or other medical examinations*, shows evidence of the development of pneumoconiosis shall be afforded the option of transferring from his position to another position in any area of the mine, for such period or periods as may be necessary to prevent further development of disease. (12, emphasis added)

Policy Meddles with Social Fabric of Work

The transfer program is a form of medical removal protection. Wage retention is provided. This type of policy (which exists for only a few workers under OSHA's jurisdiction), though promoted on relatively sound public health grounds, is not without its problems. It relies on identifying and transferring potentially sick workers rather than on identifying and controlling job hazards. It implicitly tolerates health hazards. It also meddles with the social fabric of work by limiting a worker's employability (since pneumoconiosis is irreversible) and wages and by limiting an employer's latitude in assigning work. It also limits the applicability of job seniority as a criterion for promotion, one of the most strongly defended rights in labor–management contracts. (13)

Nevertheless, job transfer should remain one tool of disease control because there are instances when it is appropriate. In coal mining, however, this program has failed in practice. A few numbers document this failure. In the first round of chest X-rays, in the early 1970s, more than 50 percent of eligible miners participated. In the third round, in the early '80s, participation was 32 percent. (14) Of the 9,138 miners with positive films who had become eligible by 1987 to exercise their transfer rights, only 2,119 had done so. In 1985, there were only 199 miners who had exercised their rights. Moreover, it is not known whether the program succeeds at arresting the progression of CWP.

There are several prominent problems with the medical surveillance program. By limiting the required medical examinations to the chest X-ray, miners are subject to differences of opinion among radiologists about the interpretation of films. Miners with lung disease caused by mine dust but not visible on a chest film do not receive rights to transfer that Congress intended. Neither NIOSH nor MSHA has addressed these problems. (15)

A second problem is that too many important decisions have been delegated to mine operators—for example, notifying miners of the plan, selection of a facility to make the X-ray film, and selection of a radiologist to interpret it—leaving many miners with the perception that this program is a mine operators' program and not one managed and directed according to the Mine Act. Finally, there is a section of the National Bituminous Coal Wage Agreement that provides miners who have positive films the option of exercising "superseniority" in bidding on low-dust jobs. Under this provision, miners with positive films may obtain additional wage protection, but only by compromising rights conferred by job seniority. It also results in many miners waiving their transfer rights for long periods (thus continuing to be exposed) while they wait for a desirable job to be posted.

The Worst Record of Success

It should not be surprising then that, with these problems, confusion about the purpose and mechanics of the program is common. Of the three elements of disease control, the medical monitoring program has the worst record of success.

THE BLACK LUNG PROGRAM

The Black Lung Compensation Program has been the subject of much controversy. Since its beginning in 1970, more than $20 billion, significantly more than originally estimated, has been paid out in benefits. (16) The cost of prior years' of neglect was obvious. The magnitude of payments may make this program one of the rare instances in which compensation payments, when they are combined with other preventive efforts, provide a real incentive to prevent disease.

But there are other important lessons to emphasize. In the original plan, there are imaginative administrative solutions to some common medical problems. The central medical problem is that it is difficult to precisely identify causes of chronic diseases that develop over many years of exposure and that have both occupational and nonoccupational causes, such as the collection of chronic lung diseases called black lung.

The black lung compensation program solved or addressed this problem with a series of presumptions that can be used to establish eligibility for benefits. These presumptions have status similar to the presumption of innocence in criminal law. Until the 1981 amendments to the Act, there were seven such presumptions that claimants could invoke in trying to establish eligibility for compensation.

For example, under the black lung program, a claimant must establish three things: (a) that he (or she) has pneumoconiosis as defined in the statute and regulations, (b) that it was caused by coal mining employment, and (c) that it is disabling. Since conditions caused by exposure to coal mine dust, such as chronic bronchitis and emphysema, also are caused by other exposures, primarily cigarette smoking, and are clinically indistinguishable from each other, proving causality for individual claimants is difficult.

Shifting the Burden of Proof

The black lung program handled this problem in the following manner: if a claimant had pneumoconiosis, and he had worked 10 years or more as a miner, he was allowed to invoke a rebuttable presumption that his condition was caused by coal mine employment. When this presumption was invoked, the burden of proof was shifted from the miner who had to prove what caused the disease to (most often) the mine operator who had to rebut the presumption and prove employment as a miner did not cause it.

This, along with other presumptions, was a practical solution to the problem of establishing causality. When the Act was amended in 1981, this and other presumptions were removed, which made it more difficult for claimants to establish eligibility. With the right to invoke this presumption lost, miners now have to prove that employment in the mines caused illness. For a disease with many manifestations and other possible causes, this is often difficult.

The loss of presumptions seems analagous to the changes in policy in civil rights law that have been rationalized by invoking the legal rhetoric of strict constructionism. Under this philosophy, a person must show conscious intent to discriminate in order to win a case. Under the black lung program without presumptions, a claimant must prove causality. This approach promotes the myth that science can provide certain answers and systematically ignores inherent ambiguity in determining exactly what caused a person's disease.

Reform of the black lung program has not stopped with removing presumptions. Efforts have been made to narrow the definition of compensable disease to further restrict eligibility. (17) The conventional medical definition of pneumoconiosis has been restricted in practice to the observation of opacities that can be seen with a chest X-ray. The conventional medical dogma is that disability only occurs with complicated pneumoconiosis and not with the simple form. The statutory and regulatory definition of pneumoconiosis is broader, and encompasses all the health effects of inhaling excessive amounts of coal mine dust—pneumoconiosis, bronchitis, emphysema, impaired diffusion, and shortness of breath. These definitions obviously differ. If compensation were limited only to those persons with complicated CWP, many miners disabled by breathing coal mine dust but whose X-rays did not show complicated CWP would not be eligible for compensation. (18)

The broader term is preferred for two reasons. First, chest X-ray readers disagree about the interpretation of films, including films interpreted as complicated CWP by one reader. This makes eligibility for compensation dependent on the fickleness of radiologists. Second, coal mine dust causes disabling chest diseases in addition to CWP. (19)

Reforms Undermining Disease Control

A coherent and effective public health plan to control occupational lung disease among miners was created with the Coal Mine Health and Safety Act of 1969 but is now being weakened by piecemeal reforms. Proposed changes in mine ventilation regulations would have the effect of relaxing appropriately complementary ventilation controls designed to limit exposure to respirable dust. The medical monitoring program designed to halt the progression of disease has been only halfheartedly implemented. And an imaginative solution to establishing eligibility for compensation for miners has been significantly restricted.

The combined effect of these changes is that we may lose primary, secondary, and tertiary public health controls on this disease. Miners no doubt will experience the consequences in years to come.

ACKNOWLEDGMENT

Portions of this chapter were presented at a meeting of the Coalition of Black Lung and Respiratory Disease Clinics, Lexington, Kentucky, on October 31, 1989.

NOTES

1. Weeks, JL. "Is Regulation Effective? A Case Study of Underground Coal Mining." *Ann New York Acad. Sci.* 572 (1989):189–199.

2. Jacobson M., Parobeck PS., and Hughes ME. *Effects of Coal Mine Health and Safety Act of 1969 on Respirable Dust Concentrations in Selected Underground Coal Mines*. U.S. Bureau of Mines, IC 8536. Washington, DC: U.S. Government Printing Office, 1971.

3. Social Security Administration. Social Security Bulletin. Table M-31. 52 (April, 1989):52.

4. Althouse RB. "Ten Years' Experience with the Coal Workers; Health Surveillance Program, 1970-1981," *Morb. Mort. Wkly Rept 34* (1985): 33SS-37SS.

5. Derickson A. "Down Solid: The Origin and Development of the Black Lung Insurgency." *J. Pub. Health Policy 4* (1983):25-44.

6. Boden LI and Gold M. "The Accuracy of Self-Reported Regulatory Data: The Case of Coal Mine Dust." *Am. J. Indus. Med. 6* (1984):427-440.

7. Department of Labor, Mine Safety and Health Administration. 30 CFR Part 75, Safety Standards for Underground Coal Mine Ventilation; Proposed Rule. Fed Reg 53(1988):2382-2424.

8. Bureau of Mines. *Guidelines for Dust Control in Small Underground Coal Mines*. Washington, D.C., U.S. Department of the Interior, 1987.

9. Mine Safety and Health Administration. *Belt Entry Ventilation Review: Report of Findings and Recommendations*. Washington, D.C., U.S. Department of Labor, 1989.

10. Sec. 303 (b) of 30 USC 801 et seq.

11. Hurley JF, Alexander WP, Hazledine DJ, Jacobsen M, and Maclaren WM. "Exposure to Respirable Coalmine Dust and Incidence of Progressive Massive Fibrosis." *Br J Indus Med 44*(1987):661-672.

12. Sec. 203 (b) of 30 USC 801 et seq.

13. Freeman RB and Medoff JL. *What Do Unions Do?* New York: Basic Books, 1984.

14. Hoffman J. "X-Ray Surveillance and Miner Transfer Program: Efforts to Prevent Progression of Coal Workers' Pneumoconiosis." *Ann. Am. Conf. Governmental Industrial Hygienists 14:*(1986)293-297.

15. Speiler EA. "Can Coal Miners Escape Black Lung? An Analysis of the Coal Miner Job Transfer Program and its Implications for Occupational Medical Removal Protection Programs." *WV Law Review 91* (1989):775-816.

16. Nase JP. "The Surprising Cost of Benefits: The Legislative History of the Federal Black Lung Program." *J. Min. Law Policy 4*(1988-89):277-319.

17. Renzetti AD and Richman SI. *Current Medical Methods in Diagnosing Coal Workers' Pneumoconiosis and Review of the Medical and Legal Definitions of Related Impairment and Disability*. Washington, DC, U.S. Department of Labor, 1984.

18. Weeks JL and Wagner GH. "Compensation for Occupational Disease with Multiple Causes: The Case of Coal Miners' Respiratory Diseases." *Am. J. Public Health 76*(1986):58-61.

19. Marine WM, Gurr D, and Jacobsen M. "Clinically Important Respiratory Effects of Dust Exposure and Smoking in British Coal Miners." *Am. Rev. Respir. Dis. 137*(1988):106-112.

6

The Myth of Injury Prevention Incentives in Workers' Compensation Insurance

GEOFFREY C. BECKWITH

This chapter seeks to examine workers' compensation insurance in an effort to determine whether the system provides effective financial incentives for employers to invest in the prevention of occupational injuries and illnesses.

It has long been argued that financial incentives exist that compel employers to reduce injuries and illnesses in the workplace, and that the workers' compensation system provides a classic example of such an incentive.

The typical belief is that firms ultimately pay for the costs of occupational injuries and illnesses through workers' compensation insurance. It follows that these payments, either in the form of insurance premiums or direct expenditures to disabled employees, should therefore increase in accordance with the number and severity of injuries, and with the value of the outgoing benefits to said workers. In the case of declining injuries and illnesses, employers' workers' compensation costs should decrease accordingly.

The savings in workers' compensation costs that may be achieved through the prevention of injuries and illnesses represents the financial incentive. Firms need only tally up their potential savings in order to make decisions about whether to invest in measures to reduce occupational hazards.

However, there is significant disagreement regarding this simple concept. The workers' compensation system is very complex. Benefits to injured workers, actual insurance premium costs, insurance premium rates, types of rate-making structures, the problems workers face in proving the work-relatedness of occupational diseases, and the difficulty employers face in quantifying nonevents (avoided injuries) represent major variables in the system.

These factors combine to create a tangled mass of market forces and government regulation that impact large and small businesses and risky and less-risky lines of work in dramatically different ways.

The result is a workers' compensation system that does not provide a clear financial incentive for employers to invest in occupational injury and illness prevention initiatives.

BACKGROUND

The workers' compensation system is in place to assure workers of compensation in the event of injuries or illnesses caused by work, and to limit employers' liability for work-related accidents.

Workers' compensation programs exist in every state in the nation. Quite simply, these laws provide compensation to workers who have suffered occupational injuries or illnesses. This compensation comes in the form of medical care, cash benefits (for lost wages, temporary and permanent disabilities, and death benefits), and rehabilitative services. (1)

Workers' compensation laws have created a type of no-fault insurance. Employers finance the cost of these benefits to workers through insurance premiums, payments into government-administered funds (in some states), or, in the case of a small percentage of very large firms, through self-insurance. Except in the instance of self-insureds (companies large enough to establish full programs of their own), the actual payments to workers are made by insurance carriers or the state funds.

Employers' insurance payments are administered on a state-by-state basis, and generally follow the same process. A complex system of more than 600 separate business classes has been established, based on the riskiness and likelihood of losses in each category. Employers pay premiums based on each $100 of payroll in each job category, as set by the rate for each class. This is called the manual rate. A company's premium may be adjusted above or below the manual rate in limited instances.

Benefit levels for injured workers are established by statute. Premium rates are established by the state insurance commissioner, based on rate requests submitted by the insurance industry.

Claims fall into five categories: medical benefits, temporary total disability, permanent partial disability, permanent total disability, and death benefits. (2) Benefits may come in the form of weekly or one-time "lump sum" payments. In general, the benefits are set at a level lower than the employee's original salary to provide an incentive for workers to get back on the job as soon as possible, and to provide a disincentive for fraud. (3)

Claims made by employees may be disputed by employers based on questions concerning the work-relatedness or severity of the disability or injury. Disagreements are settled by hearings adjudicated by the state workers' compensation agency.

HISTORY

Workers' compensation programs were established in the United States during the early part of this century. Following the example of European countries, most notably Great Britain and Germany, the United States began to deal with the troublesome issue of compensating the victims of occupational injuries.

The industrialization of the nineteenth century had caused an explosion of workplace injuries, and the legal systems regulating employer liability made it very difficult for those injured to receive financial relief. (4)

While it is obvious that workers were being injured well before the early 1900s, "when the problem of industrial injuries arose in the first half of the nineteenth century, the first reaction of both the British and the American courts was to protect industry." (5)

This protection came in the form of common law defenses against employee claims. The difficulties of going to trial, proving negligence and overcoming common law defenses that allowed employers to blame co-workers (the fellow-servant rule), suggest that the employee had assumed the risk as a part of the job (assumed risk), or suggest that the employee had contributed to the cause of the injury (contributory negligence), resulted in an inadequate framework for dealing fairly with injured workers. (6)

The "sheer inhumanity of this system had reached such proportions that it could no longer be borne, especially since it was obvious that industry was no longer so feeble and immature as to require such subsidies" (7) by the turn of the century.

Compensation for injured workers commenced only after long, expensive court actions, and it was clear that many of those who deserved to win in court did not. In Europe, the response was to create systems whereby employers were obliged to provide compensation and cover most, if not all, of the costs. (8)

The United States was behind Europe in moving to address the inequities of the legal system. The lack of statistics from the late 1800s and early 1900s makes it difficult to evaluate the problem in human terms. However, "reports from the period indicate that from 6 to 30 percent of industrial accidents were compensated by employers." (9)

If accurate, these figures suggest that between 70 and 94 percent of injured workers were unable to secure some compensation. Obviously, legitimate concern about the lack of care and fair treatment for disabled workers had developed.

The result of this concern was the adoption of workers' compensation laws in this country. Between 1911 and 1920, nearly all states enacted some type of workers' compensation system (10), a virtual explosion of statutes throughout the nation. Interestingly, there was no effort to adopt a nationwide law for all workers. Each state crafted its own system.

The exceptions are for maritime and railroad workers, who are covered by federal statutes. (11) These laws provide a system that allows for tort action by employees, and

> efforts to provide workers' compensation coverage to railroad workers and seamen have been resisted by their unions. They feel they are better off under the modified tort system which provides a larger percentage of actual damages than does workers' compensation *when the employer is proven to be negligent*. (12, emphasis added)

Of course, with workers' compensation there is no requirement that the employer be found negligent. The premise behind the new laws was a compromise, or trade-off, between management and labor:

> A bargain was struck between the rights of employers and employees: Employers became liable to payment without a finding of fault in return for a guarantee that their liability would be restricted to statutorily set benefit amounts; employees gave up the right to sue their employers for full damages in return for a guarantee of prompt, certain compensation for work-related injuries. (13)

In order to finance the system, employers paid insurance premiums into a pool that was drawn from by workers qualifying for benefits. (14) Injured employees could access the benefit money more quickly, although the compensation did not fully replace lost wages, and was otherwise limited to out-of-pocket expenses (primarily medical costs). No awards for pain and suffering could be obtained.

In essence, workers agreed to limit their right to seek full compensation in exchange for a more certain expectation of a lower level of benefits, and employers agreed to amend their ability to block most claims in exchange for a more stable and predictable system of compensation that avoided potentially high awards to injured workers.

The workers' compensation system was one of the first systems of no-fault insurance. At first, constitutional questions led most states to balk at requiring all firms to participate, and the laws were voluntary. But later, after the courts had clarified the issue, almost every state established a compulsory system.

By the mid-1970s, nearly 90 percent of the nation's workforce was covered by workers' compensation. (15) Workers not falling under the programs included some farm workers, domestic workers, self-employed individuals, and small groups of public employees. (16) All states have workers' compensation laws, and all but a handful of these statutes are compulsory for employers.

Today, workers' compensation continues to stand as a vital issue in American business and labor relations. Of course, labor remains interested in the system because workers continue to be injured on the job. Business is interested because costs have increased dramatically over the years.

The premiums that U.S. employers pay for workers' compensation insurance doubled during the 1980s, and benefits paid to insured workers tripled during that time as a result of increases in medical costs, cash benefits, and the number of injuries and illnesses. (17) Estimates place the total cost of workers' compensation insurance at more than $50 billion in 1990. (18)

As the workers' compensation system has become more financially burdensome, businesses are now paying close attention to ways of reducing costs. These efforts have primarily focused on reducing benefits, avoiding payments, restructuring administrative procedures to reduce the number of eligible claims, controlling medical costs, lowering employees' lost time from the job, and preventing injuries. These strategies are all on the table as viable alternatives for business to use when seeking to reduce costs.

It is difficult to judge the extent of employers' commitment to injury and illness prevention as a means of controlling costs. If one examines the trends during the 1980s, however, it is clear that there has been a steady increase in the rate and number of workdays lost to occupational injuries (see Figure 6.1). It is important to note that the progression of the injury rate has occurred while costs have been skyrocketting (see Figure 6.2). Obviously, these two factors are related. But as costs go up, in theory the financial incentives to prevent injuries, if they really exist, should become even more apparent to employers.

One can argue that, at best, employers have focused on ineffective strategies to promote prevention, or, at worst, they have failed to seriously consider prevention as a corporate priority.

POLICY

Because workers' compensation has developed as a loosely connected federation of statutes throughout the nation, no one statement of policy,

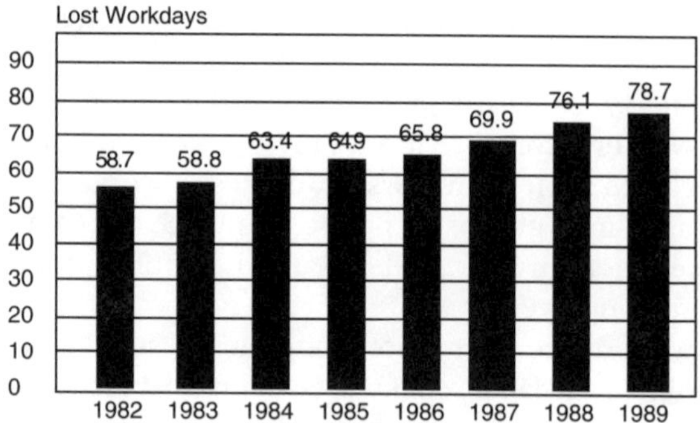

FIGURE 6.1. Occupational injuries lost workdays per 100 full-time private sector workers. *Source:* National Council on Compensation Insurance. *Issues Report, 1991*, Boca Raton, FL: NCCI, 1991, p. 34.

or rationale, serves as a common denominator in all 50 state laws (plus the District of Columbia).

However, a typical statement of policy can be found in the preamble to California's 1917 Workmen's Compensation, Insurance and Safety Act:

> A complete system of workmen's compensation contains adequate provision for the comfort, health, safety and general welfare of any and all employees and those dependent on them for support to the extent of relieving from the consequences of any injuries incurred by employees in the course of their employment, irrespective of the fault of any party; also full provision for *securing safety in places of employment*, full provision for such medical . . . treatment as is requisite to cure and relieve from the effects of such injury, full provision for adequate insurance coverage against the liability to pay or furnish compensation, full provision for regulating such insurance coverage . . . full provision . . . that *the administration* of this act shall accomplish

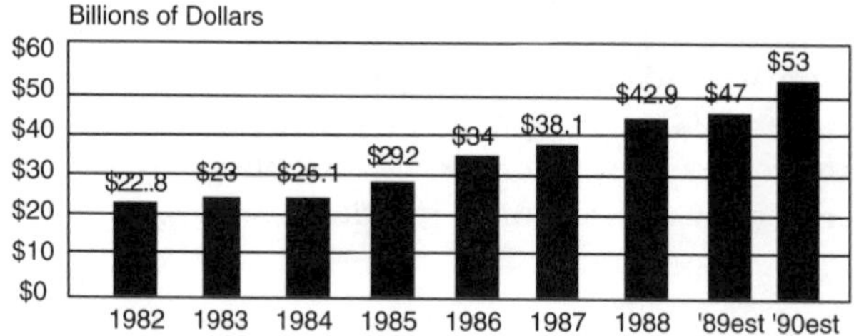

FIGURE 6.2. Workers' compensation costs. *Source: John Burton's Workers' Compensation Monitor*, Ithaca, NY: Vol 4:2, March/April, 1991, p. 1.

> substantial justice in all cases expeditiously, inexpensively and without incumbrance of any character. (19, emphasis added)

The key elements of the California law included compensation for injured workers and their families, safety in the workplace, payment for necessary medical treatments, regulations required to ensure the availability of insurance premiums to employers, and fair state administration of the system. Although it was not further specified in the legislation, "securing safety in places of employment" was one of the initiative's goals. Injury prevention was considered to be one of the benefits of enacting the law.

Six major goals for workers' compensation have been identified: internalizing the costs of industrial accidents; compensating workers without regard to fault; providing medical care for workers who have been injured; providing replacement of wage losses; providing swift and certain compensation; and limiting the liability of employers. (20)

For our purposes, the first goal, internalizing the costs of industrial accidents, requires closer analysis. The notion here is straightforward:

> Internalizing costs encourages employers to maintain a safe and healthy workplace, since the margin of profit would depend in part on how much it cost to compensate their workers for accidents. Employers with safer workplaces and hence fewer compensable accidents would have a competitive edge over employers who were so negligent as to allow their workers to be hurt. (21)

This is the system dynamic that serves as the chief financial incentive for employers to reduce injuries. Two of the remaining goals conflict somewhat with this prevention incentive.

First, replacement of lost wages generally is limited to two-thirds of the employee's salary, or the state's average weekly wage, whichever is lower. Employers do not have to compensate workers for the full value of their lost time. In other words, the injured employee subsidizes the value of his or her compensation, even when one considers that most benefits are not subject to taxation.

Second, a limit on employers' liability also serves to reduce costs to management. The full cost of injuries, including pain and suffering, cannot be won by employees. This shielding of employers from full exposure to the value of a worker's claim was a major incentive for businesses to seek workers' compensation programs.

These two components of the workers' compensation system conflict somewhat with the aim of internalizing the costs of injuries and illnesses, in that they blunt the potential impact of the full costs associated with industrial hazards.

Another way of looking at the question is to suggest that injury prevention is a very broad goal, one that is impacted by many other issues beyond workers' compensation. The remaining five goals of workers' compensation (compensating workers without regard to fault, providing medical care for workers who have been injured, providing replacement of wage losses, providing swift and certain compensation, and limiting the liability of employers) speak more to the mechanics of the system. They deal with the immediate policy needs after accidents occur, which is the primary focus of workers' compensation.

The issue of prevention goes far beyond workers' compensation:

> Actually, the cost of compensation benefits is but a fraction of the cost of a work injury to the employer and to society. The effort to prevent work injuries and to improve occupational health must continue, not only to reduce the cost of compensation insurance, a worthwhile objective in itself, but also for the more important purpose of minimizing as much as possible the loss of income to injured workers, the loss suffered by employers through decreased production and other hidden costs of work injuries, and the cost to society of the loss of productive workers. (22)

Employers and workers have joined to establish this system to alleviate many of the problems caused by the old policy. Workers were primarily concerned with the lack of certain compensation, and employers were concerned with their increasing exposure to full and unpredictable liability. As momentum began to grow behind the workers' compensation movement, the notion that injuries and illnesses could and should be prevented became part of the rationale behind adoption of this compromise.

The extent to which workers' compensation deals with injuries before the fact depends on how broadly the system can be constructed to offer incentives to business. In the early part of this century, some attention was focused on prevention in the workers' compensation system. Industry formed the National Safety Council in 1912 as a response to workers' compensation statutes. (23) Large companies discovered that by investing in simple safety equipment and enforcing basic standards, they could reduce costs caused by accidents. One early example is U.S. Steel, which invested $5 million in equipment and worker education, saw its accident rate fall 40 percent, and slashed its injury-related costs by 35 percent. (24) However, "after some early success in reducing accident rates, the corporate safety movement seemed to run out of steam. Compensation costs were stabilizing . . . the return to normalcy began a 40-year consignment to oblivion of the safety issue." (25)

Nevertheless, the desire to provide an incentive to reduce workplace accidents remains popular as a rationale behind the system. "The workers' compensation system promotes other social goals, one of them being injury reduction. As the National Commission on State Workmen's Compensation Laws stated emphatically in its 1972 report, "the encouragement of safety is one of the basic objectives of a modern workers' compensation program." (26)

Historically, injury and illness prevention has been seen as one of the goals of the workers' compensation system, a notion that remains intact today.

PREVENTION INCENTIVES

The concept behind financial incentives for business to practice injury and illness prevention in the workplace is straightforward: internalize the costs of industrial accidents and employers will respond by investing in ways to reduce the level of those losses. A common view is that "workers' compensation demands that each manufacturer enjoying the profit of a product should assume the costs of injuries. Each producer is consequently motivated to lower accident costs to achieve a competitive advantage." (27)

There are several major factors that affect the ultimate cost of workers' compensation to management. These include the cost of medical care, the level of benefits in the state, and the number and severity of claims. All of these costs, plus charges for administration of the program, are reflected in insurance premiums.

In examining financial incentives for injury and illness prevention efforts, we must look at the relevant factors that affect insurance premiums based on the number and severity of industrial accidents.

It is important to note that, as the cost of workers' compensation increases, the more likely it is that management will focus on controlling those factors. It has been argued that, when a state legislature passes a benefits increase, employers are given a greater incentive to control losses and perhaps enjoy an enhanced injury prevention incentive.

However, business does not always react accordingly. Currently, this country is facing a "workers' compensation crisis," according to the U.S. Chamber of Commerce. (28) The Chamber has editorialized in favor of the following key factors to bring down rising costs: "administrative reforms can reduce costly litigation and streamline the ways that claims are handled. Medical-fee schedules can trim doctors' bills. And payment guidelines can more clearly define what benefits are available for certain types of relatively new stress and disease claims." (29) There is little mention of the need to practice prevention in order to drive down rates. As we shall see, one reason for this may be the fact that relatively small firms, such as those belonging to the Chamber network, are not able to access much rate relief through prevention efforts.

It is true, though, that the Chamber does include "the encouragement of maximum employer interest in safety through the practice of (past safety) experience rating for the purpose of setting the price for workers' compensation premiums" (30) as a primary goal of workers' compensation. But the application of safety incentives through rate making is difficult to carry from the concept stage to the implementation stage.

In studying the application of prevention incentives to the real world of workers' compensation, we must take a look at the vehicles for cost cutting. Invariably, the focus is on the rate-making procedures used to calculate premiums.

Rate Making

Rate setting is of crucial interest because the manner in which a company's insurance premium is calculated speaks to the potential financial incentives for injury prevention for that firm.

There are several ways for insurance carriers to establish a workers' compensation premium. These include class rates (the manual rate discussed earlier), experience rating, retrospective rating, and schedule rating.

Class Rating

Separate rates have been developed for more than 600 categories of industrial activity, ranging from office workers to ironworkers. Employers pay a certain amount, or rate, for each $100 of payroll they have in each category of work. The classes have been established based on expected losses for all employers in that category. Thus the rate that is set, usually called the manual rate, reflects an industry average.

For all businesses, this is the first calculation used to determine base premiums. For smaller businesses, with annual premiums up to $5,000, this is the primary factor used to calculate premiums. Larger firms, with a higher volume of premiums, may have their payments further adjusted based on experience or retrospective rating, as we shall see.

A majority of firms are so small that they qualify only for class rating, accounting for only about 10 percent of the dollar volume of premiums underwritten. (31) This means that little incentive exists for smaller firms to engage in prevention since a higher or lower accident rate than the class average would have no impact on their premiums.

Experience Rating

Larger firms, with annual premiums exceeding a threshold, usually about $5,000, qualify for rate adjustments based on their past claims losses. Called experience rating, this system allows insurers to examine a firm's losses over the previous three-year period and compare that record with the industry average during that time. (32)

If an employer's losses were one-third lower than the industry average, this would translate into a 33 percent premium discount. Similarly, if an employer's losses were 50 percent higher than the industry average, then the premium would be 50 percent higher.

As usual, though, the system is much more complex. Because the base manual rate is developed as an average, chance plays a large part in these calculations. The result is a system that has individual firm's ratings fluctuate from year to year. This is one reason behind figuring experience rating on a three-year average. The chance factor impacts smaller firms to a greater degree than larger firms. A smaller employer with no losses for most years and one serious disability case could see its rating fluctuate wildly. Yet, a larger firm in the same business with a higher payroll would tend to have more predictable losses.

In order to justify this inherent problem with experience rating, a credibility factor is calculated by insurers. For example, firms with premiums below the $5,000 threshold have no credibility, and thus are simply charged the manual, or class, rate. As firms get larger, insurers assign more credibility to their experience, ranging from 1 percent to 100 percent credible. Very few firms have enough premiums and "exposure" to be rated 100 percent. (33)

Two examples of how this system works follow: Employer A has a poor experience rating, with losses double the industry average, yet its credibility factor is only 25 percent. Employer B has a better injury record, 50 percent lower than the industry average, with the same 25 percent credibility factor. With no credibility factor used, experience rating would double Employer A's premium and give a 50 percent discount to Employer B. With the credibility factor, Employer A is charged only 25 percent above the industry average, and Employer B receives a discount of just 12½ percent.

The larger the firm, the less impact the credibility factor has on rates. However, it is important to note that the effect of the credibility factor is to reduce financial incentives for injury prevention, and to blunt financial disincentives for firms that accrue poor industrial accident records.

It has long been assumed that experience rating promotes prevention efforts. As the National Commission on State Workmen's Compensation Laws has stated, "the allocation of the costs of the workmen's compensation program to the industries responsible for the losses thus encourages the economy to make a better use of its resources." (34)

Yet the group also qualifies its endorsement of experience rating by stating that "the variations of Experience Rating Plan values to produce a significantly more direct effect of actual experience upon the employer's premium is a mechanical possibility—but not really a practical or actuarial one," especially for smaller firms. (35)

In fact, so few firms are large enough to be self-rated (100 percent credible) that in most instances the potential impact of setting insurance premiums based on past experience is blunted. The potential rewards or penalties in the form of discounts or surcharges are reduced in most cases. Additionally, since rates are based on an average of the last three years of experience, any successful investment in injury prevention initiatives takes at least several years to be reflected in rates, further undercutting the payback incentive industry looks for when making such investments.

Retrospective Rating

Generally reserved for large firms with premiums more than $100,000 (36), retrospective rating bases a firm's premium on the losses experienced during the current policy period.

Employers and insurance carriers negotiate a maximum credit for favorable loss behavior, and a maximum surcharge for poor performance, when compared with the class rate for that industry segment. Within the confines of the agreed-upon floor and ceiling, employers can work to achieve savings in a current year. The system is one of quasi-self insurance, gaining a stronger financial incentive for employers than experience rating. (37)

However, it must be stressed that the system is not practicable for most firms, and the incentives are still kept in a confined range.

Schedule Rating

Under schedule rating, insurers may give credits to employers based on a review of important aspects of their workers' compensation program. (38) For example, an employer beginning a new loss prevention program that is evaluated by the carrier as including effective initiatives would qualify for a premium discount.

Schedule rating is helpful as a minor tool to help well-intentioned smaller firms that receive no benefit from experience or retrospective rating. However, in practice the credit is not that large unless surcharges can be placed on firms with poor records in order to finance the cost to the carrier.

Self-Insurance

Self-insured employers are limited to those very few companies large enough to finance all of their losses. This is the only group that directly pays out the full cost of benefits to injured workers. This is the case because all other firms are pooled in the insurance system, and businesses with high losses are invariably underwritten by premiums paid by other firms with fewer or no losses. This is the essence of insurance.

The traditional school of thought on self-insurance is that it offers the purest financial incentive possible in the workers' compensation system:

> If workers' compensation is to contribute to efficient accident prevention, the premiums charged to a firm must reflect its prevention efforts. The ideal premium arrangement is one in which premiums are discounted by the costs of the accidents prevented . . . This finely tuned adjustment is approached only when the firm acts as its own insurer. In this situation, any costs saved by preventing an accident benefit the firm. Thus, self-insurance maximizes the firm's incentive for providing safe working conditions. (39)

However, in a recent empirical analysis of the workers' compensation system, one expert concluded that "experience-rated savings often exceed self-insured savings. This flies in the face of conventional wisdom, according to which self-insured firms receive the maximum savings—equaled only by large self-rated insured firms." (40)

The study makes a point that should underscore this discussion of self-insurance and all of workers' compensation: "Generalizations about the size of [workers' compensation] financial incentives—for either policy or research purposes—should be made with great care," because results depend on the particular state statute, benefit levels, the cost of medical care, the hazards inherent in the industry, the size of the firm, and regional wage scales. (41)

While it is successfully argued that self-insurance provides firms with a somewhat clearer understanding that injury prevention will prevent losses, the extent of the incentive is still subject to debate, discussion, and disagreement.

Benefit Levels as an Influence

It is obvious that the level of benefits given to injured employees is a major factor in establishing the level of rates and premiums. Benefits are set on a state-by-state basis and vary according to the wishes of state legislatures throughout the nation.

The higher the benefits, the greater the cost of workers' compensation. Yet, as we have seen, the benefits workers receive, with the exception of medical care, are below the full value of employees' losses. For example, replacement of lost wages is generally capped at two-thirds of the employee's salary or at the statewide average weekly wage, whichever is lower, and no pain and suffering awards are allowed, as they would be under the tort system.

At first blush, it looks as though increasing the benefit levels in a state would increase an employer's financial incentive to invest in occupational hazard prevention initiatives. However, it has been argued by many (and articulated in particular by James Chelius) that

> an increase in workers' compensation benefit levels represents an increase in the cost of an injury to the employer and a decrease in cost to the employee. Economic theory predicts that when the cost of something increases the quantity decreases, and when the cost of something decreases the quantity increases. Therefore, increases in workers' compensation benefit levels may have two opposite effects on the level of resources devoted to prevention, a greater quantity of prevention demanded by employers and a lesser quantity demanded by employees . . . no theoretical analysis can predict which force is stronger. (42)

Chelius states that there is a conflict between the employer's prevention incentives and the employee's sense of security when benefits increase. (43) His analysis suggests that there is a trade-off when benefit levels increase, in the sense that employers may have a greater incentive to look at prevention, but employees may either file more frequent claims because they would not bear as much of the cost of an injury as in the past, or they would not be as fearful of accidents because of more adequate compensation.

Others have been a little more direct in their interpretation: "the 1987 *Economic Report of the President* notes that a growing body of research found that workers' compensation benefits have unfavorable effects on safety" (44), and "there is speculation that the positive association between work-related injuries and workers' compensation benefits observed across states largely reflects that incentive for workers to report injuries rather than an economic incentive for employers to invest less in workplace safety." (45)

It has been suggested that one way to clarify the tension caused by benefit increases is to "raise injury taxes on employers, but not to make the whole tax payable to the employee . . . [although] it must be acknowledged that this proposal is probably not politically feasible." (46)

In other words, in order to make benefit levels serve an adequate role in establishing incentives, employers would have to be charged the full cost of injuries (to the extent allowed by the workers' compensation system), but employees, who are already compensated for less than the full value of their injuries, would be forced to receive even less.

The impact of benefit levels on employers' financial incentives is not certain, and may even have a neutral impact on claims rates. This is another in a series of clouding factors in our review of this issue.

The Problem of Occupational Disease

As we have examined the question of whether adequate financial incentives exist to encourage employer occupational injury and illness prevention efforts, we have done so with the tacit understanding that injuries and illnesses are (or should be) covered by the workers' compensation system. In addressing occupational disease, we have come to an area that has been identified as a critical flaw in the system. It has been persuasively argued that workers' compensation does not adequately deal with the victims of occupational disease. This is crucial, because the extent of the occupational disease problem in the United States is quite large.

Projections have placed the number of workers disabled each year as a result of occupational illness at about 400,000 and the number of deaths at up to 100,000 (this includes those suffering from illnesses with long latency periods). (47)

However, despite the magnitude of the occupational disease problem, it has been estimated that *only 5 percent* of the workers disabled by work-related illnesses receive workers' compensation benefits (48), and "data has shown that less than 3 percent of all workers' compensation cases deal with occupational diseases." (49)

Workers may pursue two avenues for compensation for work-related diseases: the tort system for product liability cases (against third parties such as chemical manufacturers) and workers' compensation.

Of course, product liability suits against third parties are also possible for occupational injuries, but these suits are in addition to the benefits that may be awarded through the workers' compensation system, and are therefore not central to the question of whether a worker's disability will be recognized and compensated in some way. But "the attributes of a disease differ greatly from those of an industrial accident, and placing disease victims into accident-oriented compensation systems has proven unsatisfactory." (50)

It is far more difficult for a worker to receive workers' compensation benefits for work-related illnesses than it is for occupational injuries. The "accident-oriented" nature of the workers' compensation system forces victims of occupational disease to move into the courts in an effort to receive benefits. Those legal actions are not focused on the worker's employer, but rather take action against a third party such as a manufacturer or supplier, although the third party eventually may seek to recover any awarded damages from the employer, a move called subrogation.

In the courts, occupational disease cases have had to fight a number of obstacles, including enormous difficulty proving causation and work-relatedness, time limits imposed by statutes of limitations, and the very expensive nature of the court system, making the exercise very inefficient, costly, and difficult for the plaintiffs. (51)

The workers' compensation system has failed to adequately deal with occupational diseases for a number of reasons.

First, many workers (and their physicians) do not realize that their maladies may be related to their work, or may not make the connection between their disease and work that they performed many years ago. (52) Long latency periods make it very difficult to separate occupational diseases from ordinary diseases of life.

Second, employers contest a very high number of claims that are filed, and prevail. (53)

> Studies have shown that it can take six times as long for the administrative system to respond to occupational disease claims as for accident claims. Sixty-three percent of all occupational disease cases were controverted [denied] as opposed to 10 percent of accident cases, and of those litigated work relatedness was the primary reason in 73 percent of the disease cases versus just 21 percent of the accident claims. (54)

As Leslie Boden has pointed out, "the apparent unfairness and inefficiency of workers' compensation of occupational diseases arises in great measure from the inherent uncertainty about whether many chronic diseases are work-related." (55)

A claimant must show that the disease was related to the work performed, and that causation is very difficult to prove, even though sound arguments may be offered. The burden of proof is on the disabled worker (56), and businesses and insurance lobbyists have succeeded in keeping the definition of occupational disease sufficiently narrow to compound that burden. (56)

Additionally, even in those few cases where benefits are won, compensation is usually very low. For example, average benefits for workers permanently suffering a total disability from an occupational disease are about $9,700, while averaging $23,400 for accident victims—and average cash benefits to occupational disease survivors are $3,700, compared to $57,500 for accident victims. (58)

When comparing the workers' compensation system to the tort system, or when simply attempting to determine the level of justice in the workers' compensation system for occupational disease victims, it is clear that very severe problems exist.

> The tort system is very costly for long latent chemical poisonings; incentives to take care in cases of probabilistic harm are not appropriate; the risk that defendants will be judgement proof is not small; the probability of bringing suit may be low. On the other hand, administrative compensation systems may not perform well as substitutes for the tort system. Problems of demonstrating cause remain. In addition, wage replacement is quite low, *reducing both compensation and incentives below appropriate levels*. Finally, only a small proportion of occupational disease victims file workers' compensation claims. *This further dilutes the effectiveness of workers' compensation in delivering both safety incentives and insurance.* (59, emphasis added)

Business is bearing very little of the true cost of occupational diseases as a result of the lack of claims, the high level of claims denial, and the relatively low benefit levels when awards are given. This seriously undercuts the entire notion that workers' compensation costs can serve as an incentive for employers to invest in prevention initiatives.

Furthermore, the failure of the tort system to serve as an alternative means that most victims of occupational disease either suffer with no recourse or win substandard benefits. A growing school of thought suggests that alternative systems of more direct regulation must be established to deal with this area of concern. A consensus seems to be shifting toward a separate type of disability insurance, or a regulatory system that deals head-on with the "fault or causation" concerns that serve to promote injustice in the current programs. (60)

To conclude, the issues raised in this section on the special problems posed by occupational disease and illnesses have a significant impact on our discussion. Workers' compensation originally was put in place to guarantee swift (if slightly less than adequate) compensation for injured and ill workers, and to place a limit on management's liability. In the case of the treatment of occupational diseases within the workers' compensation system, there is no guarantee for swift (or even close to adequate) compensation for workers, and there is an exceptionally high level of protection from liability for employers.

The lack of coverage for occupational diseases damages the argument that the workers' compensation system offers adequate financial incentives for employers to invest in prevention. Within the workers' compensation system, employers bear very little of the true cost of work-related diseases.

Even if employers were to bear a greater share of the cost, the long-term nature of the problem means that the payback period for investments in prevention may take years to develop, and few employers seem anxious to make such a commitment when other investments would bring about a higher return on the dollar. "Even with complete coverage, [workers' compensation] gives firms little economic incentive to prevent diseases with long latency periods. Private firms use discount rates of 10 percent and higher, and, at such rates, costs more than 20 years in the future shrink into insignificance." (61)

Until such time as management bears a large portion of this cost, the workers' compensation system will serve no useful purpose in promoting occupational disease prevention behavior on the part of employers, and even with full coverage under workers' compensation, any potential financial incentive would be suspect.

Summary

This section has sought to examine the features of the workers' compensation system that are purported to create financial incentives for employers to invest in the prevention of occupational injuries and diseases.

Victor put it quite clearly when he defined workers' compensation financial incentives for injury prevention as "the savings in workers' compensation costs that the employer expects to realize through prevention." (62)

In exploring the nature of workers' compensation, we have noted that the system already provides a great deal of protection to employers by shielding them from liability, and limiting their costs to levels below the true cost of the injuries to workers. The trade-off that served as a premise for creating the workers' compensation system has built-in savings for businesses. (63)

As we looked at the traditionally accepted vehicle for financial incentives, the rate or premiums paid by employers, we reviewed the rate-making framework within which employers' workers' compensation costs are determined—class or manual rates, experience rating, retrospective rating, schedule rating, and self-insurance.

We discovered that the rate-making process is so complex that it is very difficult to determine whether true incentives exist. Smaller companies are usually charged the same amounts, regardless of performance and commitment. Larger companies find most potential savings blunted by the "credibility factor," which reduces potential credits for sound performance, thereby reducing available capital to invest in prevention. And firms with poor performance records find that, based on the nature of insurance, their losses are underwritten by employers with more favorable safety records.

Very few firms are large enough to be self-rated, and even fewer are self-insured. Even in these cases, it is difficult to ascertain whether the incentives that may exist are powerful enough to promote action.

As we discussed earlier, business may prefer to reduce or control their workers' compensation costs by opposing benefit increases, fighting or streamlining claims, and reducing doctors' bills.

Even when the costs of the entire system are inflated by raising the level of benefits, we discovered that the two forces of management and labor may conflict and neutralize incentives to invest in prevention.

Finally, we examined the most glaring problem with workers' compensation: the blatantly inadequate handling of occupational disease compensation. Employers bear very little of the cost of the 400,000 annual occupational disease-related disabilities and 100,000 annual deaths. There are no real market forces to create financial incentives for prevention of occupational diseases, and therefore little incentive at all for employers to take occupational disease seriously.

Underscoring all of this analysis is a constant tension. The purpose of insurance is to spread a risk among a large population. The way to create an incentive is to focus on the consequences of one's behavior. The workers' compensation system contains a continuous conflict between the insurance system's goal of diffusing responsibility for industrial accidents, and the need to create and place the burden of responsibility for the cost of injuries clearly and surely on employers' shoulders. Put another way:

> In considering the role of liability insurance, one has to recall the dichotomy between the two aims of liability: prevention of accidents and compensation of victims. Historically and on the face of it, the purpose of the insurance is to protect the person liable from the consequences of his liability; in so far as his protection can be bought at a reasonable price, the economic deterrent effect of liability disappears. (64)

The very nature of insurance tends to discourage the potential financial incentives needed to create prevention activity. This is especially true when we consider the fact that in the workers' compensation system, employees share the cost of injuries with employers, and to an even greater degree the system places costs for illnesses on employees.

This is further placed in context by understanding that "employer motivations to improve the safety of the workplace are somewhat limited by the fact that at some point, the cost of eliminating injury and illness will exceed employer capabilities." (65) Combined with a low financial incentive, if one exists at all, and knowing that "there will always be an optimal level of work injuries and illnesses under any workers' compensation program," (66) it is difficult to feel confident that the system is capable of offering enough incentive to compel employers to invest in prevention efforts.

At the beginning of this section, we cited a common view of workers' compensation claiming that the system "demands that each manufacturer enjoying the profit of a product should assume the costs of injuries. Each producer is consequently motivated to lower accident costs to achieve a competitive advantage." (67) Our discussion leads to a conclusion that suggests that this principle is more appropriate as a theory and is not supported when one looks at its application.

In short, the workers' compensation system does not offer significant financial incentives for employers to invest in prevention efforts. Further, the system's framework is too stifling to allow a clear vision of these issues, and that in itself allows the exaggerated notion of the existence of market-driven incentives to continue. This myth diverts attention away from efforts to bring about meaningful change.

INDUSTRY'S RESPONSE TO INCREASED PREMIUMS

Earlier it was stated that, in theory, when the costs of premiums increase, the financial incentives for employers to invest in prevention increase as well. Yet, we also have seen the broad range of constrictive policies that tend to blunt the impact of rate incentives.

It has also been recognized that industry has alternative ways of controlling costs other than looking to injury prevention. A good example of the political strategy to control costs can be found in Massachusetts industry's response to skyrocketing premium costs. (See this chapter's addendum, "Massachusetts as an Example.")

Reflecting a national business agenda, Associated Industries of Massachusetts (AIM) targeted workers' compensation reform as a top priority in 1990. (68) AIM, the state's leading business lobbying association, has filed legislation to control costs by: reducing lawyers' fees; limiting allowable medical costs and capping physicians' charges; changing the administration of the system, making it easier to discontinue benefits to workers, and limiting the scope of certain hearings; and reducing benefits to injured workers. (69)

This approach is well represented by the U.S. Chamber of Commerce, which emphasizes administrative reforms to streamline claims, medical fee changes, and "guidelines" for benefits to injured workers.

While consultants to industry still push for injury prevention and loss control as components of their workers' compensation cost control plans (70), prevention is a small part. Other areas of emphasis include pressuring employees to return to work earlier, documenting accidents to prepare better defenses against claims, maintaining safety programs because "statistics indicate that people—not workplace conditions—cause 80 percent of all injuries. Carelessness, daydreaming, bad attitudes, problems at home and physical ailments are all root problems." (71)

Other recommendations include keeping employees happy and finding clerical errors.

Overall, in Massachusetts, and probably around the nation, industry's response to increasing workers' compensation costs has been to hold down the number of eligible claims, reduce benefits, and control medical costs. All of these steps place a greater share of the cost of injuries on disabled employees.

IMPROVEMENTS OR ALTERNATIVES

Workers' compensation remains at the center of a significant debate. Is it possible to amend the workers' compensation system to adequately enhance the financial incentives for employers to invest in prevention initiatives on their own?

Of course, this question is very hard to answer. And it is impossible to achieve a consensus when addressing the issue. However, it is instructive to review a sampling of current thoughts regarding the adequacy of workers' compensation as a model for prevention. The following are a few of the ideas that receive the most discussion.

Keeping the System Intact

Many argue that the workers' compensation system has performed reasonably well in introducing the no-fault aspect of compensation. In terms of developing reforms that work within the workers' compensation system, James Chelius has offered the most straightforward.

Chelius, noting that "the change from the tort system to workers' compensation is . . . a positive one," for both employers and employees (72), has offered a model for dealing with the question of inadequacy of incentives in the system.

He recommends that both the employee and the employer participate in the reintroduction of a modified liability system. Employees would bear a greater share of the cost of less serious injuries through reduced benefits and medical costs. Employers would be subject to pain and suffering claims (within the workers' compensation system, not the courts), although he would make it more difficult for employees to prevail by adding a gross negligence test on employers, which is greater than being simply at fault. (73)

As cited earlier, an additional injury tax on employers could serve to increase the costs on high-risk employers, although it has been suggested that the money should not be put into increased benefits for fear that claims would increase. It is argued that the only incentive should be for firms to reduce injuries. (This proposal becomes more attractive if the funds were to be used to enhance OSHA programs, or promote prevention-based strategies.)

Other recommendations for working within the system include the adoption of deductibles for firms, so that at least part of the cost of a workplace injury would not be spread to other employers. (74) While this proposal might serve as an incentive for small firms, its impact would not be significant on larger employers.

The general discussion in this examination of workers' compensation has been highly critical of the system, and has pointed to the absence of financial incentives for prevention initiatives. It would be a large leap of faith to embrace the workers' compensation system as adequate with changes as custodial in nature as these. Given the analysis that we have developed, one can argue that these changes would not add a dramatic amount of clarity to the issue of prevention incentives.

Government Regulation

As the workers' compensation crisis has blossomed across the nation, laws to "reform" the system have been enacted in a number of states. While the major thrust of these statutes has been to control medical costs, regulate benefits, expedite the resolution of claims disputes, and so on, there have been a number of injury prevention and safety requirements.

In the past, some states have created programs to provide voluntary safety education and training to industry. These initiatives have ranged from New York's Safety and Health Training and Education Program (it has provided grants totalling more than $13 million to more than 270 employers and organizations), to Connecticut's agency-based technical assistance, training, and education program (funded at more than $1 million a year), to the Office of Safety in Massachusetts.

Since 1989, a number of states have moved one step further, mandating injury prevention programs for firms with particularly poor loss performance records within their industrial class.

In 1989, Texas adopted a reform law that establishes an extra-hazardous employer program to identify firms whose injury frequencies substantially exceed those expected in their industry. (75) Extra-hazardous employers are required to have an outside consultant conduct an audit of unsafe practices, and develop a detailed accident prevention program. The Texas Division of Workers' Health and Safety is empowered to monitor and enforce the implementation of the injury prevention plan. The law also creates health and safety training programs, requires insurers to offer health and safety consulting services, and establishes a back injury education program.

As a part of Oregon's 1990 reform measure, that state enacted a provision requiring hazardous businesses with 10 or more employees to establish joint health and safety committees. (76) Firms with lost workday case incidence rates in the top 10 percent of their industrial class, or the top 25 percent of premium rates for all classes, come under this mandate. Although the health and safety committees are not given specific resources, the effort is an attempt to inject prevention issues into the workplace.

Maine's workers' compensation reform law, passed in the summer of 1991 (77), requires employers who are experience-rated at twice the manual rate to develop workplace health and safety plans approved by the state's Department of Labor.

This new trend in state-mandated health and safety programs is interesting for two reasons. First, these states seem to recognize that past reliance on any financial incentives in the workers' compensation system has resulted in few serious injury prevention programs at the company level. Second, the mandated programs are relying on the insurance industry's experience rating system to identify hazardous employers, creating another incentive for firms to avoid poor loss performances. However, as we noted previously, a minority of firms are large enough to be experience rated. These programs do not impact most businesses.

A NEW MODEL FOR OCCUPATIONAL DISEASES

Earlier, we discussed the lack of coverage for occupational illnesses in the workers' compensation system, and discovered a great deal of information that led us to reject the notion that work-related diseases are dealt with adequately in workers' compensation. In fact, there is general agreement around the development of an alternative plan for occupational diseases. Even Chelius concedes that there may be no satisfactory resolution to the problems of compensating occupational disease within the traditional workers' compensation framework. Since the limitations of the work-relatedness criterion are so great, more serious attention should be paid to reforms that attempt to remove occupational disease from the workers' compensation umbrella. (78)

Others have argued that, while workers' compensation has "poorly served those with work-related illnesses," and a new system is needed, given the nature of regulation and the nature of occupational disease, any "compensation system will inherently be flawed because of the complexity of the disease phenomenon." (79) Generally, many people suggest that an initiative similar to Social Security's Disability Income with a presumption of work-relatedness may offer the best model available. Of course, this model would completely abandon even the pretense of placing a burden on the employer.

The broadest possible change would be to merge a number of our insurance programs together into one 24-hour-a-day program.

Certainly, this initiative would represent complete change in our system. It would recognize the weak incentives on the part of employers to engage in prevention, and develop alternative approaches.

CONCLUSION

There are grave doubts regarding the adequacy of workers' compensation as a vehicle to drive prevention efforts.

Some labor advocates have called workers' compensation "a swindle" (80), because it limits an employer's liability while transferring much of the cost of injuries and illnesses to the worker.

Other labor advocates have called the system "workers' noncompensation," and state that the system fails nearly all its original goals except one: limiting the liability of employers, and insulating them from the true costs of work-related illnesses and injuries. (81)

From our analysis, it is possible to draw several conclusions:

First, the workers' compensation system has not demonstrated an ability to serve as a consistent incentive for investment in prevention initiatives. There is a constant struggle within the workers' compensation system to resolve a fundamental conflict: the nature of insurance is to spread and diffuse risk, and the nature of financial incentives requires focused costs and easily identified savings. From our analysis, it seems as though the power of insurance has prevailed, and thwarted the rise of any significant financial incentive for employers.

Second, the workers' compensation system does not merely limit an employer's liability, it transfers much of the cost of injuries and illnesses onto the employee, and institutionalizes those costs in ways that make it virtually impossible for injured workers to recover them. As long as this dynamic prevails, there will be little incentive for employers to support significant change in the system, regardless of the relative expense of insurance premiums.

And third, the problem of occupational disease is a major blow to the credibility of the workers' compensation system. With 400,000 people suffering occupational illnesses each year, and with 100,000 deaths each year, the workers' compensation system deals with only 3 to 5 percent of these cases. With no incentive for employers to invest in preventative measures for occupational diseases, a gaping hole remains in the fabric of society's compensation policy.

Given this harsh analysis of the workers' compensation system, there naturally are a number of limits placed on policy recommendations that may be offered.

We should not expect too much from the workers' compensation system. Putting aside the question of prevention for a moment, the system can still serve a useful purpose in providing benefits to injured workers in a no-fault system. Under the old tort system, there were no guarantees of benefits or compensation. Common law is not acceptable, unless relaxed statutes of limitations, rules of evidence, court costs, interim benefits, and speedier decisions are guaranteed. At least now, many more individuals are sure of receiving something to offset part of the costs of their injuries. However, the compensation is universally low, and efforts by industry to force workers to bear an even larger burden must be opposed.

For victims of occupational disease, the workers' compensation system is not satisfactory. It seems likely that society will be forced to create new programs, such as disability insurance, to deal with this serious gap in current policies.

If we accept the system for what it is—not what it pretends to be—then perhaps we can develop a mindset that will allow advocates to come together to deal with prevention in other more effective ways.

And as we look for a truer means of encouraging or even mandating prevention, we must work to improve the workers' compensation system. Government should recognize the primacy of prevention, education and training programs by increasing and financing them through premium surcharges. We can take other steps to encourage prevention in the public sector, where the audience may be more receptive.

As Lawrence White says, "the great scandal of industrial injury, occupational disease, and workers' compensation, is that the life of an American worker is so cheap that it costs industry less to injure and kill than to create safe working conditions." (82)

In its current form, workers' compensation contributes to the neglect of injury and illness prevention strategies. It is part of the problem, not part of the solution.

MASSACHUSETTS AS AN EXAMPLE

Let us examine the major aspects of one state's workers' compensation program—Massachusetts's—in the context of our discussion. Important items for us to review include the rate structure, the state's investment in safety and prevention, and the business community's response to very high insurance premiums.

Massachusetts first passed its workers' compensation law in 1912. It was a voluntary statute, allowing employers to adopt the provisions of the law, and avoid tort liability. Thirty years later, in 1943, the legislation was updated and made compulsory for all businesses.

The fundamentals of the system are quite similar to most states'. Benefit levels are set by statute; insurance premium rates are proposed by the insurance industry and set by the state's commissioner of insurance, with premiums based on manual rates per $100 of payroll for more than 600 classes of industrial activity; the Department of Industrial Accidents administers the program; there is no state fund—all of the insurance coverage is underwritten by private carriers; and there is a great deal of concern in the business and political communities around the issue of workers' compensation because of dramatically increasing premium costs.

In 1985, Massachusetts passed its first significant workers' compensation legislation since World War II. Benefits were increased, claims backlogs were targeted for reduction, case adjudication procedures were streamlined (lawyers fees were reduced by removing incentives for attorneys to 'lump sum' cases, that is go for one-time settlements from which they would receive a standard fee of 33 percent, leaving the disabled worker with fewer benefits in the long-run, and attorneys with higher pay in the short- and long-run).

In 1985, the total volume of workers' compensation premiums paid by industry was approximately $800 million. Despite the 1985 reform law, which was intended to stabilize rates, premiums soared to nearly $2.4 billion in 1991. Insurers filed for a 46 percent rate hike request for 1992 premiums. (a) To put this figure into perspective, Bay State companies pay only a little more than $500 million in corporate excise taxes. Not only have rates more than doubled in the last five years, but employers pay much more in workers' compensation costs than they do in taxes to the state.

Of course, this is part of a national trend. (b) However, as we shall see, the reaction to these increased costs speaks to industry's perception of the value of financial incentives to prevent injuries.

The Massachusetts Rate Structure

If a financial incentive in workers' compensation is to be found, one must examine the system that establishes premium levels for employers. In Massachusetts, approximately 110,000 firms purchase workers' compensation insurance premiums or manage their own self-insurance programs. (c)

The system allows for experience rating, retrospective rating, and scheduled (or merit) rating, and these are the major factors in rate adjustment above or below the class rating that is established using the industry segment average as a base.

The same issues that raised earlier concerns regarding the impact of the rate-making process on potential financial incentives are relevant in Massachusetts.

In experience rating, only firms above a $5,500 premium threshold qualify for rate adjustment. One-third of the businesses in the state are large enough to be experience rated, and only 100 to 150 employers are self-insured. (d)

Retrospective rating, a voluntarily negotiated range of either surcharges or credits based on the current year's losses, is not used frequently because insurers do not believe that rates are adequate, and therefore there is not too much incentive on the part of insurers to offer credits—the same can be said of any potential dividend plan, where any credits that are offered are not offset by potential surcharges on poor-performance firms. (e)

Two-thirds of the employers in Massachusetts receive their workers' compensation premiums based on the manual rate, with one-half of those eligible for a small adjustment from a schedule rating-based initiative by the insurance industry called the Merit Rating Program. (f)

The Merit Rating Program sets a simple policy. If an employer has one lost-time claim, the firm stays at the manual or class rate. If the employer has no lost-time claims, the firm receives a 5 percent credit. If the business has two or more lost-time claims, the employer pays a 5 percent surcharge.

Employers eligible for the Merit Rating Program owe annual premiums of between $501 and $5,500. This raises an interesting question regarding the value of the financial incentive: if an employer has a premium costing $5,000, a 5 percent savings translates into $250 a year—the investment of $250 a year in prevention would have to be stretched far in order to make a difference.

One-third of the businesses in the state are so small that they pay premiums of less than $500, and thus are underwritten based on the manual rate regardless of their safety records. (g) These firms have no financial incentive to invest in prevention.

Any analysis of the Massachusetts system must also include a review of the voluntary and involuntary insurance pools in the state. For it is here that we see many of the central issues of insurance collide with the notion of financial incentives in the workers' compensation system.

Massachusetts has a compulsory workers' compensation law. This means that every employer is required to participate in the system, either with private insurance or as a self-insured entity. The concurrent requirement on insurers is that collectively they must provide insurance coverage to all firms in the state.

For firms with little risk, or with a solid record of loss control, insurers voluntarily offer coverage, and in a few cases even compete for business. In order to ensure full coverage for all the state's companies, each insurer that does business in the state (approximately 230 carriers) must also underwrite a certain number of firms with whom no carrier wishes to contract.

In practice, employers are placed in two categories—one is the voluntary pool, and the other is the residual market, called the Assigned Risk Pool (ARP). If a carrier underwrites 10 percent of the premiums in the voluntary market, that insurer is required to underwrite 10 percent of the ARP premiums.

In 1980, 10 percent of the premiums in Massachusetts were in the ARP. However, during the next decade, as a result of dramatic increases in the premium costs and the insurance industry's belief that rates were not adequate to meet costs, the percentage of premiums in the ARP increased to more than 50 percent, representing nearly 70 percent of the state's employers. (h)

It may be helpful to think of insurance writing as a zero sum game. In a zero sum game there are always winners and losers, and adding up the score at the end, the sum should be zero. In order for one player in a zero sum game to win 10 points, another player must lose 10 points.

When writing insurance premiums, in order to offer a bonus of 10 points to one firm, an underwriter feels that the bonus (in the case of workers' compensation it would be a credit) must be offset by a surcharge of an equal amount on another firm. This is necessary when the bonus (in this case experience rating and retrospective rating savings) is based on performance against the industry average.

The ARP incurs large losses each year, and until recently those losses were eventually financed through higher rates on the entire market. In other words, the firms that have performed better than the industry average were partially subsidizing the state's worst risks.

The cost to firms in the voluntary market was very high, as the losses in the ARP reached hundreds of millions of dollars in recent years.

Employers with good loss performances saw a very high fixed part of their premiums go to employers who, once in the ARP, had little incentive to act, and were in fact subsidized by firms that may take prevention more seriously.

Carriers have taken steps to reverse the trend toward even higher ARP costs. In 1990, the Workers' Compensation Rating and Inspection Bureau (WCRIB), the group that represents the workers' compensation insurance industry's interests in Massachusetts, launched an innovative program to initiate surcharges on firms with poor loss experience.

The All Risk Adjustment Program (ARAP) was constructed to apply retrospective surcharges of up to 49 percent on ARP and some voluntary-market firms. Based on a formula designed to rate loss performance, WCRIB seeks to force a system whereby the greater-risk employers pay a greater share of their costs. Although this program is limited to the one-third of employers who are experience rated, this revenue-neutral initiative has reduced premiums to non-ARAP employers. Approximately one-half of the experience-rated firms face ARAP surcharges, averaging about 7 percent. The remaining qualifying firms enjoy a projected savings of 7 percent. (i)

In addition, the WCRIB has commenced another schedule-rating measure, the Qualified Loss Management Program (QLMP). The effort is designed to give experience-rated firms an incentive of up to 10 percent (although it is usually 5 percent or lower) discounts if they institute comprehensive loss control programs. (j) The initiative's primary focus is on business practices after injuries occur, such as medical care and rehabilitation services, early return to work or light duty incentives, claims control, and improved communication with injured workers. A safety and training program is required.

The premium discounts last a maximum of three years. In order to qualify for the full 10 percent credit, employers must hire an appropriate certified consulting firm, implement a comprehensive loss management program, and reduce their losses by 20 percent in the first year. Based on the loss experience, and the nature of a firm's program, carriers can reduce or even deny the discount. Employers are not guaranteed to achieve an immediate savings in their premiums, especially when the consultant's costs are included, although it is almost certain that long-term savings can be achieved.

The fact that the initiative is even being offered suggests that insurers recognize a perception by employers that stronger incentives are necessary to compel action.

Summary on Rates

It is important for us to examine a system such as the Massachusetts workers' compensation program to more fully understand the forces at work.

Earlier, we concluded that workers' compensation does not provide adequate financial incentives for employers to invest in prevention. Now, as we look closely at the Massachusetts model, and reasonably expect that other states are experiencing the same issues, we see that, in practice, there is even more confusion around incentives.

One Massachusetts official noted that "although theoretically the Assigned Risk Pool is the pool of last resort, it shouldn't be an experience where I [as an employer] can cut my costs." (k) In spite of the ARAP initiative, the eventual subsidy that ARP firms receive serves as a disincentive for investment in prevention.

State Prevention Efforts

In 1917, when discussing the need to further develop and reform the state's workers' compensation system, a legislative panel was formed and made a presentation to the Commonwealth. The Joint Special Recess Committee on Workmen's Compensation Insurance Rates and Accident Prevention made the following statement:

> While certain and prompt payment of compensation to injured workmen and their dependents has very properly been made the subject of thorough study, and has been secured by legislative enactment, the Commonwealth has not yet given adequate attention to the problem of preventing accidents. Statistics show that the number of accidents, and consequently the burden upon employees and society at large, is steadily increasing. It is our opinion that the Legislature should no longer delay the adoption of such changes in the present system and administration as are necessary to make them effective. There can be no dispute of the statement that every accident is simply so much economic waste, the amount of which is only measured by the severity of the accident. Moreover, the waste is threefold, in that the employee and his dependents must be cared for while he is incapacitated; his earning power is suspended for a time, and in the more severe cases is decreased for all time. (l)

Seventy-four years later, in 1991, Massachusetts had a Department of Industrial Accidents (DIA), which administers the workers' compensation system. DIA had an annual budget of $13 million, and spent $400,000 on accident prevention, virtually all of the funding going to finance a series of grants to labor, industry, and government groups to engage in occupational safety and health educational and training programs.

According to a former director of DIA's Office of Safety, these grants serve a useful purpose. They are given to unions, companies, nonprofit organizations, hospitals, and other similar interests, and the funds are used to focus on the "need for public education, training in prevention of illnesses and injuries. It is a modest grant program with a focus on making employers and employees realize that a proactive strategy to reduce injuries is a benefit to all parties." (m)

DIA officials saw this program as a first step in the state's activity promoting prevention. "The 1985 reform act [which created the Office of Safety, in addition to revamping the entire workers' compensation statute] infused into the flow of our comp system the issue of injury prevention. It remains a small part of the department, but it has been introduced." (n)

The Office of Safety's creation was a positive step in the state's effort to begin a program to promote the prevention of occupational injuries and illnesses. Yet the administration that came into office in 1991 began to approach this issue with an unsuccessful attempt to reduce DIA's budget by $300,000, even though the department is financed through workers' compensation premiums, and costs average taxpayers nothing. Even the future of the Office of Safety's modest program was unclear.

EDITORS' NOTE

While this chapter was originally in press, the Commonwealth of Massachusetts enacted a sweeping reform of its workers' compensation system. As the legislation made its way through the process, proposals to increase the focus on injury prevention were rejected, and the major cost savings in the initiative were achieved through dramatic cuts in benefits to injured workers and efforts to remove an unspecified number of "fraudulent" cases from the system. In short, the state blamed injured workers, and not the occurrence of injuries, for the problem of runaway workers' compensation costs.

NOTES

1. Worrall, John D., ed. *Safety and the Workforce: Incentives and Disincentives in Workers' Compensation*, Ithaca, NY: ILR Press,1983, p. 2.
2. Worrall, *Safety and the Workforce*, pp. 4–5.
3. Baram, Michael S. *Alternatives to Regulation*, Lexington, MA: Lexington Books (Heath), 19, p. 80.
4. Darling-Hammond, Linda, and Thomas Kniesner. *The Law and Economics of Workers' Compensation*, Santa Monica, CA: The Institute for Civil Justice, Rand Corporation, 1980, p. ix.
5. Cheit, Earl, and Margaret Gordon, eds. *Occupational Disability and Public Policy*, New York: Wiley and Sons, Inc., 1963, p. 78.
6. Chelius, James R. *Workplace Safety and Health: the Role of Workers' Compensation*, Washington, D.C.: American Enterprise Institute for Public Policy Research, 1977, pp. 17–18.
7. Cheit and Gordon, *Occupational Disability and Public Policy*, p. 78.
8. Darling-Hammond, *The Law and Economics of Workers' Compensation*, p. ix.
9. Chelius, *Workplace Safety and Health*, p. 19.
10. Darling-Hammond, *The Law and Economics of Workers' Compensation*, p. ix.
11. Nackley, Jeffrey V. *Primer on Workers' Compensation*. 2nd ed, Washington, DC: The Bureau of National Affairs, 1989, p. 1.
12. Chelius, *Workplace Health and Safety*, p. 21.

13. Darling-Hammond, *The Law and Economics of Workers' Compensation*, p. ix.

14. The Commonwealth of Massachusetts had passed this country's first law regulating employer liability for industrial accidents in 1887, the Employers' Liability Law, and that action had helped to plant the seeds for the workers' compensation insurance business, as underwriters subsequently offered premiums to insure employers against losses incurred in the courts.

15. Chelius, *Workplace Safety and Health*, p. 20.

16. Darling-Hammond, *The Law and Economics of Workers' Compensation*, p. x.

17. Thompson, Roger. "Fighting the High Cost of Workers' Comp," *Nation's Business*, Washington, DC, March, 1990, p. 20.

18. Burton, John F., Jr. "Workers' Compensation Costs in 1990," *John Burton's Workers' Compensation Monitor*, Vol. 4:2, Ithaca, NY, April, 1991, p. 1.

19. Cheit and Gordon, *Occupational Disability and Public Policy*, p. 281.

20. White, Lawrence. *Human Debris: the Injured Worker in America*, New York: Seaview/Putnam, 1983, pp. 74–75.

21. White, *Human Debris*, p. 75.

22. Cheit and Gordon, *Occupational Disability and Public Policy*, p. 101.

23. Berman, Daniel M. "Death on the Job: Occupational Health Struggles in the United States," New York: Monthly Review Press, 1978, p. 77.

24. Berman, "Death on the Job," p. 77.

25. Berman, "Death on the Job, "p. 77.

26. Victor, Richard B. *Workers' Compensation and Workplace Safety: The Nature of Employer Financial Incentives*, Santa Monica, CA. The Institute for Civil Justice, Rand Corporation, 1982, p. vii

27. Darling-Hammond, *The Law and Economics of Workers' Compensation*, p. xiii.

28. See cover story and editorial, *Nation's Business* (published by the U.S. Chamber of Commerce), March, 1990.

29. Editorial, *Nation's Business*, March, 1990, p. 3.

30. Baram, *Alternatives to Regulation*, p. 79.

31. Chelius, James, ed. *Current Issues in Workers' Compensation*, Kalamazoo, MI: W. E. Upjohn Institute for Employment Research, 1986, pp. 200–211.

32. Chelius, *Current Issues in Workers' Compensation*, pp. 211–212.

33. Chelius, *Current Issues in Workers' Compensation*, p. 212.

34. Interdepartmental Workers' Compensation Task Force. *Research Report*, Volume 4, Washington, DC: U.S. Department of Labor, 1979, p. 81.

35. Interdepartmental Workers' Compensation Task Force. *Research Report*, Volume 4, p. 86.

36. Chelius, *Current Issues in Workers' Compensation*, p. 211.

37. Chelius, *Current Issues in Workers' Compensation*, pp. 213–214.

38. Chelius, *Current Issues in Workers' Compensation*, p. 213.

39. Chelius, *Workplace Safety and Health*, p. 25.

40. Victor, *Workers' Compensation and Workplace Safety*, p. x.

41. Victor, *Workers' Compensation and Workplace Safety*, p. x.

42. Worrall, *Safety and the Workforce*, p. 154.

43. Worrall, *Safety and the Workforce*, p. 159.

44. Kniesner, Thomas J., and John D. Leeth. "Separating the Reporting Effects from the Injury Rate Effects of Workers' Compensation Insurance: A Hedonic Simulation," *Industrial and Labor Relations Review*, Vol. 42, No. 2, Ithaca, NY: Cornell University, January, 1989, p. 280.

45. Kniesner, *Industrial and Labor Relations Review*, p. 292.

46. Chelius, James R. "The Influence of Workers' Compensation on Safety Incentives," *Industrial and Labor Relations Review*, Vol. 35, No. 2, Ithaca, NY: Cornell University, January, 1982, p. 241.

47. "Notes: Compensating Victims of Occupational Disease," *Harvard Law Review*, Vol. 93, 1980, p. 916.

48. White, *Human Debris*, p. 79.

49. Massachusetts Workers' Compensation Advisory Council. *Report to the Legislature on Occupational Disease*, Boston: Commonwealth of Massachusetts, Department of Industrial Accidents, May, 1990, p. 48.

50. "Notes," *Harvard Law Review*, p. 917.

51. "Notes," *Harvard Law Review*, pp. 926-928.

52. White, *Human Debris*, pp. 79-80.

53. White, *Human Debris*, p. 79.

54. Mass. Workers' Compensation Advisory Council. *Report to the Legislature*, p. 47.

55. Chelius, *Current Issues in Workers' Compensation*, p. 323.

56. White, *Human Debris*, p. 80.

57. Berman, *Death on the Job*, p. 69.

58. Mass. Workers' Compensation Advisory Council, *Report to the Legislature on Occupational Disease*, p. 48.

59. Boden, Leslie I. "Comment on Epstein," *Journal of Legal Studies*, Vol. 13, Chicago: University of Chicago, August, 1984, p. 516.

60. See Boden in *Journal of Legal Studies*, Cheit and Gordon in *Disability and Public Policy*, Notes in *Harvard Law Review*, and Chelius in *Current Issues in Workers' Compensation*.

61. Mendeloff, John. *Regulating Safety: An Economic and Political Analysis of Occupational Safety and Health Policy*, Cambridge, MA: The MIT Press, 1979, p. 12.

62. Victor, *Workers' Compensation and Workplace Safety*, p. 3.

63. Not part of our discussion, but worthy of note, is the fact that businesses can pass costs on to consumers. This very simple economic fact of life underscores the fact that employers do not bear the full cost of injuries and illnesses. One of this region's first workers' compensation advocates, William W. Kennard, Chairman of the Commonwealth of Massachusetts Industrial Accident Board in the early part of this century, said, in 1918, that "the Workmen's Compensation Act is not a regulation of any substantive duty; it is exclusively an economic readjustment of the burdens of industrial accident from the shoulders of the employees to the shoulders of the consuming public." (Letter to Governor Samuel McCall, included

in the *Report of the Special Recess Committee on Workmen's Compensation*, Massachusetts General Court, Boston, February, 1919, p. 23).

64. Morgenstern, Felice. *Deterrence and Compensation: Legal Liability in Occupational Safety and Health*, Geneva, Switzerland: International Labour Organisation, 1982, p. 65.

65. Baram, *Alternatives to Regulation*, p. 81.

66. Baram, *Alternatives to Regulation*, p. 81.

67. Darling-Hammond, *The Law and Economics of Workers' Compensation*, p. xiii.

68. Associated Industries of Massachusetts. *Legislative Bulletin*, Vol. 30, No. 8, March 30, 1990, pp. 1-3.

69. AIM, *Legislative Bulletin*, Vol. 30, No. 8, p. 2.

70. See *Nation's Business*, March, 1990, p. 26, and "Squeeze the Cost Out of Comp.," *Industry*, Boston, November, 1989, p. 32, for two '10 point plans' offered by industry consultants.

71. *Nation's Business*, p. 26.

72. Chelius, *Current Issues in Workers' Compensation*, p. 324.

73. Worrall, *Safety and the Workforce*, p. 129.

74. Worrall, *Safety and the Workforce*, p. 129.

75. See Senate Bill 1 from the 71st Texas Legislature, Second Called Session, 1989.

76. See Chapter 2, Oregon Laws 1990.

77. See State of Maine Chapter 615 of the Acts of 1991.

78. Chelius, *Current Issues in Workers' Compensation*, p. 324.

79. Notes, *Harvard Law Review*, p. 937.

80. Berman, *Death on the Job*, p. 186.

81. White, *Human Debris*, pp. 93-95.

82. White, *Human Debris*, p. 23.

NOTES TO "MASSACHUSETTS AS AN EXAMPLE"

a. Workers' Compensation Rating and Inspection Bureau (WCRIB), 40 Broad Street, Boston, MA. WCRIB is the organization that represents insurance carriers in Massachusetts, compiles statistics on the workers' compensation system, submits rate requests to the Commissioner of Insurance, and recommends policy changes on behalf of the insurance industry. The information provided by WCRIB and used in this discussion is based on interviews with agency officials conducted in 1990 and 1991.

b. Thompson, *Nation's Business*, p. 20.; and the *National Council on Compensation Insurance, Issues Report, 1991*: A Summary of Issues Facing Workers Compensation, Boca Raton, Florida: National Council on Compensation Insurance, 1991, pp. 26-42.

c. WCRIB.

d. WCRIB.

e. WCRIB.

f. WCRIB.

g. WCRIB.

h. WCRIB.

i. WCRIB.

j. WCRIB.

k. Day, Stephen. Executive Director of the Massachusetts Workers' Compensation Advisory Council, Department of Industrial Accidents, Boston, MA, May, 1990 [Interview].

l. Joint Special Recess Committee on Workmen's Compensation Insurance Rates and Accident Prevention. *Report of the Committee*, Senate No. 370, Boston, February, 1917, page 35.

m. Russell, William. Interview with Former Director of the Office of Safety, Comm. of Mass. Department of Industrial Accidents, Boston, MA, May, 1990.

n. Linsky, Steve. General Counsel, Comm. of Mass. Department of Industrial Accidents, Boston, May, 1990 [Interview].

SECTION B

CASE STUDIES: INDUSTRIAL DISEASE

7

The Occupational Lead Standard

A GOAL UNACHIEVED, A PROCESS IN NEED OF REPAIR

ELLEN K. SILBERGELD
PHILIP J. LANDRIGAN
JOHN R. FROINES
RICHARD M. PFEFFER

Lead is an ancient metal. It has been recognized since antiquity to cause disease in industrial workers. Cases of occupational lead poisoning were described by Hippocrates (1), Ramazzini (2), Thackrah (3), and Hamilton. (4) Efforts to regulate the exposure of workers to lead also extend far back in time. The ancients recognized the importance of inhalation in causing occupational lead poisoning, and Pliny recommended that miners in the Roman Empire protect themselves against lead by wearing an animal bladder tied tightly across the face. (1)

In the United States, the most recent effort to reduce occupational lead exposure and prevent toxicity was the adoption in 1978 of a comprehensive lead standard by the Occupational Safety and Health Administration (OSHA). This standard has been very influential in reducing exposures in many workplaces and in reducing the incidence of occupational lead poisoning. Nevertheless, despite many successes, there are problems with the OSHA lead standard—problems in the extent of coverage, in the implementation and enforcement of the standard, and in the proliferation of exclusionary "fetal protection policies" that result in evasion of the intent of the Occupational Safety and Health Act to provide a safe workplace for all workers. In addition, recent biomedical research on lead has shown that lead is toxic to adults at levels that previously were thought to be safe, and that are below the thresholds established on that assumption in the standard more than 10 years ago. This review will address some current problems associated with the OSHA lead standard and will explore their implications for public policy.

The modern era of lead regulation began in Britain in the early years of this century with the work of Sir Thomas Legge, the first Medical Inspector of Factories (5), who proposed four axioms for control of occupational lead poisoning:

1. Unless and until the employer has done everything—and everything means a good deal—the workman can do next to nothing to protect himself, although he is naturally willing enough to do his share.
2. If you can bring an influence to bear external to the workman (that is, one over which he can exercise no control) you will be successful; and if you cannot or do not, you will never be wholly successful.
3. Practically all industrial lead poisoning is due to the inhalation of dust and fumes; and if you stop their inhalation you will stop the poisoning.
4. All workmen should be told something of the danger of the material with which they come into contact and not be left to find it out for themselves—sometimes at the cost of their lives.

Under Legge's influence, occupational lead poisoning was made a reportable disease in Britain in 1899. With the continuing surveillance and control that followed this action, the number of reported cases of industrial lead poisoning fell from 1,058 with 38 deaths in 1900, to 505 in 1910, and to 59 in 1973, despite the considerable increase in consumption of lead that occurred over that period. It should be noted that these figures do not cover industries utilizing tetraethyl lead, where deaths occurred particularly during the 1920s. (6)

With some modifications, these axioms have been translated into several fundamental principles of industrial hygiene that are reflected in modern health standards. First, the employer has primary responsibility for protecting workers' health. Second, engineering controls are the most effective means to control exposure levels, because they can contain and eliminate contaminants at their sources and are not so dependent as other methods upon modifying workers' behavior. Third, workers must be informed about hazards and how to control them so that they can protect themselves.

THE OSHA LEAD STANDARD

The OSHA lead standard, promulgated in 1978 (7), has been judged by many as among the most influential actions undertaken by OSHA. Thousands of workers have benefited, and many workplaces and work practices have been improved. Also the reductions in occupational exposure to lead mandated by the OSHA standard have contributed to the overall reduction in releases of lead into the American environment during the last 15 years.

Two aspects of the OSHA lead standard are distinctive. First, it is one of the few standards that requires employers to monitor both external and internal exposures to a toxic agent: it requires monitoring of both air lead and blood lead levels. Second, as discussed below, it contains a requirement for medical removal protection that also provides substantial job security.

Additionally, it is one of the occupational standards developed independently by OSHA and not derived from consensus standards established by the American Conference of Governmental Industrial Hygienists (ACGIH) or the American National Standards Institute (ANSI). (8) The lead standard was based on a comprehensive review of the biomedical literature undertaken at OSHA under the direction of then-Assistant Secretary Dr. Eula Bingham. This review, summarized in the preamble to the standard, reflects a public record of extraordinary excellence in regulatory proceedings. As a consequence of its high scientific quality, the OSHA lead standard has had influence worldwide, providing a model for policies of the World Health Organization (WHO) and the European Economic Community (EEC), as well as for occupational regulations in many nations. The successes of the OSHA standard have been based upon two principles. First, it utilized a consistent and objective definition of occupational disease linked to an objective biological marker, the level of lead in blood. This objective definition was important because it freed the worker from subjective medical judgment. Secondly, the OSHA lead standard provided workers with considerable protection from dismissal based on early signs of overexposure. This protection was provided by the medical removal protection (MRP) provision of the lead standard, which obligates employers to remove workers based upon evidence of excess exposure, while at the same time protecting most workers' rights to return to employment, wages, and seniority. Removal under MRP is triggered by a blood lead level of 50 μg/dl, and the removed worker cannot be returned to a high exposure job until the blood lead level has fallen to the "return trigger" level of 40 μg/dl. No other occupational standard in the United States embodies this enlightened concept.

EXPOSURE STILL A PROBLEM

Nevertheless, lead toxicity and excess exposures to lead in the workplace remain problems:

• Lead poisoning and elevated blood lead levels continue to be discovered with disturbing frequency among workers in the United States. (9, 10) In 1987, in New York, California, and New Jersey, more than 1,000 workers were found to have blood lead levels above 40 μg/dl, of whom approximately 200 had levels in excess of 50 μg/dl. (11) The most severe problems were seen in smelters, foundries, construction work, demolition, and automotive radiator repair. (12)

• Workers in the lead industries continue to suffer from lead-induced disease and dysfunction of multiple organ systems. Scientific advances made since promulgation of the lead standard show that these toxic effects can occur at levels of exposure to airborne lead below those permitted by the standard and at blood lead levels below the MRP trigger level of 50 μg/dl.

• There have been failures of implementation and enforcement. Because of a court-ordered stay of the OSHA standard, many of the lead industries have not yet been required by OSHA to reduce airborne lead levels. An analysis of airborne lead concentrations in U.S. industry using data from the OSHA Integrated Management Information System (IMIS) has identified 52 industries with lead exposures exceeding the OSHA standard. (12)

• There are gaps in coverage. The federal OSHA lead standard does not cover all workers exposed to lead. A particularly important exempt group is construction workers engaged in demolition, lead paint abatement, and bridge repair. Numerous cases of lead poisoning have been seen in these workers.

• Exclusionary policies have developed. Prevention of lead exposure has been achieved in part by means of exclusionary policies, such as "fetal protection policies" that exclude whole classes of workers from employment with lead—for example, pregnant women and, in some cases, all fertile women. (13)

These continuing problems with lead in the workplace occur against a background of continuing exposure to lead and widespread lead toxicity in the general environment. Although environmental exposures to lead have been reduced, largely as a result of the deleading of gasoline, millions of Americans are still exposed to residues of past uses of lead in air, dust, soil, paint, and drinking water as well as to continuing uses of lead. (14)

ADVANCES IN KNOWLEDGE OF LEAD TOXICITY

Absence of Evidence for Thresholds

Recent experimental research on lead toxicity and several large-scale epidemiologic studies have substantially changed scientific understanding of the nature of lead toxicity. It is now generally accepted that lead has no useful or beneficial role in mammalian metabolism, and that at the molecular level there may be no threshold for its toxicity. (15–17) Further, the old notion of clearly delineated categories of lead toxicity, "subclinical" versus "overt," has been replaced by a view that lead intoxication is a continuum, in which adverse effects are expressed at the cellular, organ, or whole organism level depending on the dose. The molecular approach to understanding mechanisms of lead toxicity underlies this understanding. The OSHA preamble invoked the concept of a continuum a decade ago. However, the standard was nevertheless based upon a distinction between (for instance) enzyme inhibition in red cells and "evidence of ill health or impending loss of health," such as anemia. As OSHA recognized, these distinctions are difficult to draw: higher levels of exposure to lead cause frank anemia, while lower exposures inhibit heme synthesis and inhibit the enzyme, pyrimidine-5'-nucleotidase. (18) All of these effects are now recognized to be intrinsically interrelated, and it is primarily the level of analysis that determines the definition of toxic effect. Likewise, in the nervous system, lead at high levels causes convulsions, while at lower doses it blocks calcium-dependent processes of cell:cell communication. (16)

Lead and Cardiovascular Disease

Epidemiological studies also have provided data to challenge the old assumption that there are thresholds for lead below which no adverse effects occur, most clearly in the case of lead-induced hypertension. (19) Even relatively small increases in systolic and diastolic pressure associated with lead exposure greatly increase the risk of hypertensive heart disease and stroke in the United States population. (19) Smaller studies of populations occupationally exposed to lead in the Netherlands, Canada, France, and England have corroborated this relationship between lead and hypertension. (20)

Lead Neurotoxicity

Prospective studies, in which groups of individuals are followed forward over time, beginning before their first exposure to lead, have increased the sensitivity of epidemiologic analyses to detect neurotoxic effects at blood lead levels well below 40 μg/dl. In these studies, function is assessed repeatedly, with each subject serving as his own control. Thus a prospective study conducted in Finland of new entrants to the lead industry showed a significant relationship between relatively small increases in body lead burden and decreases in nerve conduction velocity in workers with blood lead levels of 30 to 40 μg/dl. (21) Similar studies of central nervous system function also have demonstrated deficits at low levels of exposure to lead. (22–24)

Lead Nephropathy

Chronic occupational exposure to lead also has been associated with kidney disease. (25) The lowest dose at which renal toxicity is induced by lead is not certain. Most of the published epidemiological studies have studied long-term, high-level exposures, because the populations studied were employed during the 1950s and 1960s, before the OSHA lead standard went into effect. Recent toxicologic research has shown, however, that cellular manifestations of toxicity can be induced in the kidneys of experimental animals by very low levels of lead. (26) Moreover, in recent studies of lead workers never diagnosed with lead poisoning, there is evidence for higher burdens of lead stored in those workers with end-stage renal disease as compared with those without kidney disease. (27) This finding suggests an association between renal failure and chronic asymptomatic increased absorption of lead. Two recent studies in England and the Netherlands reported evidence of kidney dysfunction in workers whose blood lead levels never exceeded current occupational standards. (28, 29)

Carcinogenicity of Lead

In animals, inorganic lead has been shown to cause kidney cancer following lifetime exposures to relatively high doses. (30–32) Also in experimental studies, lead has been shown to damage DNA and to affect chromosomal integrity (33–35), changes which may be related to the induction of cancer. (36, 37) There are few epidemiologic studies of cancer in lead workers, but two case reports have noted kidney cancer in lead workers with chronic high-dose exposure to lead and renal impairment. (38, 39) The U.S. Environmental Protection Agency (EPA) recently has concluded that lead is a probable human carcinogen, although the data are not considered sufficient to calculate an estimated potency. (40)

Reproductive Toxicity of Lead

The toxicity of lead to reproduction is a very controversial topic because of its implications for policy. Experimental evidence indicates that lead at high doses is toxic to reproductive function in both male and female animals. (41) Clinical reports, most of them from the first half of this century, described reproductive toxicity in workers of both sexes with high-dose exposure to lead; the incidence of spontaneous abortion was reported to be increased in female lead workers, as well as in the wives of male lead workers. (42, 43)

In male workers heavily exposed to lead (mean blood lead level, 74.5 μg/dl), and also in males with moderately increased lead absorption (mean blood lead level, 52.8 μg/dl), decreased sperm counts and an increased prevalence of abnormal sperm have been reported. (44) Corroboration of these findings is provided by recent American and Italian studies, which also observed sperm count depression at relatively high blood lead levels (>60 μg/dl). (45, 46) The lowest dose at which spermatotoxic effects are induced is not known. Studies are needed of the effect of lead particularly on male reproductive function at lower levels of exposure.

A most difficult policy issue is raised by recently reported findings that lead causes neurological damage to the fetus at blood lead concentrations in the mother as low as 10–15 μg/dl—levels substantially below current workplace exposure standards. (17) Lead passes virtually unimpeded across the placenta, and the neurological impairment produced in the fetus by lead appears to be irreversible. Lead also may be mobilized from maternal bone stores during pregnancy and then transferred to the fetus. (47) The implications of these findings have been cited by industry in support of "fetal protection policies" that exclude women from employment in the lead trades.

LIMITATIONS IN IMPLEMENTATION OF THE OSHA STANDARD

The OSHA lead standard was deliberately limited in its application. Workers in the construction and agricultural industries, as well as workers in most industries using organolead compounds, are not covered. While organolead is of diminishing importance in the United States as a result of EPA regulations limiting lead additives to gasoline, the failure to provide protection to construction and other trades has resulted in many cases of toxicity. (48) To resolve this problem, the State of Maryland has developed a state lead standard extending coverage to construction workers; no other state has to date taken this step. Moreover, with increasing employment in waste clean-up and in lead paint abatement, two occupations where employment has grown dramatically since 1978, new jobs with potential for high levels of exposure to lead have been created. Because these operations frequently employ nonunion workers and have high turnover, regular medical surveillance is generally absent; surveillance studies of members of the laborers' union in Baltimore indicate a need for closer examination of this workforce.

The OSHA standard has been inadequately implemented. Between 1979 and 1985, OSHA conducted approximately 4,000 inspections in which lead in air was sampled. More than 32 percent of these inspections found at least one air lead level exceeding the standard, and there was no improvement in air lead concentrations over this period. In 97 different types of industries, at least 20 percent of the median air lead concentrations were at values greater than the standard. These data indicate that the problem of excessive occupational lead exposure is widespread and continuing. (12)

This failure of implementation reflects the fact that a central requirement of the lead standard—that employers comply with the PEL of 50 μg/m^3 of air exclusively by means of engineering and work practice controls—has never been in force in most of the lead industries, because implementation of this requirement has for years been judicially stayed. OSHA until recently has failed to press the courts to lift this stay. Thus, those industries have been allowed to control air lead exposures only down to the previous PEL of 200 μg/m^3 and have been allowed to comply with the much lower PEL of the lead standard by any combination of compliance methods they may choose, including respirators.

FETAL PROTECTION POLICIES

"Fetal protection policies" are an unforeseen consequence of the OSHA lead standard. These policies are in part an industry response to the entrance of women into jobs with potential lead exposure; they are an alternative to reducing occupational lead exposures to levels low enough to protect all workers, including fertile women. Under these policies, industry intends to exclude reproductively active women from the workplace for two stated reasons: first, that employers can be held liable for damage to the fetus of employees, and second, that the fetus is more sensitive than either the adult male or female to lead. (13) It is argued in these policies that the period of special sensitivity of the fetus begins in the early days of intrauterine development, before a woman may know that she is pregnant. On this rationale many fetal protection policies extend not only to pregnant women but to all women of childbearing capacity. The legal justification advanced by industry for these policies is that the risk of lead intoxication to the fetus could result in employer liability for damages expressed in the child, who is not covered by workers' compensation. (49)

Fetal protection policies raise delicate and complex questions for medical science, law, and public policy. Because they exclude workers rather than limit workplace exposure, they appear to embody the tactic of blaming the victim. The polarization of opinion that has arisen over the legal status of the fetus and the importance of civil rights issues in these policies makes it difficult to develop consensus policy alternatives in this area.

A fundamental biomedical point that must be acknowledged in developing a consensus is that the fetus is highly vulnerable to serious neurological consequences induced by lead. These effects appear to occur in the fetus—and perhaps also during later development as well—at lower levels of exposure to lead than are associated with neurological toxicity in either the adult male or female. Whether the fetotoxic dose of lead is lower than the dose associated with cardiovascular toxicity or reproductive toxicity in the adult male or female is not known, although recent data suggest that it may not be.

As a matter of law and public policy, if industry can reduce the air lead concentration through engineering and work practice controls to levels low enough to protect the fetus and to adequately protect workers of both sexes, it should be required by OSHA to clean up the workplace rather than exclude half the workforce. To date, the lead industries have hardly done all they can to reduce air lead levels. However, in some high-exposure industries (for example, smelters and foundries), it may be impossible to reduce air lead levels sufficiently by engineering controls and work practices alone. If such reductions are not feasible, then the best alternative to current exclusionary policies is only partially available because OSHA standards are limited by feasibility, according to the OSH Act. Removing pregnant women temporarily from occupational lead exposure, even in the unlikely event that removal could be achieved soon after conception, may also not protect the fetus, because mobilization during pregnancy of the woman's previously accumulated lead body burden could still pose a risk to the fetus. (47)

CHALLENGES TO BIOLOGICAL BASIS FOR MEDICAL REMOVAL

The medical removal protection (MRP) policy in the OSHA lead standard is based on the assumption that short-term removal of workers with elevated blood lead levels from heavily lead-contaminated work areas will protect these workers from lead toxicity, especially from acute neurological toxicity. MRP provides up to 18 months of removal of workers with blood lead levels greater than 50 μg/dl from job situations with excessive air lead exposures (30 μg/m^3) until the worker's blood lead level has fallen to the trigger level (40 μg/dl) for return to work. At that point, the employer is free to return the worker to a position with higher exposure to lead.

The OSHA standard assumed that MRP would provide some protection against lead toxicity to veteran workers with high body burdens and also would protect the anticipated small number of workers hired after the standard became effective whose blood lead levels would exceed the 50 μg/dl removal trigger. Thus MRP represented, in part, an attempt to address the problem of the high outlier blood lead levels that it was anticipated would arise infrequently as a result of the heterogeneity of the population.

Biologically, MRP assumes that blood lead is an adequate marker for both acute and chronic risks of lead toxicity. It assumes that short-term changes in the blood lead level reflect changes in concentrations of lead at critical target sites and in overall body burden. However, the assumption that blood lead levels accurately predict the risks of chronic injury appears to be incorrect in light of recent data, as discussed above. Blood is the compartment in the body with the most rapid turnover of lead, approximately 30 to 40 days (50), and it is not a target for chronic or irreversible lead toxicity. (51)

Under conditions of variable, repeated exposures to lead, as occur when the MRP provision is used to rotate workers in and out of heavily exposed jobs, the relationship between blood lead and tissue lead is not at steady state. Such intermittent lead exposures—dosing followed by a brief unexposed period only long enough for blood lead levels to fall—are likely to result over time in overall increases in body lead burden. These increases will occur in compartments with longer half-lives for lead than blood, particularly bone, where lead has a half-life in excess of 25 years. (50, 52) These storage compartments may be sites of toxicity, for example, the kidneys, or sources of internal redistribution, for example, bone. Internal redistribution is observed in workers who have had chronic exposure to lead, whose blood lead levels remain higher for years after retirement than those of unexposed persons. (53) Bone lead levels in these workers can be 10–100 times those found in unexposed controls. (52, 53) Release of lead from bone has been suggested to be associated with nephropathy in older men. (27) In post-menopausal women, release of lead from bone may be substantial over the first four to five years after menopause. (47)

These data argue against the assumption that blood lead is an adequate biological marker for risk of chronic lead toxicity. (54, 55) Further research will be required to develop appropriate biological markers to address this issue. For example, it may be necessary to supplement blood lead monitoring with monitoring of bone lead by X-ray fluorescence (XRF) analysis. (55) This would provide a much sounder biological basis for assessing risks of chronic lead toxicity. Most importantly, in the absence of validated markers of body and organ lead burdens, it will remain necessary to monitor airborne lead levels in the workplace and also to assess acute and chronic lead absorption in urine and blood of lead-exposed workers. It is important to obtain whatever information is available on both internal and external dosimetry for lead.

THE CONTINUING PROBLEM OF LEAD IN PUBLIC HEALTH

Lead toxicity is a significant problem not only in the workplace. It is also one of the most prevalent diseases of environmental origin in the United States today. (14) Using a blood lead criterion of 15 μg/dl as an index of increased lead absorption, a value currently under review by the Centers for Disease Control, more than 67 percent of black inner-city children and nearly 17 percent of all children under the age of 5 in the United States—a total of 2.3 million children—have excessive internal exposures to lead. (14)

Many adults also have elevated body lead burdens as a result of general environmental exposures that put them at increased risk for neurotoxicity, renal toxicity, and hypertensive heart disease. Occupational exposures represent an additional burden over and above this background for that segment of the population working with lead. EPA's Toxics Release Inventory confirms that industry remains a significant source of lead releases into the environment. For instance, in 1987, more than 33 million pounds of lead and lead compounds were released in land disposal sites. Primary and secondary lead smelters in the United States currently emit more than 17 million tons of lead per year into the environment. (56) Reducing exposures to lead in the occupational setting will, if properly implemented, have the additional benefit of reducing releases of lead into the general environment.

THE OPTIMAL SOLUTION

For many reasons, as discussed above, the blood lead level triggers for MRP removal and return in the lead standard must be reduced substantially. The optimal solution to the problem of fetal protection is to prevent the accumulation of lead body burden in both men and women. Using blood lead as the biological marker of exposure must be adjusted to prevent the accumulation of a body burden that is toxic to adults of either sex or to the fetus. As discussed below, it is not clear what blood lead level would correspond to this goal, or even if blood is the appropriate compartment to monitor exposure to achieve this goal. Clearly, the current removal and return triggers should be lowered from 50 μg/dl and 40 μg/dl, respectively, possibly to levels no higher than 20 μg/dl and 10 μg/dl. Reducing MRP trigger levels would significantly improve protection for all workers. It should not present serious feasibility problems when applied to new workers. However, for veteran workers with substantially elevated body burdens and persistently high blood lead levels, the lower trigger levels may need to be phased in over a period of time.

In addition to lowering MRP trigger levels, there are two other steps OSHA should take to reduce occupational exposures to lead. OSHA's permissible exposure limit (PEL) for lead must be set as low as current medical evidence dictates necessary to protect health over both the long and short term. This action would constitute a radical departure from OSHA's traditional approach to setting PELs. At present, regardless of continuing significant risks to health, OSHA sets PELs only as low as can feasibly be achieved by engineering and work practice controls. The PEL for lead was set at 50 $\mu g/m^3$, because that was the lowest level in OSHA's opinion that could feasibly be achieved by engineering and work practice controls as of 1978.

If OSHA were to establish a PEL for lead at a level lower than is achievable by engineering and work practice controls, it would be necessary to place greater reliance than heretofore on respirators. That step must be undertaken with great care and with the explicit recognition by all parties that it would not constitute a retreat from OSHA's long-standing commitment to a hierarchy of controls that quite properly values engineering and work practice controls above personal protective equipment such as respirators. Accepting a greater reliance on respirators would be only supplementary protection, and would be justified by the lack of other options to reduce occupational lead exposure to very low levels to protect the health of workers of both sexes as well as of the fetus.

It makes considerable sense to require industry to implement both kinds of controls: to implement engineering and work practice controls, and then additionally to rely on respirator programs for supplementary protection where engineering controls are incapable of achieving appropriately low levels. If this approach were to be adopted by OSHA, close regulatory oversight of the balance between control technologies would be required. The movement already evident within some sectors of industry to rely excessively on respirators must aggressively be checked.

BAN UNNECESSARY USES

Strong consideration should be given to banning all unnecessary uses of lead. This step would be the most reliable method of reducing both occupational and nonoccupational exposures to lead. This could be accomplished by OSHA or by the EPA using its powers under Section 6 of the Toxic Substances Control Act, as has been done for asbestos and polychlorinated biphenyls. EPA has specific statutory powers to ban the production and importation of toxic substances.

As a first step toward that goal, OSHA or the National Institute for Occupational Safety and Health (NIOSH) should conduct a national survey of all lead-using industries. In conjunction with EPA, those in which the use of lead is not essential and where suitable substitutes are available could be identified. Then, to encourage restriction of the use of lead to essential needs, phaseout of these uses should be introduced either through legislation or regulation. An alternative to regulation was proposed recently by the Environmental Defense Fund, a heavy excise fee, perhaps four times the current price of lead, on all originally produced lead, that is, on all lead produced by primary smelters as well as on all imported lead. Such a fee would provide a strong market incentive to find substitutes for lead whenever substitution is feasible and cost-effective. This action likely would result in the elimination or reduction of many occupational lead exposures. Such a fee also would encourage more recycling and therefore greater care in disposing of lead wastes, since no fee would be imposed on domestically recycled lead. However, if recycling is to be encouraged, it is critical to assure that workers in all lead-exposed jobs in waste handling and recovery operations are well protected by OSHA.

In summary, these strategies for reducing lead exposure and body lead burdens—(a) changing the way PELs are set and thereby lowering the PEL for lead; (b) lowering MRP trigger levels for lead to deal with the problem of lead body burdens; and (c) attempting to find suitable substitutes for lead by limiting use and by imposing a heavy excise fee on the production and import of all lead, except domestically recycled lead—would largely eliminate the need for exclusionary fetal protection policies.

CONCLUSIONS

This review of the current status of the occupational lead standard produces the following conclusions:

- The lead standard must be vigorously and consistently enforced. No standard is better than its implementation. The laudable goal of the lead standard, to eliminate occupational lead toxicity, has not been fully met because of failures of oversight and commitment within OSHA, the courts, the states, and the Congress.
- The PEL for lead in the workplace air and the trigger level of lead in blood for MRP are both too high in light of current medical knowledge. Toxicity is known to be occurring in workers in multiple organ systems at air and blood lead levels below currently permissible levels of exposure.

• The concept of MRP based solely on blood lead levels needs to be reexamined. MRP may not be effective as currently utilized in preventing chronic lead toxicity. Periodic monitoring of body lead burden, perhaps by XRF analysis of the lead content in bone, may have to be used to supplement blood lead monitoring. In the interim, we suggest that the medical removal and return trigger blood lead levels be reduced to 20 and 10 μg/dl, respectively.

• The creation by the lead industries of fetal protection policies excluding women from employment in the lead trades is a fundamentally unacceptable approach to control of lead exposure. To counter these exclusionary policies, OSHA must make an explicit commitment to protect the civil rights of both male and female lead workers to employment in a safe and healthful workplace. Reduction of the PEL and of the trigger levels for blood lead as well as critical reexamination of the necessity for lead in commerce would constitute components of that commitment.

• More information about lead use will be required to systematically track lead poisoning. Lead poisoning and elevated blood lead levels should be reportable on a state and national basis, and registries should be maintained to track trends and effects of regulation on disease prevention. Data collected under the biological and environmental monitoring provisions of the lead standard should be transmitted routinely by industry to OSHA and to NIOSH and used for surveillance of trends and for identification of problems, as has been the case with the mining industry. State programs for reporting cases of lead poisoning and elevated blood lead levels should be implemented nationwide and coordinated into a NIOSH registry. Epidemiological studies of such registry data could be used to resolve such major issues as the possible carcinogenicity of lead and the long-term effects of lead in populations as they age and potentially release lead from bone stores.

• All industries and occupations with potential for exposure to lead should be covered by the occupational lead standard. This coverage should extend to all forms and all uses of lead from extraction through waste disposal to recycling. Training workers for abating environmental lead hazards—such as lead paint removal and hazardous waste clean-up—is a high priority for cooperative research and development similar to training projects currently mandated by the Superfund legislation.

• Occupational and environmental health criteria should be consistent with each other. The notion that workers may be permitted to be exposed to much higher levels of risk than the general population is inherently objectionable. The OSHA standard for lead exposure is now seriously inconsistent with the health-based criteria recently expressed by EPA and CDC to bring blood lead levels of all Americans below 10–15 μg/dl. For this reason and to eliminate the justification that has been advanced for fetal protection policies, OSHA should amend the lead standard to bring occupational standards into line with the broad national health goal of preventing lead toxicity in the United States population.

- All nonessential uses of lead must be eliminated. A national study is required to identify all uses of lead and to target as many as possible for phaseout. A recent proposal by the Environmental Defense Fund for an excise tax on newly mined and imported lead deserves serious consideration.

AUTHORS' NOTE

The opinions expressed in this chapter by its authors are their own and do not necessarily reflect the views of any organization or agency with which they may be affiliated.

NOTES

1. Hunter, Sir D. *The Diseases of Occupations*. Boston: Little, Brown (5th Edition), 1973.

2. Ramazzini, B. *De Morbis Artificum Diatriba*. Transl. Wright, W.C. Chicago: University of Chicago Press, 1913.

3. Thackrah, C.T. *The Effects of Arts, Trades and Professions and Civic States, and Habits of Living on Health and Longevity with Suggestions for the Removal of Many of the Agents which Produce Disease, and Shorten the Duration of Life*. London: Longman, Rees, Orme, Brown, Green and Longman, 2nd ed., 1832.

4. Hamilton, A. *Exploring the Dangerous Trades*. Boston: Little, Brown, 1943.

5. Legge, Sir T. *Industrial Maladies*. London: Oxford University Press, 1934.

6. Rosner, S. and Markowitz, D. "A gift of God." *Amer. J. Public Health*.

7. Occupational Safety and Health Administration. U.S. Department of Labor. "Occupational Exposure to Lead." *Federal Register* 43:54353-54616, 1978.

8. Castleman, B., Ziem, G.E. "Corporate influence on threshold limit values." *Amer. J. Ind. Med.* 3:530-559, 1988.

9. Cullen, M.R., Robins, J.M. and Eskanazi, B. "Adult inorganic lead intoxication: Presentation of 31 new cases and review of recent advances in the literature." *Medicine* 62:221-247, 1983.

10. Rudolph, L., Maizlish, N., Tepper, A., Gerwel, A., Wenzel, T., Melius, J.M., Stone, R., Martin, J., Pichette, J. and Montopoli, M. "Surveillance for occupational lead exposure—United States, 1987." *MMWR* 38:642-646, 1989.

11. Goldman, R.H., Baker, E.L., Hannan, M. and Kamerow, D.B. Lead poisoning in automobile radiator mechanics. *New Engl. J. Med.* 317:214-218, 1987.

12. Froines, J.R., Baron, S., Wegman, D. and O'Rourke, S. "Characterization of the airborne concentrations of lead in U.S. industry." *Amer. J. Industr. Med.*, 18:1-17, 1990.

13. Curran, W.J. "Dangers for pregnant women in the workplace." *New Engl. J. Med.* 312:164–165, 1985.

14. Mushak, P. and Crocetti, A.F. "Determination of numbers of lead-exposed American children as a function of lead source: integrated summary of a Report to the U.S. Congress on Childhood Lead Poisoning." *Environ. Res.* 44:210–229, 1989.

15. Landrigan, P.J. "Toxicity of lead at low dose." *Br. J. Industr. Med.* 46:593–596, 1989.

16. Silbergeld, E.K. "A neurobiological perspective on lead toxicity." In Needleman, H.L. (ed.) *Human Lead Exposure*. New York: Raven Press, 1991.

17. Bellinger, D., Leviton, A., Waternaux, C., Needleman, H.L. and Rabinowitz, M. "Longitudinal analyses of prenatal and postnatal lead exposure and early cognitive development." *New Engl. J. Med.* 316:1037–1043, 1987.

18. Angle, C.R., and McIntire, M.S. "Low level lead and inhibition of erythrocyte pyrimidine nucleotidase." *Environ. Res.* 17: 296–302.

19. Pirkle, J.L., Schwartz, J., Landis, R. and Harlan, W.R. "The relationship between blood lead levels and blood pressure and its cardiovascular risk implications." *Amer. J. Epidemiol.* 121:246–258, 1985.

20. Victery, W., Tyroler, H.A., Volpe, R., and Grant, L.D. "Summary of discussion sessions: symposium on lead–blood pressure relationships." *Environ. Health Persp.* 78:139–155.

21. Seppalainen, A.M., Hernberg, S., Vesanto, R. and Kock, B. "Early neurotoxic effects of lead exposure: a prospective study." *Neurotoxicology* 4:181–192, 1983.

22. Hanninen, H., Hernberg, S., Mantere, P., Vesanto, R. and Jalkanen, M. "Psychological performance of subjects with low exposure to lead." *J. Occup. Med.* 20:683–689, 1979.

23. Valciukas, J.A., Lilis, R., Fischbein, A., Selikoff, I.J., Eisinger, J. and Blumberg, W. "Central nervous system dysfunction due to lead exposure." *Science* 201:465–467, 1978.

24. Baker, E.L. Jr., Landrigan, P.J., Barbour, A.G., Cox, D.H., Folland, D.S., Ligo, R.N. and Throckmorton, J. "Occupational lead poisoning in the United States: clinical and biochemical findings related to blood lead levels." *Brit. J. Industr. Med.* 36:314–322, 1979.

25. Goyer, R.A. and Rhyne, B. "Pathologic effects of lead." *Interntl. Rev. Exper. Path.* 12:1–77, 1973.

26. Fowler, B.A. "Mechanisms of metal-induced renal cell injury: roles of high affinity metal-binding problems." *Contrib. Nephrol.* 64:63–92, 1989.

27. Van de Vyver, F.L., D'Haese, P.C., Visser. W.J., Elseviers, M.M., Krippenberg, L.J., Lamperts, L.V., Wedeen, R. and DeBroe, M. "Bone lead in dialysis patients." *Kidney Interntl.* 33:601–607, 1987.

28. Heiber, R.F.M., Verplanke, A.J.W. and Verschoor, M.A. "Early kidney effects by exposure to xenobiotics." Interntl. Conf. Heavy Metals in the Environment. I:314–317, 1989.

29. Mason, H., Somervaille, L.J., Wright, A., Chettle, D.R. and Scott, M.C. "The renin-angiotensin axis and kallikrein excretion in lead exposed workers." Interntl. Conf. on Heavy Metals in the Environment. II:254–257, 1989.

30. Boyland, E., Dukes, C.D., Grover, P.L. and Mitchley, B.C.V. "The induction of renal tumours by feeding lead acetate to rats." *Br. J. Cancer* 16:282–288, 1962.

31. Mao, P. and Molnar, J.J. "The fine structure and histochemistry of lead-induced renal tumors in rats." *Am. J. Pathol.* 50:571–603, 1967.

32. Van Esch, G.J. and Kroes, R. "The induction of renal tumors by feeding basic lead acetate to mice and hamsters." *Br. J. Cancer* 23:765–771, 1969.

33. Nordenson, I., Beckman, G., Backman, L. and Nordstrom, S. "Occupational and environmental risks in and around a smelter in northern Sweden. IV. Chromosomal aberration in workers exposed to lead." *Hereditas* 88:263–267, 1978.

34. Forni, A., Sciame, A., Bertazzi, P.A. and Alessio, L. "Chromosome and biochemical studies in women occupationally exposed to lead." *Arch. Environ. Health* 35:139–146, 1980.

35. Sarto, F., Stella, M. and Acqua, A. "Studio citogenetico su un gruppo di lavoratori con indici di aumentato assorbimento di piombo." *Med. Lav.* 69:172–180, 1978.

36. DiPaolo, J.A., Nelson, R.L. and Casto, B.C. "In vitro neoplastic transformation of Syrian hamster cells by lead acetate and its relevance to environmental carcinogenesis." *Br. J. Cancer* 38:452–455, 1978.

37. Casto, B.C., DiPaolo, J.A. and Meyers, J. "Enhancement of viral transformation for evaluation of the carcinogenic or mutagenic potential of inorganic metal salts." *Cancer Res.* 39:193–198, 1979.

38. Lilis, R. "Long-term occupational lead exposure, chronic nephropathy, and renal cancer: A case report." *Am. J. Ind. Med.* 2:293–297, 1981.

39. Baker, E.L. Jr., Goyer, R.A., Fowler, B.A., Khettry, U., Bernard, D.B., Adler, S., White, R.D., Babayan, R. and Feldman, R.G. "Occupational lead exposure, nephropathy and renal cancer." *Am. J. Ind. Med.* 1:139–148, 1980.

40. U.S. Environmental Protection Agency. "Review of lead carcinogenicity and EPA Scientific Policy on Lead." (EPA-SAB-EHC-90-001), Washington, DC: EPA, December 1989.

41. Rom, W.N. "Effects of lead on reproduction." In: Infante, P.F. and Legator, M.S. eds. *Proceedings of a Workshop on Methodology for Assessing Reproductive Hazards in the Workplace.* Washington, DC; National Institute for Occupational Safety and Health, 1980.

42. Oliver, Sir T. *Lead Poisoning: From the Industrial, Medical and Social Points of View.* Lectures delivered at the Royal Institute of Public Health. New York: Hoeber, 1914.

43. Hamilton, A. and Hardy, H.L. *Industrial Toxicology.* Acton, MA: Publishing Sciences Group, 1974.

44. Lancranjan, I., Popescu, H.I., Gavenescu, O., Klepsch, I. and Serbanescu, M. "Reproductive ability of workmen occupationally exposed to lead." *Arch. Environ. Health* 30:396–401, 1975.

45. Cullen, M.R., Kayne, R.D. and Robins, J.M. "Endocrine and reproductive dysfunction in men associated with occupational inorganic lead intoxication." *Arch. Environ. Health* 39:431–440, 1984.

46. Assennato, G., Paci, C., Baser, M.E., Molinini, R., Candela, R.B., Altamura, B.M. and Giorgino, R. "Sperm count suppression without endocrine dysfunction in lead-exposed men." *Arch. Environ. Health* 4:387–390, 1986.

47. Silbergeld, E.K., Schwartz, J., and Mahaffey, K.R. "Lead and osteoporosis: Mobilization of lead from bone in post-menopausal women." *Env. Research* 47:79-94, 1988.

48. Marino, P., Franzblau, A., Lilis, R. and Landrigan, P.J. "Acute lead poisoning in construction workers: the failure of current protective standards." *Arch. Environ. Health* 44:140-145, 1989.

49. *International United Auto Workers vs. Johnson Controls*, 886 F.2d 871 (7 Cir. 1989) (en banc). Cert. granted, 58 U.S.L.W. 3609 (U.S. March 27, 1990), (No. 89-1215).

50. Rabinowitz, M.B., Wetherill, G.W. and Kopple, J.D. "Kinetic analysis of lead metabolism in healthy humans." *J. Clin. Investigation* 58:260-270, 1977.

51. Hernberg, S. "Biochemical and clinical effects and responses as indicated by blood lead concentration." In: Singal, R.L., Thomas, J.A., eds. *Lead Toxicity.* Baltimore: Urban and Schwarzenburg, pp. 367-399, 1980.

52. Somervaille, L.J., Chettle, D.R., and Scott, M.C. "In vivo measurement of lead in bone using X-ray fluorescence." *Phys. Med. Biol.* 30:929-943, 1985.

53. Skerfving, S., Nilsson, U., Attewell, R., Schutz, A., Christofersson, J.D., Mattson, A., Tell, I., and Ahlgren, L. "Bone lead in lead workers." *Environ. Hlth. Persp.*, 1990.

54. Hattis, D. "Dynamics of medical removal protection for lead—a reappraisal." Massachusetts Institute of Technology, Center for Policy Alternatives (Report No. CPA-81-245) Cambridge, MA, 1981.

55. Landrigan, P.J. "Strategies for epidemiologic studies of lead in bone in occupationally exposed populations." *Environ. Hlth. Persp.*, 91:81-86, 1991.

56. U.S. Environmental Protection Agency. *The Toxics Release Inventory: A National Perspective.* (EPA 560/4-84-0005) Washington: EPA, June 1989.

8

Multiple Chemical Sensitivities

DEVELOPMENT OF PUBLIC POLICY IN THE FACE OF SCIENTIFIC UNCERTAINTY

MARK CULLEN

THE NUMBERS STORY

Although there are no data available from population surveys, the syndrome of Multiple Chemical Sensitivity (MCS) in its full-blown clinical form is probably uncommon. For example, among more than 2,700 patients treated at Yale's Occupational and Environmental Medicine Clinics since 1986, only 49, or 1.8 percent, had MCS. Of course, it is likely that this figure represents only a fraction of Connecticut cases—many less severe or misclassified cases must exist. But MCS remains an unusual syndrome among the millions exposed regularly to chemical irritants and intoxicants.

The belief that some people might be exquisitely sensitive to chemicals in air, water, and food is not new. Early allergists proffered this idea during the 1950s, and case descriptions date back probably to antiquity. (1–3) However, it was not until the last decade, when public concern about chemicals and other toxic exposures skyrocketed, that this problem of apparent intolerance to chemicals in very low doses has begun to be taken quite seriously. As our society has begun to control exposures to chemicals to reduce scientifically established risks for morbid and fatal diseases, the special problems of the "sensitive" portion of the population have challenged the concept and basis for defining "safe" limits for exposure. There has been the perception, too, that a larger number of people are experiencing symptoms attributed to low-level chemical exposures than was previously imagined. (4)

Early in the postregulatory era, attention to this issue was limited to a small subgroup of medical practitioners who called themselves clinical ecologists. Applying a theoretical framework adapted from the earlier work of clinical allergists such as Randolph and others (5, 6), these practitioners subscribed to the view that a wide range of clinical maladies was caused by an accumulation of chemical agents in air, water, food and medications. Not only patients who had complaints directly referable to chemical exposures but also many others with ill-defined constitutional, respiratory, gastrointestinal, and musculo-skeletal problems were treated using novel diagnostic and therapeutic techniques that involved isolation from contact with any man-made chemicals while attempting to enhance the patient's tolerance to those to which exposure would be inevitable in our society. Calling the illness "Environmental Illness" or "20th-Century Disease," these practitioners at once endeared themselves to a faithful patient population that had failed to get satisfaction from more traditional practitioners and at the same time inspired the wrath of the community of allergists and immunologists and other mainstream groups who were skeptical of the theory and methods of ecology practice. (7) But in any event, the numbers of both patients and clinical ecologists remained relatively small; most physicians were entirely unaware of the clinical problem involved or the medical practices at issue.

In the 1980s, however, three interrelated phenomena occurred that brought the problem of chemically sensitive people "out of the closet." First, for uncertain reasons, the number of recognized "cases" started to grow, enhancing the likelihood that medical practitioners of every description would start seeing these patients in their practices. Second, as a response to the larger societal interest in chemically induced disease and occupational health, clinics and centers arose throughout the country specializing in the treatment of individuals with health problems suspected to be of workplace or related environmental origin. (8) The overwhelming majority of these care facilities were based in very traditional medical settings such as medical schools, hospitals, and multispecialty clinics, and were staffed by physicians with generally recent, scientifically oriented training. Finally, the growing awareness of work-related disease created by regulation and societal orientation and the legal successes of some early victims of asbestos-related disease combined to create a highly litigious climate around all aspects of occupational and environmental health. From the corporate perspective, employers and manufacturers were at once being required to reduce noxious exposures and provide those exposed with detailed information about risk, that information in turn raising concern and apparently the likelihood of workers' compensation or liability suit. From the perspective of the workforce and population, chemicals were increasingly felt to represent an unacceptable threat to health for which legal action would be appropriate as redress. The net effect was a marked upsurge in claims brought for chemical injuries of all kinds and an increasingly circumspect response by the defendants of these claims. The explosive rise in health care and workers' compensation costs during the decade further fueled both action and reaction.

The combined impact of these phenomena was an unexpected upswell of patients with symptomatic problems around low-level chemicals presenting to the new occupational health clinics. No longer could the problem be quietly shunted by word of mouth to a small group of clinical ecologists whose methods and costs were increasingly unacceptable to health care payors. So, too, the clinical ecologists failed to be sufficiently convincing in the legal process to serve the needs of patients who they felt were unable to work or had been permanently injured by chemicals. Out of the observations at these clinics came a new name for the disorder—Multiple Chemical Sensitivities (9)—and the first serious discussions of the problem in mainstream medical journals. In just a few years, debate, both scientific and policy-related, moved from the "closet" to the nation's most august scientific and medical organizations and from the courthouse to the halls of legislatures and policymakers.

Unfortunately, what did not happen quite so quickly was discovery of the pathogenesis of Multiple Chemical Sensitivities (MCS) or how to prevent, diagnose, or manage the problem. Not surprisingly, scientific uncertainty has, in turn, crippled effective legal resolution of "cases" and opened a pandora's box of policy questions. In the sections which follow, I will attempt to summarize the current scientific debate as a prelude to a discussion about some legal and policy considerations.

SCIENTIFIC CONTROVERSIES

Two interrelated questions lie at the heart of present scientific controversy over MCS: (a) how to define it distinctively, and (b) how to explain it pathophysiologically. Although the issue of pathogenesis generates most of the heat in debate and has been the focus of the several research activities that are now under way to study MCS, the definitional issue is very important from a scientific perspective, all the more so because there is so little agreement about the mechanism or mechanisms that cause the manifestations of the syndrome. Indeed, until there is an agreed-upon case definition, the possibility of performing reproducible and interpretable clinical studies is nil. Precisely for this reason, a growing number of interesting case series, ostensibly demonstrating the presence or absence of certain key findings in MCS patients, must be treated as anecdotal for lack of careful or relevant definition of the populations studied. (10–12)

MCS CASE DEFINITION

Almost everyone agrees on certain essential aspects. First, the disorder appears to be acquired, typically after the occurrence of a toxic reaction to some solvent, pesticide, or respiratory irritant. Subsequently, patients report the onset of "reactions"—acute, multiorgan system symptoms—that recur after exposure to various excitants or precipitants, usually chemicals in air, water, or food. These excitants are typically present in the environment in which symptoms occur at levels far below those which typically bother most people. Further, the excitants are multiple and unrelated chemically. In general, reactions are predictable after exposures to the various excitants that the patient has discovered; removal from those environments brings at least short-term relief from reaction. Most patients not only suffer the acute reactions to chemicals but also feel chronically ill with multiple symptoms between attacks. Often, this more chronic symptomatology dominates the clinical picture, acute reactions being rare because of the patient's modified lifestyle. Finally, no accepted test of organ dysfunction or disease, including psychological disease, can adequately explain the symptoms.

Unfortunately, within this broad case definition and description lies extraordinary room for interpretation, especially since all but the normal results on lab tests rely primarily on patient reports and are not readily documentable or verifiable. Three points have raised the greatest controversy. The first relates to the relative importance of the circumstances of onset. Many, including myself (13), have argued for a rigid view of this criterion, limiting the scientific designation of MCS to the most typical patients who develop the syndrome after an otherwise uneventful recovery from a single or multiple episodes of chemical injury or intoxication such as acute solvent poisoning, pesticide poisoning, or smoke inhalation. Failing such an historic event, the question arises how to construe the onset of MCS. Is MCS excluded by a workup for inexplicable symptoms before the reactions to chemicals became apparent, or, rather, should the date of onset of MCS be pushed back to include such periods as well? On the other hand, if one eliminates this criterion altogether, as suggested by the panel at a recent National Academy of Sciences conference on the subject, MCS might be difficult to distinguish from other conditions characterized by multiple somatic complaints, such as chronic fatigue syndrome.

Another criterion that evokes debate is the issue of predictability. If any historic feature distinguishes MCS from other ill-defined illnesses, such as chronic fatigue syndrome, it is the temporal association between exposures to the various low-dose excitants and the occurrence of symptoms. Yet many patients relate that sometimes they start to have a reaction and then discover the "culprit." Of course, for these individuals who react to so many different things, it is hardly diagnostically meaningful that when forced to find a "culprit" to explain symptoms, they succeed. On the other hand, certainly allergic individuals of the more classic kind can develop a wheeze, hive, or sniffle only to then "find" the source, an ex post facto association we never question. Precisely for this reason some now advocate that the criterion of predictability must be verified with blinded challenge testing to excitants. (14) Perhaps ironically, this idea for verification has been proposed both by skeptics who think that few if any patients will really meet the challenge as well as by those who, believing the disease to be biologically mediated, view objective challenge as means for vindication of their views. Unfortunately, the agreement is more virtual than real. Few centers have the capability to deliver safe, blinded exposures reproducibly to allow consistency in diagnosis. Blinding is particularly hard to achieve since many of the offending substances in question have very low odor or irritation thresholds from the norm. As if this were not problem enough, those investigators who subscribe to the immunologic view of MCS do not feel that any challenge can be legitimately interpreted until the patient has been "de-adapted," which means nothing less than days to weeks of isolation in specially prepared chemical-free environments; this would seriously limit the applicability of a validation requirement for predictability of response.

The third area of controversy over definition relates to the final point, namely the absence of alternative diagnoses or lab testing abnormalities that explain symptoms. Not surprisingly, a substantial fraction of MCS patients are atopic, that is, they have typical allergies (like 25–30 percent of the "healthy" population). Similarly, given the consequences of recurring symptoms and often restricted social and economic opportunities that result for many with MCS, it is not surprising that many are anxious and/or depressed. The problem arises, therefore, as to how to interpret these conditions vis-à-vis this criterion. Some investigators have used the frequent occurrence of gastrointestinal (GI), respiratory, or psychological dysfunction as evidence that MCS criteria are rarely filled (that is, MCS is NOT a distinctive diagnostic group, but an expression of another disorder) (10, 11, 15), while others have tried to demonstrate the high occurrence of abnormalities such as autoantibodies (16) as proof that MCS must be biologically mediated.

MCS PATHOGENESIS

If the medical community is having difficulty arriving at a consensus on diagnostic criteria, views on pathogenesis are so divergent that consensus is not a realistic short-term goal. Broadly speaking, there are two polar camps. On the one hand are those who believe that MCS represents a biologically mediated toxicity of man-made chemicals. Although there is not uniformity of view regarding the target of this toxic effect—some continue to support the immune theory of the clinical ecologists while others suggest that the Central Nervous System or mucosal surfaces of the GI and/or respiratory tracts are the primary site of injury (14)—proponents of this polar view believe that ultimately MCS will be as biologically decipherable as occupational asthma or other idiosyncratic disorders caused by low levels of chemicals.

At the other pole are those who feel that MCS is primarily a psychological disorder. Although there is a range of perspectives on pathogenesis within this group, the unifying theme is the belief that the relationship between chemicals and symptoms is symbolic. Many of the proponents of this general theory are reluctant to accept that MCS represents more than a modern presentation of well-characterized anxiety or affective disorders such as panic disorder, somatization disorder, or depression. (15, 17) Others have suggested that MCS may represent a variant of posttraumatic stress disorder or behavioral sensitization. (18, 19)

Unfortunately, neither group can currently cite credible scientific data to support its contention. The "biologists" rely most heavily on the collective anecdotes of those who have seen large numbers of "cases," especially those who appear to react to very low level exposures predictably in environmental chambers and appear to improve when isolated from chemicals over protracted time periods. Unfortunately, neither careful case definition nor rigorous methods for blinding have been employed, rendering interpretation difficult. On the other hand, the "psychologists" have reproduced more of these same errors in their zeal to refute the biological view. Although there has been evidence presented that large numbers of "cases" are free of measurable abnormalities or organ system dysfunction while harboring strong measurable tendencies toward psychological distress, these investigators, too, have failed to define their cases carefully or apply proper control groups for the various tests performed, especially the psychological tests. It would appear that those who dispute the biological view feel that the burden of proof rests entirely with its proponents—that there is reason to expose what MCS isn't but no need to determine scientifically what it is.

A large number of clinicians, especially those in the academic occupational/environmental medicine community that brought discussions of MCS into the mainstream, are unwilling to choose between these unsupported views. In an effort to maintain sufficient confidence among the patients (who tend to be strongly inclined toward the biologic view) and sufficient impartiality to continue to make reasoned observations, these physicians tend to accept both biological and psychological perspectives, viewing the disorder as some form of toxic injury amplified by personal psychological factors. This middle ground may be viewed either as conservative scientific wisdom or as a cop-out from engagement in the debate, but many, including this author, believe that the pathogenesis of MCS will ultimately be proven to involve both biologic and psychologic factors.

POLICY CONSIDERATIONS

There are three dimensions to the public debate over MCS. The first has to do with the regulation and control of nontraditional, unproven medical practices by the clinical ecologists and other "environmental" physicians. The second issue is the appropriate assignment of benefits and/or other forms of legal redress to MCS patients. The third and broadest issue relates to the development of environmental exposure standards and practices, given the existence of an apparently hypersusceptible subgroup in the population. In each case, the divergence of scientific opinion precludes straightforward application of existing approaches. What might seem fair and reasonable to those of the "biologic" persuasion is viewed as dangerous, unreasonable, and even counterproductive by those in the "psychologic" camp. Complicating the issue are the economic stakes involved, which would appear to be clouding everyone's objectivity regarding the true nature of MCS.

CLINICAL ECOLOGY PRACTICE

While the MCS problem remained confined to a small number of patients ostensibly dissatisfied with the care they received from mainstream practitioners who had little knowledge or interest in low-dose chemicals, the practice of clinical ecology went largely unrecognized, uncontested, and undiscussed. However, as the number of such practices proliferated, initially in California, practitioners in the allergy/immunology community began to take notice and offense. (7) Criticism of the unproven and highly expensive methods such as the use of isolation chambers and novel "challenge tests" for diagnosis and treatment came forward from several major medical societies and organizations. (20, 21)

Viewed in its most humanitarian context, these efforts could be construed as part of the normal vigilance that the medical profession exercises over its own to limit the proliferation of quackery and potentially dangerous practices. In this context, the actions of the medical bodies were neither surprising nor new. Further, although no objective evidence of harm from clinical ecology techniques has been proven either, it is certainly fair to state that many of the techniques used by the ecologists and billed to third-party payors have not been sufficiently tested or validated to be effective.

However, it is naive to think that the increasing attacks on clinical ecologists are driven entirely by humanitarian considerations. For one thing, it is apparent that medical societies tolerate a vast array of practices that could be considered nonstandard or unproven such as homeopathy and chiropractic. Further, many of the activities conducted in private offices daily, such as the use of vitamin injections, hormonal supplements, executive physicals, and others that have not been clearly proven to be effective remain well outside the call for regulatory intervention or withholding of reimbursement. Indeed, many of the usual (and expensive) practices of traditional allergists remain speculative, such as routine desensitization in atopic adults. Quite evidently, there are turf and economic issues that have impelled usually conservative physician groups to seek the intervention of government in regulating the behavior of their "colleagues."

From the patients' perspective, the call for regulation of the ecologists appears doubly cruel. Almost all MCS patients started their quest for medical care with traditional practitioners who tended to scoff them off or demean their complaints as unfounded. Many, if not most, have been told at least once and often more frequently that there is "nothing wrong" with them or that the problem is "all in their heads" (perhaps the reasons that the psychologic view of MCS is so unpopular among patients). Having sought and found practitioners who offered them both a dignifying theory for their illness, as well as some prospect for treatment, MCS patients are now being told that they may not continue getting such care because the ecologists are "quacks," according to the same physicians who scoffed them off.

There is, unfortunately, no trivial resolution of this debate. Clearly, dismantling the practice of clinical ecology or its variants will not make the MCS problem go away, as some might have supposed. On the other hand, it seems unlikely that continued uncritical espousal of emotionally appealing but scientifically unjustified theories and practices of ecology will cure or prevent MCS either. Ultimately, the solution rests with scientific progress to understand the basis of MCS and delineate optimal strategies for its management. In the meantime, dialogue between differing groups with acceptance of the economic reality that unproven treatments cannot be financed by our constrained health care system or the government seems to be the only solution.

DISABILITY, COMPENSATION, AND LIABILITY

Despite the absence of significant impairments on standard objective tests, many MCS patients feel too sick to work at all, while others have frequent enough reactions in various environments to preclude working in factories or even offices without diffculties. Although the basis for this disability is far from clear, it is certain that malingering, the conscious manufacturing of complaints for financial or other gain, is rarely a consideration—MCS patients are miserable. Contrary to some facetious views, failure to obtain benefits appears to have little impact on successful return of MCS patients to productivity, nor does the obtaining of economic awards result in unanticipated improvement.

Again, there are two polar views and a broad middle ground. The hard-line view is that MCS patients without demonstrable impairments are not entitled to any form of disability, save possibly psychiatric benefits if a psychiatrist is willing to declare them disabled from that perspective. At the other extreme, held largely by those who subscribe to the "biologic" view, MCS patients are victims of chemical injury, and although the mechanism of injury remains unclear, as victims they are entitled to such benefits as workers' compensation and even third-party injury suits against the manufacturers or purveyors of the chemicals associated with the onset of their problem or its perpetuation. In between are many who accept that some MCS patients are disabled in a meaningful sense and therefore entitled to some form of disability benefits, at least until they can be rehabilitated to some productive state. The problem is determining which benefit. Do we know enough about the pathogenesis of MCS to be able to ascribe the onset or exacerbation to work exposures with reasonable medical probability? Can a chemical used in its normal and presumably safe way be implicated as the cause of longstanding injury in these "susceptible" hosts, analogous to the situation with occupational asthma or allergic alveolitis?

The benefits question can be dissected most easily by reference to the underlying scientific controversies that give rise to debate. The first issue is determination of disability per se. Here, the problem relates directly to the absence of an objective basis for making the diagnosis of MCS and rating its severity. Skeptics, especially potential benefits payors, are inclined to notice that the entire process is heavily weighted by patient report, and that the objective measurement by physicians is limited to the absence of abnormalities on examination and testing, not the presence of some characteristic pattern. It is hard to trivially refute the contention that anyone claiming that chemical exposures make them feel too sick to work could qualify for some benefits under the current state of practice. On the other hand, most practitioners with experience dealing with MCS patients believe that clinical judgment can be used both to make a positive diagnosis of MCS and to determine whether a patient is able to work, either at all or in a particular environment. But such judgments are extremely taxing, the more so because of the unending pressure from payors to define and delineate the basis for the clinical judgment in each case and at every point in time. In practice, since many benefit systems intrinsically rely on treating physicians' judgment anyway, the debate is reduced to a form of trench warfare in which the clinicians are effectively pressured to minimize their ratings. In any event, independent medical examiners with "tougher" standards are easy for the payors to find.

The question of attribution, which underlies which benefit system or liability action may be entertained, relates to the debate over pathogenesis. If indeed the cause of MCS were clearly biologic, there would be little scientific debate about the appropriateness of workers' compensation benefits or the legitimacy of tort actions as with other chemical injuries. It would then become a legal issue as to how to apportion responsibility. On the other hand, if MCS proves to be primarily psychologic in origin, it may be hard to assign attribution at all. Are chemical manufacturers and those who use chemicals responsible for the symbolic impact of their products on users or those who are exposed?

Given the scientific uncertainty it is hard to be dogmatic about these issues. It does seem clear that some form of disability benefit is both appropriate and humane for the sickest MCS patients, even if determination of eligibility must reside in the hands of experienced physicians, not laboratory tests. This could be regulated if necessary by the use of panels or other normalizing strategies. Regarding assignment, it seems that present knowledge is probably insufficient to justify attribution of MCS to a particular chemical exposure in most cases. However, in patients who meet the more rigorous standard for MCS that we have previously proposed, with a clear onset in relation to specific exposure circumstances (13), there may be appropriate indications for workers' compensation benefits viewing MCS as an idiosyncratic complication of a defined toxic exposure. In such cases, compensation could be justified whether the development of MCS is construed as a biologic sequel of exposure or a psychological one.

REGULATION AND CONTROL OF CHEMICALS IN OUR SOCIETY

From an economic perspective, the biggest issue raised by the MCS problem is not compensation of victims but considerations of the control of chemical exposures. If one takes the preamble of the Occupational Safety and Health Act at face value, namely that no worker shall be harmed by the effects of workplace exposures, and add the presumption that MCS is biologically mediated by chemicals, the ramifications of MCS would be very great indeed. However, since the Supreme Court already has interpreted the Act to conform to modern realities of cost and benefit, and the chemical origins of MCS are far from proven, there remains large room for debate. For the ambient environment, the question is to what degree the MCS population should be included in calculations of risk assessment, the current basis for standard setting.

As with the compensation issue, lines tend to be drawn in parallel with beliefs about the pathogenesis of MCS, not coincidentally closely aligned with economic interests. Those espousing the biologic view of MCS believe that the long-term strategy for control of MCS must require substantial strengthening of environmental standards both within the workplace and without. Those who believe that MCS is psychological argue that strengthening of indoor and outdoor standards would be unlikely to yield benefit. Even granting that some portion of MCS might be directly attributable to biological effects of chemical exposure, critics of invoking MCS in the standard-setting process argue that cost-benefit considerations would always dictate that the most susceptible fraction of the population would not be protected by economically feasible standards.

As always with MCS, the issue is not so straightforward. The cost-benefit enthusiasts would lump MCS into the lower tail of some theoretical susceptibility curve, as they would cancer or any other chemical risk. However, whatever MCS may be, it is apparent that it does not represent a mere continuum within the population of sensitivity to chemicals such as the 1 percent with the lowest odor threshold or the lowest irritation thresholds. Patients with MCS are qualitatively as well as quantitatively different from the rest of the population in terms of their reactivities to chemicals. They don't get predictable toxic effects occurring at lower doses—they experience different effects. Unfortunately, the effects they experience may be sufficiently disabling as to badly confound any model of dose response for risk assessment, which is based on the usual continuous dose-response assumptions. Until the epidemiology of MCS is established and until the costs associated with it can be correctly calculated, the cost-benefit rationale for maintaining current standards and practices will rest on a precarious foundation.

Those who would argue for strict regulation and control of chemical usage as the solution to the MCS problem stumble on the same problem. They have not dealt adequately with the argument that most MCS reactors develop symptoms at levels of exposure orders of magnitude lower than regulated values and often effectively unmeasurable, hence unregulatable even if costs were not a factor. Further, little evidence has been provided by these advocates of the actual benefit in terms of risk or severity of MCS if exposure levels were changed. It is not obvious, for example, how to explain the apparent increases in the rate and severity of MCS during the last decade during which air and water quality have improved, albeit not enough for some. Also unexplained is the apparently low rate of MCS among the lowest paid, most heavily exposed industrial workers compared to lesser exposed white-collar workers and professionals. (4) If it turns out that MCS is mediated primarily by psychological factors, strict regulation may be no more valuable for MCS patients than providing more available sanitary facilities would be for patients with obsessive-compulsive disorder!

Given the enormous uncertainties of present knowledge, neither of these approaches appears entirely rational. It does not make sense to impose economic hardship on the entire society based primarily on a putative but highly uncertain health benefit to present and future MCS patients. On the other hand, maintenance of environmental policies that are indifferent to the MCS population is contrary to our national law and inimical to sound public policy.

I would propose as the best solution one that focused environmental efforts on provision of high-quality, "safe" environments for MCS patients, much as we currently provide special facilities for the physically handicapped. Large employers, service establishments, and even communities could be urged or required to set aside space and facilities that are acceptable to those with limitations in dealing with the chemical environment, much as smoke-free environments are now mandated in many jurisdictions. Where such special alternatives are impossible, a higher level of environmental control should be provided for all. These efforts should be coupled with a more flexible and sensitive strategy for guaranteeing that MCS patients who experience hardship in the environment are not economically punished. In return for this, the general standards for the environment should probably not be adjusted based on the goal to provide "safety" to this subpopulation. To do so would be extremely costly, highly impractical if not infeasible altogether, and would provide a highly uncertain overall benefit. Until the nature of MCS is better understood, it would seem preferable to focus resources rather than diffuse them.

CONCLUSIONS

Although it is generally cast in these terms, it is clear that the "villain" in the Multiple Chemical Sensitivities story is neither the chemical industry, its customers, and their insurers, on the one hand, nor the clinical ecologists and their faithful supporters on the other, nor the insensitive community of mainstream medical practitioners. The problem is largely one of incomplete science, which has created a circumstance in which patients with MCS may appear to be victimized by all of these players. Until there is at least an adequate understanding of the origins of MCS, there is little chance for a rational policy that optimizes resources and our humanitarian principles.

In the meantime, however, choices must be made, and the patients cannot be expected to suffer doubly because they were "born too soon." Provisions must be made for them to have sympathetic and knowledgeable health care, reasonable economic supports, and tolerable environments in which to work and function socially. Although many choices made on the basis of current conjecture will surely prove to be inadequate and perhaps misguided when a fuller understanding of MCS has been achieved, there is little basis for continuation of the status quo—to ignore, disparage, and litigate as the primary responses of our society to the unknown.

NOTES

1. Randolph, TG. *Human ecology and susceptibility to the chemical environment.* Springfield, IL, Charles C. Thomas, 1962.

2. Dickey, LD. *Clinical ecology*. Springfield, IL, Charles C. Thomas, 1976.

3. Beard, GM. *American nervousness, its causes and consequences: a supplement to nervous exhaustion.* New York, G.P. Putnam's Sons, 1881.

4. Mooser, SM. "The epidemiology of multiple chemical sensitivities." *Occupational Medicine: State of the Art Reviews*, 2:663-668, 1987.

5. Randolph, TG, Moss, RW. *An alternative approach to allergies.* New York, Lippincott and Cromwell, 1980.

6. Bell, IR. *Clinical ecology: A new medical approach to environmental illness*. Bolinas, CA, Common Knowledge Press, 1982.

7. American Academy of Allergy. "Position statements—controversial techniques." *J. Allergy Clin. Immunol.* 67:333, 1981.

8. Rosenstock, L, Heyer, NH. Emergence of occupational medicine services outside the workplace. *Am. J. Indust. Med.* 3:217-273, 1982.

9. Cullen, MR, ed. *Occupational Medicine: State of the Art Reviews*. "Workers with multiple chemical sensitivities." Philadelphia, Hanley and Belfus, 1987.

10. Black DW, Rath A, Goldstein RB. "Environmental Illness: A controlled study of 26 subjects with 20th Century disease." *JAMA* 264:3166-3170, 1990.

11. Terr, AI. "Environmental illness: a clinical review of 50 cases." *Arch. Intern. Med.* 146:195, 1986.

12. Jewett DL, Greenberg MR. "Placebo response in intradermal provocation testing with food extracts." *J. Allergy Clin. Immunol.* 75:205, 1985.

13. Cullen, MR. "The worker with multiple chemical sensitivities: an overview." *Occupational Medicine: State of the Art Reviews* 2:655-661, 1987.

14. Ashford NA, Miller CS. *Chemical Exposures: Low levels and high stakes*. New York, Van Nostrand Reinhold, 1991.

9

New Cancer Theories

POLICY IMPLICATIONS FOR CANCER PREVENTION

JOEL SWARTZ
RICHARD CLAPP

A long-running controversy in the theory of chemical- and radiation-induced cancer causation involves the threshold theory. This controversy has profound implications for cancer prevention and environmental and occupational regulations. For most toxicological endpoints, a threshold dose exists; that is, there is a level of exposure, the threshold, such that exposures to doses below this value cause no disease. According to the threshold theory of cancer, a similar no-effect level can be found for carcinogens. However, the view that has been generally accepted by scientists since the 1970s is that there are no thresholds for cancer-causing agents, and that exposure to low doses of these substances produces some increased risk of developing cancer, with the risk being approximately proportional to dose. (1)

Recently, a number of scientists have proposed a new theory of cancer causation that is a new justification for the threshold concept. This theory, which we refer to as "The New Threshold Theory," has been most consistently promoted in the work of Ames and Gold. (2) According to The New Threshold Theory, the main action of cancer-causing substances is to kill cells in target organs, thus inducing cells to divide to replace the killed cells. It is this cell division that is considered the most important step in cancer development. A consequence of this theory is that there is really no difference between cancer-causing substances, known as carcinogens, and any other toxic substance. Therefore, it follows from this theory that cancer-causing substances should not be regulated in a manner different from other toxic substances.

The threshold controversy has serious implications for public health and regulatory policy. The public is exposed to thousands of cancer-causing substances, generally at doses that are far lower than the doses at which their ability to cause cancer has been established. Under the no-threshold theory, exposure to lower doses will cause an increase in the risk of developing cancer. The New Threshold Theory, on the contrary, implies that exposure at low doses carries little or no danger of developing cancer.

The reason for the importance of the threshold controversy is that virtually all our direct knowledge of the ability of a substance to cause cancer comes from high-dose exposures to animals or humans. For example, animal cancer studies generally involve several hundred rodents. In order to show statistically significant differences between the exposed and control groups in the portion of animals with tumors, the chemical must produce tumors in about 10 percent of the animals. So, for example, 13 tumors where 10 were expected would not be a significant difference, but 20 would be significant. In turn, this requires that the animals receive the largest possible dose that does not shorten life. This dose typically ranges from 100 to 10,000 times the dose to which humans are exposed. Similarly, human epidemiologic studies generally concentrate on people with very high exposures, and very high cancer rates. For example, among asbestos workers and coke plant workers, more than 10 percent of the population developed lung cancer. For public health and regulatory purposes, we are generally interested in doses from 100 to 10,000 times lower.

PUTTING IT IN CONTEXT

To place this controversy in context, we should note that during the twentieth century many substances have been identified as inducing a substantial number of human cancers. Synthesis, production, and use of many such substances have increased dramatically, sometimes by 1,000-fold, during the postwar period in many industrial societies. (3) Rising cancer rates also have been seen in these industrial societies 10 to 25 years after the rise in use of these carcinogens. (4)

During the postwar period, some of these substances have been shown to contribute substantially to the cancer burden among exposed persons. (3) Many of these carcinogens were identified in the workplace because exposures were high and to an identifiable group of workers. (5) The list of occupationally induced cancers was already large by 1980 (see Table 9.1) and has grown considerably during the last decade.

Furthermore, there is evidence that these substances have contributed substantially to the cancer burden among the general population. One well-known example is the cluster of childhood leukemia in which the cancer rate increased with the availability of contaminated drinking water. (6) Epidemiologic studies have found significantly elevated cancer rates among the general population exposed to carcinogens at levels hundreds of times lower than workplace levels. Included among this group are persons exposed to benzpyrene and related materials in the vicinity of facilities burning fossil fuels, persons exposed to arsenic from smelters, and persons living in the vicinity of asbestos mines. (7, 8) Conservative estimates have placed the number of environmentally induced cancer deaths in the United States as high as 25,000 per year (9), while others give estimates that are considerably higher. (8)

THE NEW THRESHOLD THEORY

The New Threshold Theory has been supported most consistently in the work of Ames and Gold, although it has been promoted by a number of authors. Here we refer specifically to the perspective by Ames and Gold in "Too Many Rodent Carcinogens: Mitogenesis Increases Carcinogenesis." (2)

The underlying theme of The New Threshold Theory is that cell division is the crucial, rate-limiting step for most human carcinogenesis. (2) In particular, it is argued that tumor induction at high doses, that is, near the maximum tolerated dose, results from cell division. Further, it is argued that this cell division is induced by the acute, toxic effects of the carcinogen, that is, cells are killed by the substances, and other cells divide to replace the killed ones. Therefore, at low doses, the substance would induce no tumors, or orders of magnitude fewer tumors than would be predicted by making a linear, downward extrapolation through the origin; for example, if the substance caused tumors in 10 percent of the animals at the maximum tolerated dose, it would cause no tumors, or tumors in no more than say .001 percent of the animals at a dose of one hundredth of the maximum tolerated dose.

Further, it is argued that increased mutagenesis in humans from background exposure to environmental carcinogens is small compared to the background mutagenicity. Mutagenesis is a process in which permanent changes are made in the genes, the program for the cell. Many cancers result, at least in part, from mutations in specific genes called oncogenes. Many human and animal carcinogens are mutagens, and for some classes of chemicals the abilities to cause mutations and to induce cancer are closely related. In fact, the Ames test, developed by Ames and collaborators, was created to identify carcinogens by their mutagenicity. According

TABLE 9.1. Risk Factors Associated with Workplace Exposures to Five High-Volume Carcinogens

Chemical	Sites of primary cancers	Other chronic health effects	Occupations at risk	Latency period for cancer (years)	Risk ratios for cancer	1997 NOHS estimated numbers of workers exposed full- and part-time	1981 NIOSH estimated numbers of workers exposed[a]	
							Full- + part-time	Full-time
Arsenic	Skin, lung, liver, lymphatic system	Gastrointestinal disturbances, hyperpigmentation, peripheral neuropathy, hemolytic anemia, dermatitis, bronchitis, nasal system ulceration	Miners, smelters, insecticide makers and sprayers, chemical workers, oil refiners, vintners	10+	3–8	1,500,000	255,277 432,017 (arsenic oxides)	5,926 596
Asbestos	Lung, pleural and peritoneal mesothelioma, gastrointestinal tract	Asbestosis (pulmonary fibrosis, pleural plaques, and pleural calcification), anorexia, weight loss	Miners, millers, textile, insulation and shipyard workers	4–40	1.5–12	1,600,000 2,522,000	1,280,202	449,960
Benzene	Bone marrow (leukemia)	Central nervous system and gastrointestinal effects blood abnormalities (anemia, leukopenia, and thrombocytopenia)	Explosives, benzene and rubber cement workers, distillers, dye users, printers, shoemakers	6–14	2–3	2,000,000 1,900,000	1,495,706	147,604
Chromium	Nasal cavity and sinuses, lung, larynx	Dermatitis, skin ulceration, nasal system ulceration, bronchitis, bronchopneumonia, inflammation of the larynx and liver	Producers, processors, and users of Cr: acetylene and aniline workers; bleachers; glass, pottery, and linoleum workers; battery makers	5–15	3–4	1,500,000 (chromium oxides) 175,000 (chromium VI)	1,451,631 (oxides)	59,946
Nickel	Nasal cavity and sinuses, lung	Dermatitis	Nickel smelters, mixers, and roasters, electrolysis workers	3–30	5–10 (lung) 100+ (nasal, sinuses)	1,400,000 (oxides) 25,000 (inorganic nickel)	1,369,278 (oxides)	51,840

Source: D.L. Davis and D.P. Rall, "Risk Assessment for Disease Prevention," in L.K.Y. Ng and D.L. Davis, eds., *Strategies for Public Health* (New York: Van Nostrand Reinhold, 1981).
[a]National Institute for Occupational Safety and Health, Interim Estimate, 1981.

to The New Threshold Theory, exposure to environmental mutagens has minuscule effect on the cancer rate.

THE CASE AGAINST THE NEW THRESHOLD THEORY

There is strong evidence against The New Threshold Theory. First, cancer development is a multistage process. A cell goes through several transitions between a normal cell and a fully cancerous cell. A number of factors can contribute to this process including: point mutations (that is, a change in a single base in DNA), large-scale damage to chromosomes, movement of genes from one chromosome to another, defective cell regulation often induced by hormonal abnormalities, deficient immune protection against tumors, and cell division. All the genetic changes can be induced by mutagenic substances, and the type of genetic change often is specific to a given mutagen.

There are many model systems in which cell division can substitute for the application of a particular carcinogen. However, rapid cell division by itself does not induce cancer. Extensive cell division occurs in fetal development, and in the continuous renewal of bone marrow, the small intestine, and skin. Yet cancer is rare in children, and cancers of the blood, small intestine, and malignant skin cancer are rare compared with cancer of other organs. (10)

If the major factor in cancer induction were cell division induced by acute toxicity, then there would be strong quantitative associations between toxicity and carcinogenicity. But this is not the case. Extensive analysis of the National Toxicology Program rodent carcinogen bioassay databases indicates that there is no such consistent association between carcinogenicity and organ toxicity. (11) Furthermore, there are many situations in which the acute toxicity of a substance is totally unrelated to its carcinogenicity in animal systems. (7) The same has been demonstrated in cell culture. (12) For example, monouron causes liver degeneration, but no liver tumors at low doses. Low doses of asbestos cause mesothelioma (a cancer of the lining of the lung), but not asbestosis.

One important prediction of The New Threshold Theory is that the cancer potency is much lower at low doses than at high doses. This is because at low doses the substance would not be toxic, and hence would not induce cell division to replace killed cells. For the same reason, the theory predicts that dividing a dose into many smaller doses, with the same total dose, fractionation, causes a decrease in the cancer potency. However, it is typical that fractionation brings about an increase in the cancer potency. For example, vinyl chloride has been shown to induce tumors over a range of five orders of magnitude, and at the lowest dose the cancer potency is 30 times that at the highest dose. (12) Analysis of a large number of bioassays showed that the cancer potency increased as dose was lowered in about 75 percent of the cases. (13)

FRACTIONATION GIVES OPPOSITE RESULTS

It is also true that fractionation of the dose usually gives results opposite to those predicted by The New Threshold Theory. For example, in Druckrey's classic experiments, in which mice were fed a range of daily doses of diethylaminostilbene, a 100-fold decrease in daily dose brought about a 20-fold increase in the cancer potency. (14) The increase in cancer potency with fractionation is also general for high linear energy transfer (LET) radiation (15), and has been successfully modeled in the case of bone tumors induced by bone-seeking radionuclides (16), and that of lung tumors induced by radon. (14) For low LET radiation, fractionation has the opposite effect. (17)

Another prediction of The New Threshold Theory is that many carcinogens are interchangeable, since the key step in tumor induction is deemed to be the ability to cause cell division, and this may be caused by a large number of substances that have the same function. So according to The New Threshold Theory, the types of tumors induced by different agents would be the same, and the tumors induced by one would be indistinguishable from the tumors induced by another.

In fact, it is now possible to link specific carcinogens with specific genetic changes and with specific tumors. Recent studies have shown that carcinogens often activate genes called oncogenes, and that the activation of these genes represents one step in the transition from normal to malignant cells. The particular mutation and the particular oncogene activated may be specific to the applied carcinogens. (18) For example, rat mammary tumors induced by N-nitroso-methyl-urea almost always have an oncogene called Ha-ras activated, while few of the same tumors induced by dimethylbenzanthracene have this oncogene activated. Moreover, the site of the mutation varies as the carcinogen changes.

Another implication of The New Threshold Theory is that carcinogens would produce an undetectable number of tumors in humans or animals when the exposure is at subtoxic doses. Numerous studies have indicated that this is not the case. For example, workers exposed to less than 10 parts per million of benzene were shown to be at increased risk of leukemia. (19) Many animal cancer tests (for example, benzene, 1,3-butadiene, vinyl chloride, ethylene dibromide) were conducted at exposure levels close to levels at which humans are exposed. (19) Also, 1,3-butadiene was found to be carcinogenic at doses well below the toxic level.

CONCLUSION

So The New Threshold Theory is an attempt to support the old threshold theory with some new arguments and data. But the new arguments do not meet previous objections to the threshold theory. Moreover, we have argued that The New Threshold Theory is seriously flawed, and is contradicted by important examples of known human carcinogens, and animal models. Despite this, many corporate spokespersons have touted this theory, and argued that it casts doubt on all current regulatory policy based on animal testing.

The impact of the latest challenge to regulation of carcinogens has not yet been felt. Union and community toxic activists should be aware of the challenge, and counterarguments made by ourselves and others: for example, Rall (7), Hoel (11), Moolgavkar (20), Perera. (21) We should continue our efforts to prevent human cancer regardless of arguments by corporations and their spokespersons that claim to show that environmental and occupational carcinogens have little or no health impact. In particular, carcinogens should be treated as serious potential health hazards, and carcinogenicity should continue to be treated as a health hazard separate from acute toxicity. High-dose animal studies should continue to be used to predict human health hazards at lower doses.

NOTES

1. Pitot, H. *Fundamentals of Oncology* (New York: Marcel Dekker, 1986), Chapter 6.
2. Ames, B.N., Gold, L.S. "Too Many Rodent Carcinogens: Mitogenesis Increases Mutagenesis." *Science* 249 (1990): 970–971.
3. Davis, D., Magee, B.H. "Cancer and Industrial Chemical Production." *Science* 206 (1979): 1356, 1358.
4. Davis, D., Hoel, D., Fox, J., Lopez, A. "International Trends in Cancer Mortality in France, West Germany, Italy, Japan, England and Wales, and the U.S.A." *Lancet* 336 (1990): 474–481.
5. Davis, D. "Cancer in the Workplace: The Case for Prevention." *Environment* 22 (1981): 25–37.
6. Lagakos, S.W., Wessen, B.J., Zelen, M. "An Analysis of Contaminated Well Water and Health Effects in Woburn, Massachusetts." *Journal of the American Statistical Assn.* 81 (1986): 583–614.
7. Rall, D.P. "Carcinogens and Human Health: Part 2." Letter, *Science* 251 (1991): 10–12.
8. Epstein, S.S., Swartz, J.B. "Carcinogenic Risk Estimation." *Science* 240 (1988): 1043–1045.
9. Doll, R., Peto, R. "The Cause of Cancer Quantitative Estimates of Avoidable Risks of Cancer in the United States Today." *Journal of the National Cancer Institute* 66 (1981): 1193–1308.

10. Weinstein, I.B. "Mitogenesis Is Only One Factor in Carcinogenesis." *Science* 251 (1991): 387-388.

11. Hoel, D.G., Haseman, J.K., Hogan, M.D., Huff, J., McConnell, E.E. "The Impact of Toxicity on Carcinogenicity Studies: Implications for the Risk Assessment." *Carcinogenesis* 9 (1988): 2045-2052.

12. Peterson, A., et al. "Oncogenesis, Mutagenesis, and DNA Damage, and Cytotoxicity in Cultured Mammalian Cells Treated with Alkylating Agents." *Cancer Research* 39 (1979): 131-138.

13. Swartz, J., Riddiough, C., Epstein, S. "Analysis of Carcinogenesis Dose-Response Relations with Dichotomous Data: Implications for Carcinogenic Risk Assessment." *Teratogenesis Carcinogenesis Mutagenesis* 2 (1982): 179-204.

14. Druckrey, H., Schmahl, D., Dischler, W. "Dosis-Wirkungs-Beziehungen bei der Krebserzeugung durch4-Dimethylamino-Stilben bei Ratten." *Z Krebsforsch.* 65 (1963): 272-288.

15. Moolgavkar, S., Luebeck, G., de Gunst, M. "Two Mutation Model for Carcinogenesis: Relative Roles of Somatic Mutation and Cell Proliferation in Determining Risk" in *Scientific Issues in Quantitative Cancer Risk Assessment*, Moolgavkar, S., ed. (Boston: Birkhauser, 1990) pp. 136-152.

16. Marshall, J., Groer, P. "A Theory of the Induction of Bone Cancer by Alpha Radiation." *Radiation Research*, 71 (1977): 149-192.

17. Thomas, D. "A Model for Dose Rate and Duration of Exposure Effects in Radiation Carcinogenesis." *Environmental Health Perspectives* 87 (1991): 163-171.

18. Barbacid, M. "Mutagens, Oncogenes and Cancer." *Trends in Genetics* 2 (1986): 188-192.

19. Rinsky, R.A., Smith, A.B., Hornung, R., Filloon, T.G., Young, R.J., Okun, A.H., Landrigan, P.J. "Benzene and Leukemia, An Epidemiologic Risk Assessment." *New England Journal of Medicine*, 316 (1987): 1044-1050.

20. Moolgavkar, S.H. "Carcinogenesis Models." Letter, *Science* 251 (1991): 143.

21. Perera, F.P., Rall, D.P., Weinstein, I.B. "Carcinogenesis Mechanisms: The Debate Continues." Letter, *Science* 252 (1991): 903-904.

10

Job Stress and Heart Disease

EVIDENCE AND STRATEGIES FOR PREVENTION

PAUL A. LANDISBERGIS
SUSAN J. SCHURMAN
BARBARA A. ISRAEL
PETER L. SCHNALL
MARGRIT K. HUGENTOBLER
JANET CAHILL
DEAN BAKER

Job stress has proven to be a difficult issue for the occupational health community and the labor movement to tackle. Unlike physical or chemical hazards, there is not an obvious tangible hazardous agent. This issue has also been preempted by corporate stress management, health promotion, or employee assistance programs, which explain stress as a purely personal reaction and often treat the symptoms, not the causes, of job stress. (1, 2) There has been legitimate resistance to this "stress management" model, which "blames the victim" and ignores the objective basis of job stress. The occupational stress field also has been plagued by a variety of definitions and difficulties in measurement of stress. In addition, changes in job design or work organization are often inherently more "systems challenging" and require more radical restructuring of workplaces than reducing levels of exposure to toxic substances or ergonomic hazards. This chapter addresses such concerns. (3)

A number of specific stressful working conditions, such as repetitive work, assembly-line work, electronic monitoring or surveillance, involuntary overtime, piece-rate work, inflexible hours, arbitrary supervision, and de-skilled work, have been studied and recently reviewed. (4) Since 1979, a new model of job stress (Figure 10.l) developed by Robert Karasek (5), has highlighted two key elements of these stressors, and has been supported by a growing body of evidence. Karasek's "job strain" model states that the greatest risk to physical and mental health from stress occurs to workers facing high psychological workload demands or pressures combined with low control or decision latitude in meeting those demands. (6) Job demands are defined by questions such as "working very fast," "working very hard," and not "enough time to get the job done." Job decision latitude is defined as both the ability to use skills on the job and the decision-making authority available to the worker. In some recent studies, this model was expanded to include a third factor—the beneficial effects of workplace social support. (7)

While there are a variety of models of job stress, the "job strain" model (which is the focus of this paper) emphasizes the interaction between demands and control in causing stress, and objective constraints on action in the work environment, rather than individual perceptions or "person-environment fit." In addition, other important work-related and "social" stressors exist that are less directly connected to the concept of "job strain," but may also have significant health consequences. These include increasing work hours (8, 9), conflict between work and family roles (8, 9), and sexual (10, 11) or racial harassment or discrimination.

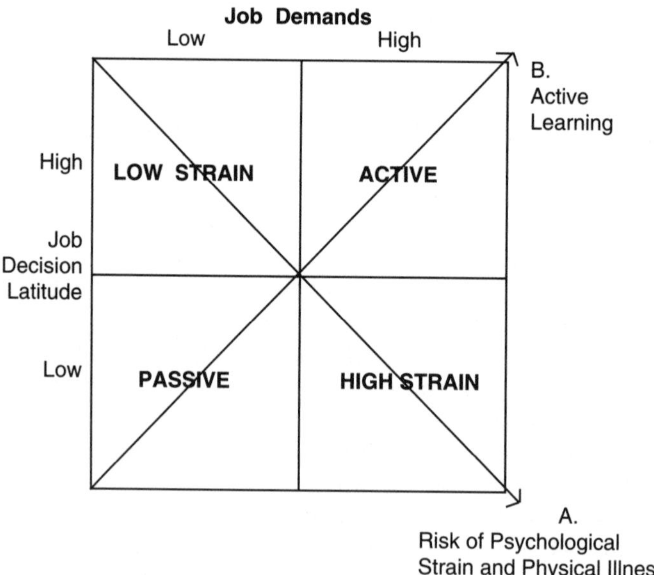

FIGURE 10.1. The Karasek "job strain" model.

Most studies of the "job strain" model have focused on outcomes such as cardiovascular disease (high blood pressure, heart disease) and psychological distress (anxiety, depression). Since heart disease is the most common cause of death in industrialized countries, and because the "job strain" and heart disease studies have been recently reviewed (12), this article will summarize the evidence that "job strain" is linked to heart disease. We then review the evidence demonstrating that interventions targeted at altering objective working conditions can produce beneficial effects on job demands, decision latitude, and social support. (13) We also review collective bargaining, legislative and other efforts to reduce "job strain" specifically and job stress in general. Together, these studies and programs provide convincing evidence that job design and work organization are risk factors for cardiovascular illness and that these can be modified.

SIGNIFICANCE OF JOB STRESS

The issue of job stress is of major importance to the occupational health community and the labor movement for two compelling reasons:

First, there is the potential for preventing much illness and death. More than 50 million Americans have high blood pressure, and, in 95 percent of cases, the cause is unknown. (14) (This type is called "essential" hypertension, as opposed to cases where there is a recognized cause, such as an adrenal gland disorder or kidney disease.) High blood pressure is a major cause of heart disease and stroke. (14) While estimates of the proportion of heart disease possibly due to "job strain" vary greatly between studies, Karasek and Theorell (5, p. 167) calculate that up to 23 percent of heart disease could potentially be prevented (*over 150,000 deaths prevented per year in the United States*) if we reduced the level of "job strain" in jobs with the worst strain levels to the average of other occupations. (15) The economic costs of job stress in general (absenteeism, lost productivity) are difficult to estimate but could be as high as several hundred billion dollars (5, p. 167–168).

Second, Karasek's model emphasizes another major negative consequence of work organization—how the assembly line and the principles of Taylorism, with its focus on reducing workers' skills and influence, can produce passivity, learned helplessness, and lack of participation (at work, in the community, and in politics). The "job strain" model (Figure 10.1) has two components—increasing risk of heart disease following arrow A, but increasing activity, participation, self-esteem, motivation to learn, and sense of accomplishment following arrow B. Thus, this model provides a justification and a public health foundation for efforts to achieve workplace democracy. Democracy at work should be promoted as not only just and fair, but also as a method to reduce ill health, and to allow for fuller development of people's emotional, intellectual, and social capacities.

EVIDENCE LINKING "JOB STRAIN" TO HEART DISEASE

During the 1980s, 14 studies on "job strain" and heart disease including two of all-cause mortality and 20 studies on "job strain" and heart disease risk factors have been published. These have been conducted in Sweden, the United States, Finland, Denmark, Australia, and Japan, and they provide strong evidence that "job strain" is a risk factor for heart disease. (12)

Of the 14 heart disease studies, 12 showed clear associations between "job strain" and heart disease. Most of these studies controlled for other (potentially confounding) heart disease risk factors. More importantly, of the eight cohort studies of heart disease and all-cause mortality, seven showed strong positive associations (7, 16–21). (The cohort studies followed people over time and therefore are less likely to be biased.)

In the 20 studies of "job strain" and independent risk factors (for heart disease), the following patterns were seen. Three studies found no association between "job strain" and serum cholesterol. Two studies found a link between "job strain" and smoking (22, 23), while two did not. (One of the negative studies [24], however, did show higher rates of smoking at lower levels of job decision latitude.) Of nine studies of blood pressure measured in a clinical setting, only one found a significant association. However, of the eight studies where an ambulatory (portable) blood pressure monitor was worn during a workday, five showed strong positive associations between "job strain" and blood pressure (26–30), while the other three provided mixed results. (25, 31, 32) Since ambulatory blood pressure is both more reliable (since there is no observer bias and the number of readings is greatly increased) and more valid (since blood pressure is measured during a person's normal daily activities) than casual measures of blood pressure, we feel confident in placing more emphasis on the ambulatory blood pressure results. We conclude that one pathway between "job strain" and heart disease is elevated blood pressure (12), possibly mediated by increases in catecholamines and cortisol (33), increased autonomic nervous system activity (for example, increased heart rate), and/or increased mass of the heart's left ventricle. (34)

CURRENT ISSUES IN RESEARCH ABOUT "JOB STRAIN" AND HEART DISEASE

A variety of issues were raised in these 34 studies, including methodological concerns, and factors that may modify the impact of "job strain" on heart disease (for example, gender, race, and socioeconomic status [SES]). For example, the studies used various methods and survey questions to measure the concept of "job strain," one limitation of the research. However, Karasek's Job Content Questionnaire (35), which contains 14 basic questions on task-level demands, skill use, and authority, was frequently used in these studies. The concept of task-level job demands was primarily measured by questions about workload demands. However, it is important to expand the concept to include other stressors such as responsibility for people, role conflict, role ambiguity, and threat of violence or injury. In addition, while questions on task-level control and demands were available in U.S. and Swedish national surveys, questions on control, either individually or collectively (36, 37), over departmental or organization-level policies and decisions (38) were not. Strengths of the "job strain" model are its simplicity and clarity, its prediction of both health and behavioral outcomes, and its emphasis on classifying features of the work environment into the categories of demands or control. However, the model would benefit from the inclusion of various dimensions of demands and control (such as those listed above) used in the more complex Michigan job stress model (39) and the NIOSH Generic Job Stress questionnaire. (40)

In addition, in 10 studies, a technique was used to develop more "objective" measures of job characteristics. National averages of job characteristics for a particular job title were assigned to individuals having that job title, ignoring the fact that job characteristics vary for people even within the same job title. Six of these 10 studies provided positive results. Thus, despite errors in measurement that make it more likely that links between "job strain" and illness will not be found even when they do exist, it is remarkable that consistent positive patterns have been found.

In seven of the 10 studies where comparisons could be made, the effects of "job strain" were similar for men and women. (However, a higher proportion of U.S. women workers face "job strain." (5, pp. 45–46)) As the Framingham Heart Study (41) and other research (42) make clear, many women face a dual set of demands from work outside and within the home. "High strain" work (that is, "job strain") may interact with home demands to increase heart disease risk for certain subgroups of workers. In addition, "job strain," as currently measured, may not adequately "capture" other stresses faced by women workers such as salary and promotion inequities or sexual harassment. (10, 11)

Another factor is race or ethnic group. Two of three studies that reported this information showed a higher proportion of African American workers in the "high strain" group (27, 31), suggesting that the increased risk of hypertension faced by African Americans in the United States may be, in part, a result of more "job strain." Since only three studies have been conducted with a predominantly non-Caucasian population, further research is needed to determine to what extent the concept is appropriate and still a risk factor among other racial/ethnic groups.

The relation between "job strain" and SES has also been debated. Many studies statistically control for education (as a measure of social class), and still find that "job strain" increases the risk of heart disease. Conventional social status scales are poorly correlated with "job strain" (5, p. 77), indicating that the "job strain" findings are not simply due to the association between lower SES and heart disease. (43) There are "high strain" white-collar clerical jobs in the low-to-middle part of the job status hierarchy, and some blue-collar craft jobs that allow for a high level of skill development and autonomy. Low status and low income occur in "low strain," "passive," and "high strain" jobs (Figure 10.1).

Only four studies compared social class groups—finding that blue-collar workers, workers with less education, or female clerical workers had a substantially stronger association between "job strain" and heart disease than higher SES groups. Lower SES groups also have higher rates of heart disease and heart disease risk factors (43), and "job strain" may interact with these risk factors. (44) Limited economic resources may play a role. "Job strain" may also interact with chemical and physical health hazards on these jobs (carbon monoxide, solvents, lead, noise, shiftwork), or other psychosocial hazards such as fear of job loss. (44) Such occupational risk factors may cluster together, and "job strain" may be increased by automation, or by increased work quotas due to budget cuts.

While scientific proof for the "job strain" model is not yet conclusive, preventive action can be undertaken to reduce potential health risks. In the United States, efforts to reduce occupational stress continue to focus primarily on changing the individual behavior of employees (for example, relaxation techniques, exercise, diet, cognitive/behavioral skills). (45) However, a growing number of programs and interventions are attempting to change various workplace sources of stress. While none of these interventions was specifically designed to reduce "job strain," many focused on changing components of "job strain" (for example, reducing demands, increasing control, enhancing support). In addition, these programs have rarely included objective measures of heart disease risk. However, their lessons provide a valuable guide to future illness prevention and job redesign efforts, and to broader efforts to increase workplace democracy. Efforts to reduce or prevent job stress have been work site-based, community-based, industry-wide (in some cases of collective bargaining) or statewide or national in scope, in the case of legislation or regulations. U.S. work site programs have mainly been the result of both social science-based organizational reform efforts (known as Action Research) and collective bargaining. These programs are reviewed in the following sections.

SOCIAL SCIENCE-BASED INTERVENTIONS

The most well-developed applied research tradition on bringing about planned change in organizations is the field of Organization Development

(OD). OD has its roots in the "human relations" management and social theorists of the 1940s–'50s, who were reacting to the dehumanization, alienation, and bureaucracy characteristic of scientific management (Taylorism). (46, 47) OD practitioners conducted innovative work reform experiments during the 1950s–'70s, including early joint labor-management Quality of Work Life (QWL) programs. These focused primarily on social relationships (for example, a sense of belonging, supportive supervision, participation in decision making) rather than the technical features of production and work organization. In the 1980s, OD practitioners "discovered" the importance of technology, especially European Socio-Technical Systems (STS) theory, which promotes semiautonomous work teams. More importantly, by the 1980s, many OD professionals lost sight of their original stated mission—to attempt to serve both employer interests and employee needs—and applied their trade primarily on behalf of employers. (48, 49)

Scandinavian work reform experiments in the 1960s–1970s, while influenced by the same human relations research (and also reacting against the dehumanizing effects of scientific management), placed a greater emphasis on technical aspects of production (for example, piece-rate, shiftwork, technology) as well as an understanding that physical illness and injury is an outcome of work organization (50)—an outcome that has been largely ignored by OD. These different emphases, along with a progressive political climate and a highly unionized workforce, led eventually to work environment legislation in the 1970s in Scandinavia and continuing job redesign and work reform efforts today. (5, 51) These experiments, and the emphasis on health as an outcome of work, also laid the foundations for Karasek's model, and much stress research both in Scandinavia and the United States.

Many OD and QWL efforts have failed, however, because of factors such as lack of support by top management or supervisors, failure to delegate authority, a bureaucratic, authoritarian climate, and rigid job descriptions and personnel practices. (52, 53) Some interventions have led to increased workload or "speed-up" (54, 55), workforce reductions (46), or were initiated as attempts to avoid unionization (56, 57) or weaken the existing union. (49, 58) However, positive experiences with cooperative programs have also been reported by some unions (59, 60), and the debate continues in the labor movement over the potential value of these programs in specific situations.

Recognizing these limitations, unions and occupational health professionals have much to gain by adopting the valuable set of techniques and processes (intervention research methods) developed by OD, and using them on behalf of workers. One of these methods is known as Action Research (AR). AR involves a partnership between outside experts (usually social scientists) and members of organizations in defining problems, developing intervention tactics, introducing changes that benefit organization members, and measuring outcomes. (38) Issues and changes that this approach typically involves include decision-making structures and processes, task and role demands, information and communication practices, work schedules, and training policies. AR can be classified into "expert-dominated" approaches (also called "weak" AR), in contrast to "strong" versions where there is relative equality among researchers and organization members in all aspects of the intervention and research process—also termed Participatory Action Research (PAR). (61) While few studies have compared these approaches, one review suggests that PAR generates more positive outcomes. (62) Several key examples of "expert" AR and PAR interventions, that focused on improving workers' physical or mental health are briefly summarized below, followed by a discussion of policy and research issues.

Expert-Dominated Action Research

In a classic example, Jackson took advantage of a state legislative mandate for more frequent staff meetings in hospitals to measure the effects of participation in decision making on job stress, job satisfaction, absenteeism, and turnover. (63) Units where the intervention was implemented held twice as many staff meetings as in nonintervention units. Workers in participating units reported greater influence, less role conflict and ambiguity, less emotional strain, and greater job satisfaction at three-month and six-month follow-up.

In another example, Golembiewski and colleagues worked with 31 "burned out" and overworked Human Resources (HR) staff of a corporation in the midst of rapid growth. (64) Four action planning groups developed recommendations, and the entire staff prioritized them and prepared implementation plans that were presented to a corporate oversight committee. As a result, an HR career ladder was introduced as well as a change in reporting structure. Effects included a 50 percent reduction in reported "burnout" that remained low four months after the last intervention, a turnover decline from 37 percent to 17 percent, and a significant increase in reports of "innovativeness."

Participatory Action Research (PAR)

An example of PAR was a six-year study by Israel, Schurman, and colleagues in a components parts plant of a major unionized automobile company. (38, 65) With agreement from local union leadership and plant management, and working with union and management representatives, they set up a representative employee committee primarily composed of shop-floor employees—the Stress and Wellness Committee (SWC)—to implement the project. Using the PAR process of iterative cycles of diagnosis, action taking, and evaluation, the committee identified four primary sources of stress and designed interventions (through subcommittees) for each: lack of participation and influence, hassles with supervisors, lack of information/ communication, and "production vs. quality." Interventions included establishment of a pilot cross-functional team in one department to address quality issues, convincing factory management to conduct state-of-the-business meetings in each department, and creation of a weekly plant newsletter. Overall, SWC members reported high levels of trust in and influence over the committee process. In addition, other employees who were more involved in and knowledgeable about the PAR project reported greater increases in participation, perceived participative climate, and co-worker support than others with less exposure. (66)

Another example of PAR in a unionized setting began with a survey by Cahill of "burn-out" and symptoms of stress among employees of the New Jersey child protection agency. (67) The survey, which found significantly higher levels of "burn-out" than in national samples of social workers, was presented by the employees' union in a legislative hearing. One result of the hearing was the formation of a labor-management stress committee, which identified the agency's existing mainframe computer system as a major source of stress. The system included repetitive de-skilled work for clericals, lack of control of data for administrators and social workers, hard-to-interpret monthly reports, and ergonomically poor workstations. The stress committee recruited a computer programmer to design software jointly with the local employees who would use a new PC-based system. Once the new system was in operation, workers reported significantly higher levels of job satisfaction, decision latitude, skill discretion, control over equipment, a more streamlined information flow between local and central offices, and improved ergonomic conditions.

A final example of PAR to reduce job stress was developed by Lerner and colleagues at the Institute for Labor and Mental Health, and was based outside the workplace. (68) Strategies for raising awareness of the social and workplace sources of stress included: meeting with unions; organizing a conference on job stress where workers told their story to government, public health officials, and the media; a "family day" with workshops on stress of family and work life; and Occupational Stress Groups (OSGs). OSGs of 10 workers, led by shop stewards, met for eight to 12 weeks to discuss stress at work, develop social support, discuss the dangers of self-blame for feelings of powerlessness or stress, and to develop strategies for collective action. At follow-up, OSG participants showed significant improvements on virtually all measures of psychological well-being in comparison to controls. Behavioral changes and initiatives taken to improve the workplace were also reported in group interviews.

Other union-sponsored and work site–based initiatives, the OCAW Work and Family Program (69) and the District 65 UAW Stress Project (70), build on the OSG format. Both employ group meetings to raise awareness of stressful working conditions (and their impact on family life) and then develop collective bargaining proposals to improve working conditions.

Discussion

PAR approaches with strong union involvement have significant advantages over weaker expert-dominated or management-dominated AR programs. Strong union involvement can ensure that the potential dangers of OD are minimized and that interventions genuinely improve the work environment. Unions played important roles in initiating and sustaining structural change in the auto parts factory and in the New Jersey state agency, as well as, of course, in developing the OSG, OCAW, and District 65 programs. However, such programs are limited by the low unionization rate in the United States. The community-based approach used by Lerner can be especially useful in nonunion settings (such as COSH group efforts to educate and help organize nonunion workers), or where unionized employers refuse to cooperate or commit required support and resources.

PAR is a flexible set of intervention processes and methods, not a prepackaged canned program. This allows it to be effective in different contexts, with different occupational groups, and with resulting different strategies and tactics. It is also an innovative social research method, which makes it valuable for occupational health research. PAR is an effective tool for the evaluation of change because both quantitative and qualitative data are included, and process, impact, and outcome are assessed (thus requiring multidisciplinary teams skilled in these techniques). For example, the intervention in the auto parts factory included three administrations of a plantwide survey (including standardized survey scales), focus group interviews and five surveys of committee members, in-depth interviews of all committee members and plant union and management leaders, and verbatim field notes from committee meetings. Other studies included standardized surveys and objective records such as frequency of staff meetings, absenteeism, and turnover. Such multimethod approaches permit "triangulation," that is, cross-validation of and increased confidence in the results. (38, p. 148) Process data enable participants and researchers to assess not just what happened but why it happened (including obstacles to change). Impact data can reveal which organizational or individual factors are affected by the intervention, and through which pathways. For example, in the auto parts factory, regression analysis of survey results indicated that the positive effects of participation were channeled through perceptions of influence. Outcome data can answer questions about health effects.

Another important research issue is the need for longitudinal designs, with adequate time for follow-up. For example, the amount of change reported by the intervention group in Jackson's study increased significantly between the first and second posttests, suggesting that participation takes time to create effects. In the auto parts factory, 1.5 years was needed to conduct organizational diagnosis and needs assessment prior to engaging in major change strategies.

Thus, PAR to reduce job stress appears to work in two main ways (corresponding to arrows A and B of Karasek's model in Figure 10.1), by: (a) modifying objective stressful conditions in the social and/or technical environment; and (b) the active (individual and collective) learning that workers experience in successfully effecting positive change (for example, enhanced perceptions of control and influence, development of skills, positive self-appraisal, strengthened relationships with co-workers).

Genuine PAR allows workers not only to problem-solve but also to, jointly, with researchers, define targets for research and intervention and evaluate change (to be involved in all aspects of the intervention). Workers bring a richness of experience that enhances problem definition and hypothesis development, as well as insights to creating change. (71, 72) For example, workers can specify the concrete manifestations of job demands or low job control in a particular workplace (not captured by standardized scales), necessary for targeting change efforts. Researchers bring a rich knowledge base, methods of questionnaire construction and research design, and other means of improving study validity. While some researchers argue that participant involvement in social research could bias results due to improper wording of questionnaires, or attempts to influence survey response, bias can also result from employees' unwillingness to participate or candidly present their opinions "when involved with conventional research projects, because they associate researchers with management and the existing hierarchical structure." In addition, PAR researchers' use of multiple methods provides limited insights from the participants' "inside" understanding of attitudes, needs, and the social environment. (38, p. 140)

Genuine PAR (as opposed to some QWL programs) increases the skills and activism of those participating in the intervention, although to date there is no evidence that it strengthens union solidarity. However, just as active and assertive union involvement in health and safety training programs strengthens the union's position and credibility in the eyes of its members (73), benefits should be expected when the union is actively involved in improving other issues of concern to workers—job design and psychosocial work environment. (74, 75)

Personal stress management and health promotion was a component in many of these programs (including the District 65 UAW stress program). By discussing personal behavior change within the context of an overall program to improve the work environment, self-blame for behaviors or feelings of stress is avoided, and the union shows it is concerned about the personal welfare of its members. It can also be an organizing tactic to help gain publicity and support for the overall program, as in the auto parts factory study. In general, multiple levels (individual, group, organization, society) need to be targeted for interventions to effectively reduce stress. (76)

Even in successful interventions, many obstacles to change remain, for example, management turnover, lack of management support, pending layoffs, and general market conditions in the auto parts factory. In the New Jersey state agency, information and technology managers were initially resistant, perceiving the new technology and software as a threat to their power. Ensuring that they received some credit for the success of the project eventually led to their strong support for the intervention.

PAR can be a valuable technique in traditional occupational health programs. (71, 77) In addition, occupational health professionals and unionists can play a critical role in the next stage of stress research and stress prevention, by: (a) adding *physical* health as an outcome in PAR programs to improve the psychosocial work environment; (b) studying the effect of the physical work environment and fear of injury, on perceived stress and psychological well-being; and (c) studying the possible interaction between physical and psychosocial hazards in the production of heart disease, hypertension, and psychological distress, and other outcomes potentially related to job stress, such as musculoskeletal disorders (78), adverse pregnancy outcomes (79), and "sick building syndrome." (80)

COLLECTIVE BARGAINING APPROACHES

In addition to more recent PAR programs, collective bargaining has been a traditional strategy to increase employee decision latitude (authority, influence, skill), and to regulate demands—through contract language on issues such as job security, overtime, seniority, discrimination, technological change, skills training, career ladders, staffing, grievance procedures, and labor-management committees. (81-83) For example, the nurses' shortage during the 1980s in the United States has been attributed to factors such as low salary and job stress. Nurses have expressed a strong desire to be treated as professionals, which can be denied through understaffing, lack of autonomy, or an authoritarian work climate. In response, unions have bargained for clinical career ladders for registered nurses in various specialties, joint physician-nurse committees, greater in-service education (84), and quality patient care and personnel committees. (82)

Many clerical workers have joined unions in the last decade, in part due to issues related to job stress: career mobility, pay equity, job security, child care, flextime, parental leave, sexual harassment, having a "voice" through union-management committees, and video display terminal (VDT) work. (85) VDT workers have bargained for better ergonomic conditions, but have also learned that adjustable equipment is not enough. For example, at a New York City newspaper, a union–management committee discovered that job design issues such as control over schedule, regular breaks, work variety, and training were as important as the purchase of new equipment. (86) The National Institute for Occupational Safety and Health (NIOSH) is conducting various studies of the role of psychosocial factors in the development of cumulative trauma disorders (CTDs) among VDT operators. (87)

At least six million U.S. workers were electronically monitored in 1987, with the number expected to grow. (88) As part of a 1992 settlement of a Communications Workers of America (CWA) lawsuit, Northern Telecom agreed to prohibit secret voice, computer, and video monitoring of employees. (89) A CWA–U.S. West contract banned monitoring in 1989 with the help of early results from a study that showed that monitored workers had higher rates not just of psychological distress but also "stiff or sore wrists," "loss of feeling in fingers or wrists," and other symptoms of CTDs. (90) Similar studies by Bell Canada and the Communications Workers of Canada led to restrictions on monitoring in 1990. (89) Recently, AT&T agreed to ban secret monitoring of the job performance of workers. (91) A new study at U.S. West by NIOSH showed that stress due to monitoring, fear of job loss, increasing work pressure, and little job decision-making opportunity contributes to injuries even when proper equipment is used. (92)

The apparent interaction between psychosocial stress and physical stress and injury and illness needs to be better understood. Monitored workers have reported aspects of "job strain" (greater workload, less job control, unfair work standards, less skill use and variety), and poorer supervisor support. Do such factors lead to fewer breaks, longer work hours, or faster typing? Does increased muscle tension play a role? While some of the 10-fold increase in reported CTDs over the last decade (93) is undoubtedly due to better reporting, these studies suggest that some may be due to work speedup, de-skilling of jobs into simpler, more repetitive tasks, lack of control, and fear of job loss.

Electronic monitoring is often used to punish, not reward (for example, by publicly displaying results), managers overrely on it, and an emphasis on quantity not quality is created. (94) However, unions have shown that there are productive alternatives to monitoring. For example, CWA members at an Arizona facility, together with AT&T management, "eliminated individual measurement and remote secret observation. AWT (average work time) was measured only for the whole group. Service observation was performed by small groups of peers by the old-fashioned "jack-in' method, where the observer sits beside the person being monitored, listens to a few calls, and then discusses the results with the employee." As a result, AWT was better than under previous methods of supervision, there were fewer customer complaints, and both the grievance rate and absenteeism were lower. (94)

The loss of the 1981 air traffic controllers' (PATCO) strike and the firing of 11,000 unionized workers was a major setback in workers' rights to organize and strike. Some argue that PATCO's biggest failure was that it could not make an effective case for job stress—a major strike issue. (95) The job of air traffic controller includes many aspects of "job strain:" (a) high demands (through understaffing, mandatory overtime, few vacations); (b) poor skill utilization (because of poor training methods, outmoded equipment, few opportunities for promotion); and (c) little authority (due to an autocratic system and military style management, where grievants are labeled as troublemakers and not promoted). (95, 96) These conditions persist and, not surprisingly, new controllers have joined a new union and stress remains a major issue.

However, medical proof of the job's hazards has remained elusive. While the major 1975–1978 health study of controllers did report prevalence of hypertension 1.5 times that of national samples, and incidence of new cases of hypertension up to four times higher (97), much analysis focused on individual and psychological differences among men in the study. In addition, the Federal Aviation Administration (FAA) emphasized only the individual differences (not the high dissatisfaction with "management policies and practices" noted in the study (97, p. 628), and never published the nontechnical summary of the study. (2, pp. 1301–1303) For years, the FAA had ordered researchers conducting their stress studies "not to make recommendations" for corrective action. (2, p. 895) The FAA's technical representative to the study later testified that if the findings of the study (and 28 other FAA studies) had been applied, "I am absolutely certain" that the 1981 strike "would have been averted." (2, p. 874) Air traffic controllers' experience of stress and desire for equity had been deflected into a debate about the quality of scientific evidence on stress and health. (98)

In 1981, PATCO's collective bargaining demands focused on ways of "escaping" rather than "confronting" job stress: reduced work hours, early retirement, and higher salary—demands that did not win public sympathy. Alternative strategies such as improving organizational climate, supervision and communication (99), or more power over the work process, for example, flow control, curbing unregulated pleasure aircraft, disciplining of authoritarian supervisors, or more new hires, were not attempted. (95, p. 187) There were, of course, other reasons why the strike was lost, such as failure to effectively build alliances with other unions (95), poor public relations (100), and, most importantly, an intransigent administration in Washington, DC. However, former PATCO officer Bill Taylor emphasized that "knowing what I know now, I think we should have tried to double our effort to inform the public what the strike was all about, which was bargaining rights, not money." (101)

A more constructive resolution to a labor-management conflict over working conditions and health was arrived at by a union of toll collectors and a New York City agency. While a specific toxin had not been identified as the cause of illness among 34 bridge toll workers in New York City in 1990, union officials had "bridled" at the suggestion that the outbreak was due to "stress." (102, 103). The union had attempted for years to improve safety and health conditions for the toll collectors, who have elevated heart disease mortality rates, due, at least in part, to documented excess exposure to carbon monoxide (CO) from automobile exhaust. After the outbreak, union officials demanded permanent air-monitoring equipment and better ventilation. Some union officials acknowledged that while the first cases in the outbreak may have been due to inhalation of toxic vapors (arising from the burning of plastic-coated wire), later cases may have been due to "anxiety." (102) The union and the agency recently bargained a substantial medical surveillance program, whose primary focus is on heart disease risk due to CO exposure. The program will also evaluate the possible role of "job strain" as an independent or interactive risk factor for heart disease.

STATEWIDE AND NATIONAL EFFORTS AND STRATEGIES TO REDUCE STRESS

Workers Compensation

Spokespersons for the insurance industry argue that claims for "mental injury" rose sharply during the 1980s, and now account for about 15 percent of all occupational disease claims nationwide (104)—figures used to justify current efforts to limit claims. However, accurate data are difficult to obtain. In California, for example, one of only six states that considers mental injuries caused by gradual mental or emotional stress to be compensable, and a state with the most liberal law, the rate of mental stress claims increased 540 percent between 1979-1988, according to state data. (105) However the 9,368 reported cases in 1988 represented only 2 percent of total disabling work injuries. According to an insurance industry institute in California, many claims are not reported to the state agency, and self-insured public employers have higher rates, suggesting that the number of stress claims is actually four times higher. (105) However, even the higher estimate does not support arguments that business "is under siege" (104), but is compatible with growing awareness of the job stress-illness link.

The California insurance institute study indicated that stress claimants are more likely to be female and older than other work-disabled employees. Sales and clerical workers filed 40 percent of stress claims. Fewer than 10 percent of the claims followed a specific incident (for example, armed robbery); rather, job pressures (69 percent) and harassment (35 percent) were the most commonly cited reasons for the claim. (105) While it is difficult to generalize from this data, since many factors influence workers' ability or intention to file for compensation, it is compatible with the model of "job strain" as cumulative exposure to job pressures and low job control. The law still generally works against the worker since the burden of proof is upon the worker to define a condition and establish work-relatedness. (106)

Recently, employers have pushed for tighter standards for stress claims. A 1990 amendment to the New York State law restricts "mental-mental" claims when stress results from a normal personnel decision (work evaluation, job transfer, demotion) when taken in "good faith" by the employer. Similarly, since 1989, in California, the law requires that workers receive a psychiatric diagnosis of mental injury, and that "actual events" in the workplace were responsible for at least 10 percent of the causation of the injury—not simply the worker's perception of stress. (105) It remains to be seen to what extent the new scientific evidence on "job strain" will be used in compensation cases to explain causation for mental injury, hypertension, or heart disease.

Legislation and Political Action

In the United States, job stressors are not covered by OSHA. There are no health standards for shift work, piece-work, machine pacing, de-skilling, job security, isolated work, or technological change (as in Scandinavia). (107) An innovative campaign, however, is being waged by the Service Employees International Union (SEIU) in Pennsylvania to reduce back injuries and stress caused by inadequate staffing in nursing homes. (OSHA has already cited several nursing homes under the General Duty Clause for insufficient staff to do person transfers.) The campaign is in support of a proposed state law that would compel nursing homes to reveal information about staffing, injuries, and profits, and set minimum staffing levels. (108) A recent SEIU national survey of nurses re-emphasized concerns about work load demands, understaffing, and stress, and called for OSHA standards for nursing (including staffing), and providing health care workers with a voice in decisions. (109)

On the national level, support by the Clinton administration for the concepts of "high-skill, high-wage strategies" and "worker participation" (110) to improve the competitiveness of U.S. businesses holds the promise for a new focus on developing healthier work environments and reducing "job strain." However, in order to genuinely promote "high skill," active and lower "strain" jobs, job training and job design programs need to: (a) go beyond basic job skills, or narrow technical skills, and include "job ladders" or "career paths;" (b) promote computer software that encourages discretion and flexibility ("system knowledge"); (c) make skill training accessible to workers' schedules; and (d) keep skilled jobs in the bargaining unit and therefore increase rather than decrease union strength. (111, 112)

In addition, a variety of current legislative proposals could help increase job control and support, for example, laws that limit electronic monitoring and regulate VDT work. Other proposals could reduce the more general burden of social stress on individuals, such as laws on parental and personal leave, day care and elder care, voluntary overtime and shift work, a limited work week to create jobs, job sharing and part-time work. (8, 9) Even the OSHA reform bill (through mandated joint committees, improved worker training and enforcement, protection against discrimination, and improved record keeping) could spur efforts to identify and reduce psychosocial risk factors, most likely through investigation of hypertension and musculoskeletal disorders. Psychosocial risk factors could be considered for inclusion in the forthcoming ergonomics standard.

The goal of all these interventions and strategies is to produce a healthy workplace—in which workers are respected, where they have the opportunity to develop their skills and abilities, and where authority is shared, in other words, workplace democracy. Therefore, it is also important to consider legislation that would strengthen workers' collective voice (that is, unions) through banning of permanent replacements for strikers, and, in general, reforming labor law, as well as other means of increasing workers' influence and economic security, such as full employment and opportunities for employee ownership.

CONCLUSIONS AND RECOMMENDATIONS

The "job strain" studies and other research support the idea that social factors play a critical role in the production of common chronic diseases, such as heart disease and hypertension. The intervention studies, and other prevention strategies, indicate that the work environment can be modified to increase employee influence, skills, authority, and support, and to regulate demands. Participatory action research, collective bargaining, and legislation can be effective tools to achieve these goals. Effective PAR requires strong union involvement, while collectively bargained programs can benefit from PAR methods to involve workers and evaluate change. While the growing evidence linking job stress with illness helps to overcome the notion that psychosocial explanations for disease are not legitimate, vigilance needs to continue against our society's dominant ideology, which uses the stress explanation to "blame the victim"—indicting those who become ill as weak.

We believe that the evidence presented supports the following actions. First, "job strain" assessment instruments should be included in workplace health surveillance and health promotion programs, and in occupational health clinic educational material. Second, unions and their allies need to further increase their emphasis on contract language, education, organizing, and legislation on issues related to their members' job design, work organization, quality of work life, schedule flexibility, and work and family concerns. Third, multidisciplinary teams (including workers, union and company officials, occupational health specialists, epidemiologists, labor and health educators, social psychologists, physicians, and nurses), using PAR methods, can design, implement, and evaluate interventions to reduce or prevent exposure to psychosocial and physical health hazards and risk of illness. Fourth, further research is needed on various health outcomes (other than cardiovascular disease) potentially related to "job strain" or stress in general, including psychological disorders (4), musculoskeletal disorders (78), adverse pregnancy outcomes (79), "sick building syndrome" (80), work injuries (113), and immune system functioning (114), and the possible synergistic effects of psychosocial and physical health hazards. Modern workplaces embrace a complex set of risk factors, including psychosocial and physical/chemical.

Research is also needed on the connection between "job strain" and heart disease risk factors such as smoking, alcohol and diet, physiological mechanisms underlying heart disease, the effects of gender, race, and social class, and time trends. Similarly, further research is needed on the mechanisms and pathways underlying the effects of participation (for example, perceived influence, skill development, social support) on improvements in satisfaction and self-esteem, as well as aspects of intervention strategy associated with genuine organizational change. The Karasek "job strain" model has contributed greatly to the field through its clarity, predictions of health and behavioral outcomes, and emphasis on the concepts of demands, control, and support. It can now benefit from the expansion of the concepts of demands and control, to include measures contained in the Michigan stress model. (39, 40)

We believe that the "high demands + low control + low support" paradigm also provides a useful working model for understanding associations between more general social stress and health. Since hypertension is prevalent in all industrialized societies (both market and state-owned economies), and since blood pressure does not typically rise with age in nonindustrial societies (for example, hunter-gatherers and agricultural communities) (115), we need to consider what aspects of industrial society (such as social class differences or "job strain") account for this effect. For example, home and family demands and lack of control may impact on health. (42) Unemployment, with its resulting health effects (116), can be conceived of as an extreme case of loss of control. Even the threat of unemployment can increase competition (demands) and lead to a decreased sense of control among remaining employees. (5, p. 307) The decline in the standard of living since the 1960s and the economic necessity for both parents to work is a major reason for increased work hours (increased demands) in the United States. (9) Finally, lower SES presents increased cardiovascular and other health risks possibly due to limited influence, resources, and opportunities, as well as a poorer physical environment. (43)

For example, rates of heart disease mortality and all-cause mortality have risen (primarily for men) in Eastern Europe since the 1960s in contrast to substantial declines in Western Europe, Canada, Japan, and the United States. (117) This has been attributed by public health officials to "lifestyle" factors such as smoking, alcohol, and a fatty diet, rather than, for example, environmental pollution. (118) However, the post–World War II period was also a period of urbanization, social migration, industrialization based on the principles of Taylorism, and introduction of and adjustment to a political system that allowed citizens limited control both in society and in the workplace. We need to consider the possible effects of these social changes not only on lifestyle behaviors, but on the prevalence of "job strain," or more directly on cardiovascular health.

Just as the elimination of infectious diseases as the major causes of mortality over the last century occurred due to social changes, improvements in sanitation and nutrition, and elimination of slum conditions (and just as the reappearance of diseases such as tuberculosis has resulted from social neglect), chronic diseases are related to the physical and social environments in which people live and work. Our social epidemiologic model of illness explicitly recognizes that work reorganization, workplace democracy, and broader societal changes (social and economic democracy) are needed to reduce the risk of cardiovascular disease and improve emotional well-being.

ACKNOWLEDGMENTS

The authors would like to thank Philip Landrigan, David LeGrande, and Dominic Tuminaro for their advice on portions of this article, as well as the suggestions of three anonymous reviewers.

NOTES

1. For example, in 1984, Dr. Robert Karasek reviewed a film on stress produced by the Federal Aviation Administration and shown to all air traffic controllers. The film stated that stress depends on demands and the individual's coping style—control over the situation was not mentioned. To cope with stress, the film advised controllers to question that real world events are causes of stress, rather interpretations of events are the major cause. The film stated that employee expectations of fair treatment are often irrational in the modern setting, and that workers' misformed expectations of fairness are the source of problems. Finally, exclusively individual solutions (for example, "visualizations" of babbling brooks) were recommended. (2)

2. U.S. Congress. House Committee on Public Works and Transportation, Hearing before the Subcommittee on Investigations and Oversight. Status of the Air Traffic Control System, 98th Congress, May 4; June 7; October 28, 1983; March 20, 21, 22, 27, 28, 29; April 4; June 26, 27, 1984; H. Rept. 98-83, pp. 785–787.

3. A number of recent developments suggest a growing awareness of the need to accord stress a major role in the occupational health agenda, including the publication of three essential books on this topic, by Karasek and Theorell (5), Johnson and Johannsson (51), and the International Labor Office. (36) Similarly, recent major conferences have focused on this topic: "Participatory Approaches to Improving Workplace Health," Labor Studies Center, University of Michigan, June 3–5, 1991 (presentation summaries available), and the 1990 and 1992 job stress conferences sponsored by the National Institute for Occupational Safety and Health (NIOSH) and the American Psychological Association (APA). (Some presentations available in "Stress and Well-Being at Work: Assessments and Interventions for Occupational Mental Health," Washington, DC: APA.)

4. Sauter, S. L., Murphy, L. R., Hurrell, J. J., Jr. "Prevention of Work-Related Psychological Disorders," *American Psychologist* 45 (1990):1146–1158.

5. Karasek, R., Theorell T. *Healthy Work*. New York: Basic Books, 1990.

6. The term "stress" is used in this paper to refer to the broad range of psychosocial factors and their resulting mechanisms that affect the worker due to impacts on behavioral, psychological, or physiological outcomes. Within the job stress research community, the term "strain" is typically used to indicate the short-term or intermediate effect of job stress (for example, alterations in the hormonal system of the body), which eventually lead to the development of disease. The terminology of "stress" leading to "strain" and then to disease is actually borrowed from the way these terms are used in engineering. In this paper, we use the term "job strain" in a more specific way—to refer to the objective workplace causes of "stress" described in the Karasek "job strain" model.

7. Johnson J. V., Hall E. M., Theorell T. "Combined effects of job strain and social isolation on cardiovascular disease morbidity and mortality in a random sample of the Swedish male working population," *Scandinavian Journal of Work, Environment and Health* 15 (1989):271–279.

8. Quinn, M. M., Buiatti, E. "Women Changing the Times," *New Solutions* 1 (1991):48–56.

9. Brandt, B. "The Problem of Overwork in America Today," *New Solutions* 2 (1991):50–65.

10. Kasinsky, R. G. "Sexual Harassment: A Health Hazard for Women Workers," *New Solutions* 2 (1992):74–83.

11. Spangler, E. "Sexual Harassment: Labor Relations by Other Means," *New Solutions* 3 (1992):23–30.

12. Schnall, P. L., Landsbergis, P. A. Baker, D. "Job Strain and Cardiovascular Disease," *Annual Review of Public Health* (1994):381–411.

13. Schurman, S. J., Israel, B. A., Hugentobler, M. K. "Changing the Work Environment to Reduce Stress: Review of Interventions in the United States and Recommendations for Research and Practice.

14. American Heart Association. *1993 Heart and Stroke Facts*. Dallas, TX: AHA, 1992.

15. This estimate of population attributable risk was derived from five studies in which multivariate models were used to calculate the association between "job strain" and heart disease. These models controlled for other risk factors (for example, age, education, cigarette smoking, serum cholesterol, and blood pressure) in many cases. To the extent that potential confounding was not adequately controlled for, the 23 percent figure may be an overestimate of the potential benefits of reducing "job strain." However, to the extent that standard heart disease risk factors are in the causal pathway between "job strain" and heart disease, controlling for them provides a conservative estimate of risk—an underestimate of the true complete effect of exposure to "job strain."

16. Alfredsson, L., Spetz, C. L., Theorell, T. "Type of Occupation and Near-future Hospitalization for Myocardial Infarction and Some Other Diagnoses," *International Journal of Epidemiology* 14 (1985):378–388.

17. Falk, A., Hanson, B. S., Isacsson, S. O., Ostergren, P-O. "Job Strain and Mortality in Elderly Men: Social Network, Support, and Influence as Buffers," *American Journal of Public Health* 82 (1992):1136–1139.

18. Astrand, N. E., Hanson, B. S., Isacsson, S. O. "Job Demands, Job Decision Latitude, Job Support, and Social Network Factors as Predictors of Mortality in a Swedish Pulp and Paper Company," *British Journal of Industrial Medicine* 46 (1989):334–340.

19. Theorell, T., Perski, A., Orth-Gomer, K., Hamsten, A., deFaire, U. "The Effect of Returning to Job Strain on Cardiac Death Risk After a First Myocardial Infarction Before Age 45," *International Journal of Cardiology* 30 (1991):61–67.

20. LaCroix, A. Z. *High Demand/Low Control Work and the Incidence of CHD in the Framingham Cohort*. Doctoral dissertation. Chapel Hill, NC: University of North Carolina, 1984.

21. Haan, M. N. "Job Strain and Ischaemic Heart Disease: An Epidemiological Study of Metal Workers," *Annals of Clinical Research* 20 (1988):143–145.

22. Green, K. L., Johnson, J. V. "The Effects of Psychosocial Work Organization on Patterns of Cigarette Smoking Among Male Chemical Plant Employees," *American Journal of Public Health* 80 (1990):1368–1371.

23. Mensch, B. S., Kandel, D. B. "Do Job Conditions Influence the Use of Drugs?" *Journal of Health and Social Behavior* 29 (1988):169–184.

24. Pieper, C., LaCroix, A. Z., Karasek, R. A. "The Relation of Psychosocial Dimensions of Work with Coronary Heart Disease Risk Factors: A Meta-analysis of Five United States Data Bases," *American Journal of Epidemiology* 129 (1989): 483-494.

25. Theorell, T., Perski, A., Akerstedt, T., Sigala, F., Ahlberg-Hulten, G., Svensson, J., Eneroth, P. "Changes in Job Strain in Relation to Changes in Physiological States," *Scandinavian Journal of Work, Environment and Health* 14 (1988):189-196.

26. Van Egeren, L. F. "The Relationship Between Job Strain and Blood Pressure at Work, at Home, and During Sleep," *Psychosomatic Medicine* 54 (1992): 337-343.

27. Blumenthal, J. A., Siegel, W. C., Phillips, B. G. "Job Strain Affects Blood Pressure in the Laboratory and During Daily Life in Patients with Mild Hypertension," *Hypertension*.

28. Theorell, T., deFaire, U., Johnson, J., Hall, E. M., Perski, A., Stewart, W. "Job Strain and Ambulatory Blood Pressure Profiles," *Scandinavian Journal of Work, Environment and Health* 17 (1991):380-385.

29. Schnall, P. L., Schwartz, J. E., Landsbergis, P. A., Warren, K., Pickering, T. G. "The Relationship Between Job Strain, Alcohol and Ambulatory Blood Pressure," *Hypertension* 19 (1992):488-494.

30. Schnall, P. L., Landsbergis, P. A., Schwartz, J. E., Warren, K., Pickering, T. G. "The Relationship Between Job Strain, Ambulatory Blood Pressure and Hypertension," Paper presented at the Ninth International Symposium on Epidemiology in Occupational Health, Cincinnati, OH, September 1992.

31. Light, K. C., Turner, J. R., Hinderliter, A. L. "Job Strain and Ambulatory Work Blood Pressure in Healthy Young Men and Women," *Hypertension* 20 (1992):214-218.

32. Theorell, T., Knox, S., Svensson, J., Waller, D. "Blood Pressure Variations During a Working Day at Age 28: Effects of Different Types of Work and Blood Pressure Level at Age 18," *Journal of Human Stress* 11 (1985):36-41.

33. Frankenhauser M, Johansson G. "Stress at Work: Psychobiological and Psychosocial Aspects," *International Review of Applied Psychology* 35 (1986): 287-299.

34. Schnall, P. L., Pieper, C., Schwartz, J. E., Karasek, R. A., Schlussel, Y., Devereux, R. B., Ganau, A., Alderman, M., Warren, K., Pickering, T. G. "The Relationship Between 'Job Strain,' Workplace Diastolic Blood Pressure, and Left Ventricular Mass Index: Results of a Case-control Study," *Journal of the American Medical Association* 263 (1990):1929-1935.

35. Karasek, R. A., Gordon, G., Pietrokovsky, C., Frese, M., Pieper, C., Schwartz, J., Fry, L., Schirer, D. Job Content Instrument: Questionnaire and User's Guide. Los Angeles, CA: University of Southern California, 1985.

36. DiMartino, V. (Ed.) *Conditions of Work Digest: Preventing Stress at Work.* Geneva: International Labor Office, 1992.

37. Johnson, J. V. "Collective Control: Strategies for Survival in the Workplace," *International Journal of Health Services* 19 (1989):469-480.

38. Israel, B. A., Schurman, S. J., House, J. S. "Action Research on Occupational Stress: Involving Workers as Researchers," *International Journal of Health Services* 19 (1989):135-155.

39. Caplan, R. D., Cobb, S., French, J. R. P., Jr., Van Harrison, R., Pinneau, S. R., Jr. *Job Demands and Worker Health.* Cincinnati, OH: National Institute for Occupational Safety and Health, 1975 (Publication No. 75-168).

40. Hurrell, J. J., McLaney, M. A. "Exposure to Job Stress—A New Psychometric Instrument," *Scandinavian Journal of Work, Environment and Health* 14 (suppl. 1) (1988):27-28.

41. Haynes, S. G., Feinleib, M. "Women, Work and Coronary Heart Disease: Prospective Findings from the Framingham Heart Study," *American Journal of Public Health* 70 (1980):133-141.

42. Hall, E. M. "Double Exposure: The Combined Impact of the Home and Work Environments on Psychosomatic Strain in Swedish Men and Women," *International Journal of Health Services* 22 (1992):239-260.

43. Marmot, M., Theorell, T. "Social Class and Cardiovascular Disease," *International Journal of Health Services* 18 (1988):659-674.

44. Siegrist, J., Peter, R., Junge, A., Cremer, P., Seidel, D. "Low Status Control, High Effort at Work and Ischemic Heart Disease: Prospective Evidence from Blue-collar Men," *Social Science and Medicine* 31 (1990):1127-1134.

45. Ivancevich, J. M., Matteson, M. T., Freedman, S. M., Phillips, J. S. "Worksite Stress Management Interventions," *American Psychologist* 45 (1990):252-261.

46. This section is based on a more detailed review in (13).

47. Examples are classic texts from this period: Lewin, K. *Field Theory in Social Science*. New York: Harper, 1951; Walker, C. R., Guest, R. H. *Man on the Assembly Line*. Cambridge: Harvard University Press, 1952; MacGregor, D. *The Human Side of Enterprise*. New York: McGraw-Hill, 1960; Kornhauser, A. *Mental Health of the Industrial Worker*. New York: Wiley, 1965.

48. For recent critical reviews, see the series Woodman, R., Pasmore, W. (Eds.) *Research in Organization Change and Development*. Greenwich, CT: JAI Press, 1987.

49. Parker, M. *Inside the Circle*. Boston: South End Press, 1985.

50. Gardell, B., Gustavsen, B. "Work Environment Research and Social Change: Current Developments in Scandinavia," *Journal of Occupational Behavior* 1 (1980):3-17.

51. Johnson, J. V., Johannsson, G. *The Psychosocial Work Environment*. Amityville, NY: Baywood, 1991.

52. Hanlon, M. D. "Reducing Hospital Costs Through Employee Involvement Strategies," *National Productivity Review* 5 (1986):22-31.

53. Kopelman, R. E. "Job Design and Productivity: A Review of the Evidence," *National Productivity Review* 4 (1985):237-255.

54. Gordon, M. E., Burt, R. E. "A History of Industrial Psychology's Relationship with American Unions: Lessons from the Past and Directions for the Future," *International Review of Applied Psychology* 30 (1981):137-156.

55. Buller, P. F., Bell, C. H. "Effects of Team Building and Goal Setting on Productivity: A Field Experiment," *Academy of Management Journal* 29 (1986): 305-328.

56. Grenier, G. J. "Twisting Quality Circles to Bust Unions," *AFL-CIO News*, May 14, 1983.

57. Jones, L. M., Bowers, D. G., Fuller, S. M. *Report of Findings 1984: Task Force on Management and Employee Relationships*. Washington, DC: Federal Aviation Administration, November 7, 1984.

58. "APWU Settles USPS Company Union Charges," *AFL-CIO News*, May 11, 1992.

59. "Partnership Sweetens Work at Nabisco Bakeries," *AFL-CIO News*, April 29, 1991.

60. "Cooperative Successes Typified at Ohio Plant," *AFL-CIO News*, June 22, 1992.

61. Whyte, W. F. *Participatory Action Research*. Newbury Park, CA: Sage, 1990.

62. Karasek, R. "Stress Prevention Through Work Reorganization: A Summary of 19 International Case Studies," in DiMartino, V. (ed.) *Conditions of Work Digest: Preventing Stress at Work*. Geneva International Labor Office, 1992: 23–41.

63. Jackson, S. E. "Participation in Decision Making as a Strategy for Reducing Job Related Strain," *Journal of Applied Psychology* 68 (1983):3–19.

64. Golembiewski, R., Hilles, R., Daly, R. "Some Effects of Multiple OD Interventions on Burnout and Worksite Features," *Journal of Applied Behavioral Science* 23 (1987):295–314.

65. Israel, B. A., Schurman, S. J., Hugentobler, M. K., House, J. "A Participatory Action Research Approach to Reducing Occupational Stress in the United States," in DiMartino, V. (Ed.) *Conditions of Work Digest: Preventing Stress at Work*. Geneva: International Labor Office, 1992: 152–163.

66. Heaney, C., Israel, B., Schurman, S., House, J., Baker, B., Hugentobler, M. "Evaluation of a Participatory Action Research Approach to Reducing Stress." Paper presented at the APA/NIOSH Conference on Occupational Stress, Washington, DC, November 1992.

67. Cahill, J. "Computers and Stress Reduction in Social Service Workers in New Jersey," in DiMartino, V. (Ed.) *Conditions of Work Digest: Preventing Stress at Work*. Geneva: International Labor Office, 1992: 197–203.

68. Lerner, M. *Occupational Stress Groups and the Psychodynamics of the World of Work*. Oakland, CA: Institute for Labor and Mental Health, 1985.

69. May, L. "A Union Programme to Reduce Work and Family Stress Factors in Unskilled and Semi-Skilled Workers on the East Coast of the United States," in DiMartino, V. (Ed.) *Conditions of Work Digest: Preventing Stress at Work.* Geneva: International Labor Office, 1992: 164–171.

70. Landsbergis, P. A., Silverman, B., Barrett, C., Schnall, P. L. "Union Stress Committees and Stress Reduction in Blue and White Collar Workers," in DiMartino, V. (Ed.) *Conditions of Work Digest: Preventing Stress at Work*. Geneva: International Labor Office, 1992: 144–151.

71. Mergler, D. "Worker Participation in Occupational Health Research: Theory and Practice," *International Journal of Health Services* 17 (1987):151–167.

72. Thus, it is analogous to the role proposed for workers in making industrial hygiene more effective, that is, involving workers in inspections, assessment of hazards and health symptoms, problem solving; and changing work practices

and organization as well as technical improvements (Senn, E. "Playing Industrial Hygiene To Win," *New Solutions* (1991):72-81).

73. Hugentobler, M. K., Robins, T. G., Schurman, S. J. "How Unions Can Improve the Outcomes of Joint Health and Safety Training Programs," *Labor Studies Journal* 15 (1990):16-38.

74. AFL-CIO Committee on the Evolution of Work. *The Changing Situation of Workers and Their Unions*. Washington, DC: AFL-CIO, 1985.

75. Cahill, J. "Economic and Non-Economic Sources of Job Satisfaction in Higher Education," Paper Presented at the American Association for Higher Education, Washington, DC, March, 1993.

76. Israel, B., Schurman, S. "Social Support, Control and the Stress Process," in Glanz, K., Lewis, F., Rimer, B. (Eds.) *Health Behavior and Health Education: Theory, Research and Practice*. San Francisco: Jossey-Bass, 1990:187-215.

77. "Moving Towards Worker-Oriented Participatory Research," Panel at the American Public Health Association, San Francisco, CA, October 1993.

78. Bongers, P. M., deWinter, R. R., Kompier, M. A. J., Hildebrandt, V. H. "Psychosocial Factors at Work and Musculoskeletal Disease: A Review of the Literature," *Scandinavian Journal of Work Environment and Health* 19 (5) (1993).

79. Brett, K., Strogatz, D., Savitz, D. "Occupational Stress and Low Birth Weight Delivery." Presented at the 24th Annual Meeting of the Society of Epidemiological Research, Buffalo, NY, June 1991. *American Journal of Epidemiology* 134 (1991): 722-723.

80. National Institute for Occupational Safety and Health. *Indoor Air Quality and Work Environment Study*. Cincinnati, OH: NIOSH, 1991 (HETA 88-364-2102).

81. Shostak, A. B. "Union Efforts to Relieve Blue-Collar Stress," in Cooper, C. L. Smith, M. J. (Eds.) *Job Stress and Blue-Collar Work*. New York: Wiley, 1985: 195-205.

82. Service Employees International Union. *Stress: Contract Provisions*. Washington, DC: SEIU, 1983.

83. Unions have also developed valuable worker education materials on job stress that use the Small Group Activity Method and the principles of empowerment education. Sources include the Communications Workers of America, District 1; District 65, UAW; and the Labor Institute, in New York City; 9 to 5, National Association of Working Women, Cleveland, OH; and the Workers Health and Safety Centre, Kingston, Ontario.

84. Landsbergis, P. "Occupational Stress Among Nurses: New Developments in Theory and Prevention," in Humphrey, J. H. (Ed.) *Human Stress: Current Selected Research* (Vol. 3). New York: AMC Press, 1989: 173-195.

85. *AFL-CIO News*, January 26, 1985; May 23,1988; July 8, 1989; February 1, 1993; Labor Notes, December 1985; October, 1989.

86. Dena Bunis, personal communication, February 1993.

87. Bernard, B. "Psychosocial Factors for Musculoskeletal Disorders," Paper Presented at the American Public Health Association, San Francisco, CA, October 1993.

88. U.S. Congress, Office of Technology Assessment, "The Electronic Supervisor: New Technology, New Tensions," OTA-CIT-333. Washington, DC: U.S. Government Printing Office, 1987.

89. "Northern Telecom Ends Worker Monitoring," *VDT News*, March/April 1992.

90. Smith, M. J., Carayon, P., Sanders, K. J., Lim, S-Y., LeGrande, D. "Employee Stress and Health Complaints in Jobs With and Without Electronic Performance Monitoring," *Applied Ergonomics* 23 (1992):17-27.

91. Ramirez, A. "A.T.&T. and Unions Praise New Pact," *The New York Times*, July 3, 1992.

92. National Institute for Occupational Safety and Health. *Health Hazard Evaluation Report*. Cincinnati, OH: NIOSH, July 1992 (HETA 89-299-2230).

93. Bureau of Labor Statistics. "Survey of Occupational Injuries and Illnesses in 1991." Washington, DC: U.S. Department of Labor, 1992.

94. "The Electronic Supervisor" (88, pp. 39, 46-48, 57); Massachusetts Coalition on New Office Technology, "Electronic Monitoring in the Workplace: Supervision or Surveillance?" Boston, 1989.

95. Shostak, A. B., Skocik, D. *The Air Controllers' Controversy*. New York: Human Sciences Press, 1986.

96. Landsbergis, P. "Is Air Traffic Control a Stressful Occupation?" *Labor Studies Journal* 11 (1986):117-134.

97. Rose, R. M., Jenkins, C. D., Hurst, M. W. *Air Traffic Controller Health Change Study*. Washington, DC: Department of Transportation, 1978.

98. Tesh, S. "The Politics of Stress: The Case of Air Traffic Control," *International Journal of Health Services* 14 (1984):569-587.

99. Bowers, D. G. "What Would Make 11,500 People Quit Their Jobs," *Organizational Dynamics* (1983):5-19.

100. Hurd, R. W., Kriesky, J. K. "The Rise and Demise of PATCO Reconstructed," *Industrial and Labor Relations Review* 40 (1986):115-121.

101. "Five Years After PATCO," *Frequent Flyer Magazine*, August 1986: 54-55, 62.

102. Golden, T. "Was Illness at Bridges in the Minds of Workers," *The New York Times* (March 12, 1990).

103. Cases of similar physical symptoms among a group of workers without an identifiable pathogen (known as "epidemic psychogenic illness," EPI) have been extensively analyzed by NIOSH and others. (For example: Colligan, M., Pennebaker, J., Murphy, L. (Eds.) *Mass Psychogenic Illness: A Social Psychological Analysis*. Hillsdale, NJ: Lawrence Erlbaum, 1982.) Work organization in workplaces where EPI has occurred has been characterized as repetitive or boring work with rigid pacing and high job pressures, strict rules, and a lack of communication and social interaction. While EPI can therefore be regarded as a desperate reaction to excessive "job strain," another explanation is that low-level chemical exposure and stress may have synergistic effects.

104. DeCarlo. D. T., Gruenfeld, D. H. *Stress in the American Workplace*. Fort Washington, PA: LRP Publications, 1989: 11.

105. California Workers' Compensation Institute, "Mental Stress Claims in California Workers' Compensation—Incidence, Costs and Trends," San Francisco: California Workers' Compensation Institute, 1990.

106. Elisburg, D. "Workers' Compensation Stress Claims: Employee Issues," Paper Presented at the APA/NIOSH Conference on Occupational Stress, Washington, DC, November 1992.

107. Deutsch, S. "Work Environment Reform and Industrial Democracy," *Sociology of Work and Occupations* 8 (1981):180–194.

108. Service Employees International Union. *The High Cost of Short Staffing*. Washington, DC: SEIU, 1992.

109. Service Employees International Union. *The National Nurse Survey*. Washington, DC: SEIU, 1993.

110. Byrne, M. "Reich, Council Explore Ways to Get Moving," *AFL-CIO News*, March 1, 1993.

111. Richardson, C. "Training to Improve Working Conditions, and Build Union Strength," Paper Presented at the International Metalworkers' Federation Conference, Helsinki, Finland, September 1991.

112. Richardson, C. "Technology, Training and Work Organization," *New Solutions* 3 (1993): 5–6.

113. Johnson, J. L. "Work Injury and Stress," Paper Presented at the APA/NIOSH Conference on Occupational Stress, Washington, DC, November 1992.

114. Henningsen, G. M., Hurrell, J. J., Baker, F., Douglas, C., MacKenzie, B. A., Robertson, S. K., Phipps, F. C. "Measurement of Salivary Immunoglobulin A as an Immunologic Biomarker of Job Stress," *Scandinavian Journal of Work, Environment and Health* 18 (suppl 2) (1992):133–136.

115. Schnall, P. L. Kern, R. "Hypertension in American Society: An Introduction to Historical Materialist Epidemiology," in P. Conrad & R. Kern (Eds.) *The Sociology of Health and Illness*. New York: St. Martin's Press, 1981: 97–122.

116. Brenner, H. *Economy, Society and Health*, Washington, DC: Economic Policy Institute, 1992.

117. Uemura, K., Pisa, Z. "Trends in Cardiovascular Disease Mortality in Industrialized Countries Since 1950," *World Health Statistics Quarterly* 41 (1988):155–178.

118. Levenstein, C. "Occupational Health in Eastern Europe During Political and Economic Transition," Paper Presented at the American Public Health Association, Washington, DC, November 1992.

11

Sexual Harassment

A HEALTH HAZARD FOR WOMEN WORKERS

RENEE GOLDSMITH KASINSKY

Anita Hill's accusations of sexual harassment against her former boss, Judge Clarence Thomas, and the public Senate Judiciary Committee hearings on them have brought the message home to millions of Americans. Sexual harassment has become a household word. The fact that 12 white males of the Judiciary Committee saw fit to contain the issue as a private trouble outraged seven Congresswomen who marched into the Senate insisting upon a public hearing. Their actions helped to elevate the problem to a public issue, and their outrage was shared by women and men across America who followed the hearings in the media.

Sexual harassment is not a new social issue; the legislation and the major federal enforcement agency, the Equal Employment Opportunity Commission (EEOC), have been in place for more than a decade. However, most ordinary Americans as well as professionals in the fields of law, politics, and industry have little clear understanding of the nature of sexual harassment and the havoc it wreaks, both emotionally and economically, on women and on the nation.

In the weeks following the committee hearings into the accusations of Anita Hill and others, there were multiple articles in the business and living sections of the print media focusing on the economic repercussions resulting from sexual harassment. The devastating health effects of sexual harassment on women employees as well as the effects on their work performance and their overall position in the workplace have been less publicized. Yet sexual harassment may be one of the most widespread occupational health hazards women face in the workplace, as well as the best guarded secret. This can be partly explained by the fact that there is an accepted norm of silence for women and those who speak out are often blamed or punished for doing so. Even more important is the acceptance of a hostile sexual workplace as part of our culture, as part of our normal environment, dominated by male norms of power. (1) "Sexual harassment is not an occupational health issue," Ronald S. Ratney, a regional administrator of the Occupational Safety and Health Administration (OSHA), maintained. "It is a personal issue. OSHA does not include sexual harassment as an occupational health hazard. It's not part of its regular mandate. There are no physiological effects." (2)

In this article, I take strong issue with this view. Sexual harassment is a major social issue in our society, hardly a personal matter. And it should be a special concern for the federal OSHA program. The stress effects of work-related sexual harassment, although often hidden, have far-reaching health consequences. Sexual harassment creates both physical and mental stress that impairs our health. To eliminate these stress effects would mean, ultimately, eliminating sexual harassment from the workplace. This is a major undertaking since sexual harassment is institutionalized in practices in the marketplace as well as in the culture at large.

SEXUAL OPPRESSION IN THE WORKPLACE AND CULTURE

Sexual harassment must be viewed as part of a larger system of the oppression of women in relation to sex and the workplace. Sexual harassment is only the "tip of the iceberg." What lies beneath is a whole system of assumptions, institutional practices, and structures through which sexuality negatively affects women's position in the marketplace. In the aftermath of the Hill–Thomas confrontation, many writers have referred to the "gender gap" in describing understandings about sexual harassment. "Women across the country understand why Anita Hill stayed at her job, why she kept phoning Thomas. They understand that you need to put up with harassment, that this is your job and your life. But men don't understand because they have options that are far better and far greater." These words were spoken by Dr. Frances Conley, a professor and one of the nation's first female neurosurgeons, who announced her resignation from Stanford University's School of Medicine last June charging that she had endured a career-long pattern of sexual harassment from colleagues and superiors. (3)

Sexual harassment grows out of the economic and cultural inequality between the sexes that is built into the workforce, as well as the whole culture. It is fostered by the organization of a workplace that is characterized by extreme segregation of jobs and occupations. In both traditional and nontraditional occupations, the position of women relative to men in the economic hierarchy reveals that men hold most of the positions with authority to hire and fire, while women are relegated to lower-level jobs. On the whole, men have higher-status jobs, higher salaries, more formal training, and more seniority than women. Men are more likely to belong to unions or to be more closely connected with informal "old boy" networks. Men can force sexual attention on women in the workplace because of the greater economic power they hold. This superior status protects men from any repercussions that could arise when a woman complains about being targeted for sexual attention. Indeed, sexual harassment is a demonstration of that status and power, and is not primarily motivated by sexual feelings. It also may be a route through which men bolster class and status differences among themselves. (4)

In addition to economic power expressed in the structures of work organizations, the confluence of sex and power that characterize "normal" male–female relationships in U.S. society may be as determinant as those power differentials in the specific workplace in sexual harassment practices. (5) There are societal rewards for domineering sexual behavior on the part of males and for a parallel passivity and acquiescence on the part of females that contribute to the existence of sexual harassment. The function of sex typing operates differently in traditional and nontraditional workplaces. In the female-dominated jobs, the origins of sexual harassment have been accounted for by suggesting that the man is acting as if the woman is a wife because her job carries that implication. The workplace is referred to as sexualized. In traditional blue-collar jobs, where men are in the majority, women are perceived as out of their gender role. Gutek (6) contends that conceptions of females as women, rather than as workers, "spill over" into the woman's work role as a major explanation for the existence of sexual harassment. Thus, status groupings and stereotypical gender attitudes that arise outside the work organization are carried into it by individual workers. Crull (7) has a different conception of the function of sex typing. She sees family roles as more key to the power configurations of harassment. The executive in the workplace has interlocking sources of power derived from both his boss status and his husband/father role. The entrance of women into nontraditional occupations threatens those sources of power and he resorts to sexual harassment. In both explanations, sexual harassment is not an aberration but rather an extension of the normal expectations for male and female behavior in the society.

How widespread is sexual harassment? Since most women never speak out or file suit, its prevalence has been underestimated. In 1976, *Redbook* magazine asked 9,000 readers whether they had been subject to unwanted sexual "attention" at work from male bosses or colleagues; 88 percent responded yes and 75 percent called the advances embarrassing, demeaning, or intimidating. (8) In a larger study of 23,000 federal government employees in 1980 and 1988, 42 percent of the women respondents reported they had been subjected to some form of sexual harassment; unwanted "sexual teasing" was identified as the most prevalent form. (9) This figure corresponds with results of a survey of *Harvard Business Review* readers. Of the 42 percent, 12 percent experienced less severe sexual harassment, including suggestive remarks, teasing, jokes; 29 percent experienced severe sexual harassment, including incidents such as receiving unwanted letters, phone calls, or materials of a sexual nature, unwanted touching or pinching, or unwanted pressure for sexual favors; and 1 percent experienced actual or attempted rape or sexual assault. (10) A Defense Department survey found that 64 percent of women in the military had suffered "uninvited and unwanted sexual attention" within the preceding year. The United Methodist Church established that 77 percent of its clergywomen had experienced incidents of sexual harassment; 41 percent named a pastor or colleague as the perpetrator, and 31 percent mentioned church social functions as the setting. (11)

Two Types of Harassment

Sexual harassment is a single label that in fact covers many different behaviors. All of them interfere with the ability of the targeted person to work, and create psychological as well as physical stress. To understand both the dynamics and underpinnings of sexual harassment, we will focus on the two distinctive patterns of harassment defined by the courts and federal policy. In 1980 the EEOC issued guidelines finding harassment on the basis of sex to be a violation of Title VII, and labeling as sexual harassment "unwelcome sexual advance, requests for sexual favors, and other verbal or physical conduct of a sexual nature."

In the first type, such cases are called *quid pro quo* in the legal literature. The type of sexual harassment first legally recognized involved women who were propositioned by a superior at work and then fired or denied some job benefit because they refused (in Latin, quid pro quo means "one thing in return for another;" in other words, "sleep with me or you are fired"). Linda Monohon's story fits this pattern of harassment.

LINDA MONOHON, INSURANCE BROKER (12)

In 1970, Linda Monohon joined a local insurance company as a secretary and rose through the ranks to become one of the few women treaty brokers in the field. Nine years later when three colleagues founded a new firm, she became the vice-president, though her salary was less than that of her male counterparts. She had had a romance with one fellow executive, John, in the old firm who was now with the new company. She had ended the affair in the early '80s over his objections and he responded by alternately ignoring her and leaving love letters on her desk. In 1983 Linda became a single parent by choice, and John responded by proposing marriage.

After her maternity leave in 1984, she had to report at work to John—who promptly shut her out of meetings, interfered with her client relationships, and even threatened to fire her. The next year, when her complaints to the firm's president went unheeded, she quit. The company then fought her claim for $164 a week in unemployment checks on grounds that she had not been fired.

She then sued in the district court charging "outrageous conduct." During the trial, other female employees of the company testified that John sometimes tore off strips of adding-machine tape in the office and compared them to the length of his penis, and had touched another woman worker's breasts. The jury awarded Linda $1.3 million for lost wages and emotional distress and $5 million in punitive damages. It was one of the highest awards ever.

In the second type of sexual harassment, there is a hostile environment or atmosphere of sexual comments, jokes, or leers directed at women. The sexual behavior is hostile and intimidating from the outset. This is the situation in which men sexually taunt women who work with them. Frequently, in atmosphere cases, the behaviors include sexual graffiti and pictures posted in work areas and the participation by several men rather than a single harasser. Similar to a pattern of harassment endured by racial and ethnic minorities, it is mixed with other actions that are clearly hostile and discriminatory, and is intended to force the woman to leave. Let's look at two case studies.

DR. FRANCES CONLEY, NEUROSURGEON (13)

Dr. Frances Conley, after 17 years as a top neurosurgeon and professor at Stanford University School of Medicine, announced her resignation in June 1991, charging that she had endured a career-long pattern of sexual harassment from colleagues and superiors within her male-dominated profession. She was protesting a general atmosphere of sexism pervading the medical school. She pointed to the cumulative effect of "microinequities" that are especially corrosive to women and that remind them they are different and less worthy. Dr. Conley said, "I was asked to go to bed . . . it was always said with four or five people around. The men wanted to see if I would put up with the rules of the club—a club that had always allowed men to use women as they wanted to. It was not harmful physically, but mentally it was disgusting. . . . Any woman is fair game" for sexual harassment. Conley said in many operating rooms the banter is so laced with sex "that the average patient would be horrified." Another example of demeaning behavior was cited by Dr. Conley: "If I'm in an examining room with a patient, and a male doctor says, "Hi Hon, can you come here for a minute?'—he's established me as being lower in the hierarchy. And what does the patient think: "Gee, are they having an affair?'"

KATHRYNE BROOKE, LONG HAUL TRUCKER (14)
Kathryne Brooke said a woman trucker has to constantly prove herself. "When you come out of a truck to get some coffee at a truck stop, you're not seen as a trucker. You're not a driver. You're a beaver. A beaver is a term given to female truck drivers. They have gross bumper stickers about beavers. One says, "Save our forest. Eat a beaver.' I don't know if it's because of the tail—piece of tail. I sure don't know. I don't have buck teeth. But what's important is to do your job and not lose your integrity. Fools are gonna talk so you just have to carry on."

The most common type of harassment that the courts as well as the EEOC deal with is the quid pro quo situation. It usually is the more extreme cases that are prosecuted. In traditional women-dominated work settings (the sexualized workplace), most women experience the quid pro quo pattern of harassment, whereas in the male-dominated occupations, the "hostile environment" pattern more often has been reported by women.

A number of researchers have used a model of the continuum to represent and compare both types of harassment. At one end of the continuum are behaviors that appear to have as their goals some sexual/social interaction, and at the other end, threatening and degrading acts that seem to serve as a substitute for patriarchal power. Corresponding to the sexual/social end of the continuum are situations where the harasser holds the greatest amount of economic power over the woman and where the sex typing of the jobs is the greatest. (15) Corresponding to the hostile end of the continuum are situations in which the man has the least amount of power over the woman and where she is violating patriarchal norms of the workplace by holding a male sex-typed job. (16)

Leading feminist theorist and practitioner Catharine MacKinnon argued in her brief in connection with the Vinson case that both quid pro quo and hostile environment forms of harassment are as closely linked as points "on a time line." A woman might remain a victim of a hostile environment for years, until she rejected it, and consequently was fired, or until she was forced to leave, as Vinson claimed she was, at which time she became a victim of a quid pro quo. For a court to require such a rejection, MacKinnon argued, amounts to forcing "the victim to bring intensified injury upon herself in order to demonstrate that she is injured at all." If a claim of hostile environment is not actionable, then as long as her harasser proceeds "with enough coerciveness, subtlety, suddenness, or onesidedness, while her job is formally undisturbed, she is not considered to have been sexually harassed." (17)

HEALTH EFFECTS OF SEXUAL HARASSMENT

The sexual harassment that prompts women to go to court, complain to their supervisors, or mail back surveys is usually a series of acts that are invasive of their personal space and/or physical privacy and escalate to the point of interfering with their work or even their safety. According to Peggy Crull, stress, physical illness, problems with job performance and self-confidence are recurring themes in working with women who have sought counseling for sexual harassment. (18) Research has shown that when the stress response is chronic, it can lead to serious health problems.

Many women's jobs are made more difficult because the woman is faced with a double bind situation. On the one hand, she fears that if she complains she will risk losing some job opportunity, yet if she tries to ignore the situation, it will only escalate. Thus, she often adopts a state of vigilance, one in which her body is always prepared to meet some physical or emotional demand. The double bind is complicated by cultural assumptions about women's roles. Many women who are the target of sexual harassment feel guilty, and these feelings are further promoted by a society that "blames the victim" and suggests that it is the woman's fault that she was victimized. These cultural attitudes make it more likely that women will internalize their problems and suffer the health consequences.

I will present some case studies of women who have been victims of sexual harassment. There are at least three different categories of stress effects that women victims experience. They are (a) effects on work performance, (b) effects on general psychological well-being, and (c) effects on physical health.

STRESS EFFECTS ON WORK PERFORMANCE

Research by the Working Women's Institute (WWI) found that even though women express pride at being able to do their jobs despite harassment, 46 percent of 518 women said sexual harassment interfered with their work performance (see Table 11.1). The most common effects on work performance were that women were distracted from their tasks, dreaded going to work, and went out of their way to avoid the harasser. (19) However, the ramifications for job performance go beyond the immediate situation. The woman's economic security can be threatened. Sexual harassment contributes to the higher rate of female unemployment, women's lower rate of continuous employment, and, therefore, results in failure to advance in a job. Emotional stress can exacerbate safety hazards that already exist on assembly lines, construction jobs, and police beats that require alertness on the job.

TABLE 11.1. Physical and Psychological Stress Symptoms

Types of stress	Harassment involving superiors	Harassment involving nonsuperiors
Psychological stress (nervousness, guilt, depression, etc.)	99.5%	88.2%
Physical stress (headaches, stomach aches, weight changes, blood presure changes, etc.)	38.1%	32.3%
Effects on work performance (distraction, decreased productivity, time off, loss of motivation, etc.)	45.5%	38.2%
Help sought for stress effects (medical, psychological, workers' compensation, disability, etc.)	31.6%	20.5%

The stress effects of sexual harassment on a woman's private life and work performance are less publicized than the economic repercussions. This table shows the proportion of women harassed by superior(s) and nonsuperior(s) who experience the types of stress indicated.
Source: Working Women's Institute, *Research Series Report No. 4*, Winter 1982.

I will present two cases of hostile environment type harassments, one of a blue-collar worker and one of a professional. Both women describe the particular way that sexual harassment has interfered with their work performance.

ANGELA SUMMER, PLUMBER (20)

Angela Summer is a plumber who described this work experience: "My supervisors—some have been real jerks and some have been real nice. This one man I worked for was a real jerk. Every joke he told was either racist or sexist. One which stuck in my mind was, 'My wife didn't mind that I was bringing home scalps, but not ones with holes in them.'—pretty gross. It really infuriated me 'cause I understood how he could get off on that—the idea of scalping a woman—a woman's cunt, as he called it.

"Another time on that job, we were working 16 hours a day and under a lot of pressure to get things done quickly. One night when we were working on the 15th floor, I was under a sink undoing a nut with a basin wrench. It was an awkward position and I was just starting to get it done when my supervisor came over and said to me, "Run down to the basement and get me a couple copper fittings.' I said, 'Okay, in a minute. I'm almost done.' He said, 'Get your cunt and go get that.' I was shocked. I wanted to cry and scream.

"What really hurt me was there was this guy standing around that I had been working with and felt really good about, and he didn't say a damn thing. It made me feel rotten. None of the other men would have said something like that to me, but with him they felt they had to stick up for another guy. Behind his back, they would tell me he was a creep, but there was this peer pressure with him around."

SURGICAL RESIDENT AND TECHNICIANS (21)

Two cases that stand out are a woman who was a surgical resident in a hospital (surgery is the most male-dominated medical specialty) and two women who were technicians in a lab where only one other woman had ever worked.

In both cases, there was harassment from co-workers that consisted of intrusive sexual comments, graffiti, and lewd sexual jokes about them, and statements that did not belong. In addition, the work environment was made inconvenient for the surgeon and unsafe for the technicians. The surgical resident's story has an extra twist to it—she also was being propositioned by her attending physician (her boss); this, in combination with harassment from fellow residents, eventually led to her being drummed out of the program for "disciplinary' reasons even though she had exceptionally high academic evaluations. The two technicians eventually quit their jobs in fear for their safety.

An Irish study of policewomen described their responses to being brushed against and touched by more-senior male colleagues or given a wolf whistle by squaddies from the army or Ulster Defense Regiment as "to cringe and suffer in silence. Their flesh creeps, but they do little other than try to avoid situations where this might recur. They ignore the sexual jokes and innuendoes, and do not participate in the sexual horseplay of the occupational (police) culture." They are outsiders and loners. To avoid being loners, some of them employ defensive humor and treat sexist remarks "as a joke" to defuse the seriousness of the incident and redefine it as nonthreatening. (22) Sexual harassment serves to increase the safety hazards that already exist as part of the job, as we have seen in previous cases.

TOTAL INVASION OF A WOMAN'S LIFE

The negative effects of sexual harassment are not limited to the work setting. They invade every aspect of a woman's life and often are manifested as general psychological stress symptoms. In the WWI research, at least one negative effect was reported by more than 94 percent of the women in the sample. The reaction most often mentioned was excessive tension. Anger and fear were other frequently reported responses. However, when these normal feelings remained unexpressed or were denied, they often led to confusion, depression, and abnormal tension. The feelings of fear and anger often interfered with relations with lovers, friends, and family, creating sexual difficulties and general estrangement. Other common symptoms of stress were depression, embarrassment, sleeplessness, and guilt.

Another study of 139 Detroit auto workers looked at the psychological health effects. The investigators concluded that sexual harassment affects feelings about work and general self worth. The researchers found correlations between frequent sexual harassment and poor relationships between the woman victim and her co-workers and supervisors, as well as an inability on her part to develop job skills. The more the woman had encountered sexual harassment, the lower was her self-esteem and general life satisfaction. As cited in the previous research, general well-being (self-esteem and life satisfaction) is disrupted even more than work performance. (23) The following is a case in point.

KATIE MURRAY, SHEET METAL WORKER (24)

Katie Murray is an African American woman. She got a job in Seattle at a company that made trucks. "When I first started working there, they gave me a hard time and wrote dirty words on the ladies room walls about what they would like me to do for them—sex and all. After a while I just got tired of looking at it and I spoke to my supervisor, who finally got the walls cleaned off. And I would walk up the aisles, they would make wisecracks about what they would like to do. I just kept walkin' and pretended I didn't hear 'em. It made me feel trampy. That's because I was a black girl . . . but the white men think that they can take advantage of a black woman.

"I can't really describe how I felt when I first walked in there. I was scared. In the plant the smell is terrible and the sounds—you have to wear ear protection. When I first started, I was in sheet metal subassembly, where it's very quiet. It was pleasant, except for my lead man. He was a Southerner from North Carolina, and he was very, very prejudiced. He made me have a lot of depression headaches. I never relaxed around him. He always put you down. It was very uncomfortable. That's why I transferred out.

"Let me tell you how that place has affected me. Because of the way it is, I had begun to drink quite a bit. When I drank, I wouldn't go to work because I knew I couldn't do my job if I was drinking. Personnel told me if I didn't go . . . for alcohol treatment, I would lose my job. The white male, who missed the same amount of time as I did, he didn't have to go, but to save my job, I had to go. Now, that's on my record. And another thing, I felt I was discriminated on when I had domestic problems and filed for a divorce. . . . And here I'm black and a woman and it's hard. So, I think sex is on one side and color's on the other. This job's changed my home life. It has made me cranky. And made me be short with my husband. When I get cranky, we don't have no sex life and I think that's what made the domestic problems that we had. 'Cause I would come home after a bad day, like today, and I didn't know what I was gonna do. You know, it really hurts, the way they discriminate against us. This job has changed me so much."

STRESS AND PHYSICAL DEBILITATION

Sexual harassment for many women has meant debilitating physical ailments, which they attributed to the harassment experience. In the WWI study, 38 percent of the women alluded to physical symptoms such as nausea, tiredness (often, a sign of depression), headaches, and drastic weight change.

DONNA PUCKETT, GAS STATION MANAGER (25)

Donna Puckett, a gas station manager in Savannah, Georgia, filed a complaint in 1986 with the company when her supervisor sexually harassed her. Under company pressure, Puckett dropped her claim, but soon, she said, another executive began making repeated unwanted advances. Puckett, who remarried in 1988, began suffering from painful stomach disorders and hair loss. She again turned to the EEOC in April 1989. "They assured me I wouldn't get fired, but that's exactly what happened," she said. Two weeks into a month-long medical leave, she received a letter that, citing her absence, terminated her employment. Puckett sank into depression. "I wouldn't leave the apartment. I stopped cleaning. I didn't want to talk to my friends or make love with my husband," she said.

Physical symptoms of harassment are not limited to mild or temporary problems, as such cases have demonstrated. Katie experienced a change of lifestyle due to alcoholism. Donna had to take a medical leave because of painful stomach disorders, and Patricia, in the next story, suffered from hair loss and an ulcer.

PATRICIA KIDD, ORGANIZER OF OFFICE SPACE FOR PUBLIC EMPLOYEES (26)

Patricia Kidd, a divorced single mother, was an organizer of office space for a public employees union. She was propositioned by her boss, Carter. "When are you going to have sex with me?" he kept asking. Day after day, she awoke with a knot in her stomach. In August, disgusted and dejected, Kidd went to personnel to request a job transfer. She was told that approval would have to come from Carter. Carter refused to give it.

His pressure for sex turned to threats. "You know, I can get rid of you," she said he told her. Carter began calling her at home, but she made excuses. She described an incident when she was at work. Carter called. "He told me to get over to the Comfort Inn. I slammed down the phone, but he called back," said Kidd, choking back tears. "He said, 'Remember, I can fire you.' I couldn't make any more excuses. I had to survive. I was numb." She joined him in bed. She said, "It was violent and demeaning." She slept with him one more time at his insistence. "He had sex with me and then sodomized me. Afterward, I felt filthy."

She swore, no matter what the consequences, she "would never subject myself to that again." When Kidd once again spurned his advances, Carter "started retaliating something awful." He took away her support staff and her computer. . . . He excluded her from meetings. He stopped assigning her work. When Kidd appealed to Carter's supervisor in 1988, she was told to improve her attitude.

Her emotional devastation took a physical toll. Her hair fell out in clumps; she developed an ulcer. "Everything was gone. No one would help." She finally took Carter and his two supervisors to court. The jury found they had intentionally inflicted emotional distress and awarded her $300,000 in damages. One year after her legal vindication, the pain and the stigma remain. Kidd's personal finances are depleted from the court battle and she had received none of the money awarded to her.

Carter retired soon after the trial. The two superiors are appealing. Even now, Kidd cringes when Carter's friends, who still work with her, glare as she passes by. "Every day I go in, I pray the Lord just to give me peace. You just close the door and cry. Carter said he would destroy me and he did."

In addition to the cases presented and the Working Women's Institute (WWI) and autoworker health studies, more recent studies have confirmed the health hazards and the high economic costs of sexual harassment.

A 1980 Merit Protections Board federal study asked some questions to assess the influence of sexual harassment on emotional and physical conditions, feelings about work, and ability to work. A large proportion of respondents indicated that their feelings about work and their emotional or physical conditions had been damaged by their experience. A smaller proportion noted that their attendance at work or their ability to work had suffered. (27) Federal researchers calculated the cost of sexual harassment in terms of job turnover, absenteeism, and the use of health benefits and concluded that it costs taxpayers $189 million over a two-year period. When the government resurveyed the costs in 1987, they had risen to $267 million over two years. In 1988, a study of sexual harassment in 160 Fortune 500 service and manufacturing companies concluded that sexual harassment typically costs each company approximately $6.7 million a year. (28)

Questions about the severity and permanence of the health problems stemming from sexual harassment require further research. However, the women whose cases have been presented obviously believe that the damage to their health has been far-reaching.

TOWARD ELIMINATING SEXUAL HARASSMENT

Making basic organizational changes would go a long way toward eliminating sexual harassment and the resultant emotional and physical stress that accompanies it from the workplace. In the long run, this means altering conditions in the marketplace and in our culture, where the economic and social power is unequally held by men, so as to more evenly distribute it between both sexes. Dr. Conley called for "an organizational culture that promotes equal opportunities for women. There must be a recognition that there is a value received" when women are part of the enterprise. (29) Women are a significant and growing part of the labor force whose needs and wishes must be taken into account. Organizations have to separate work and sex more generally, especially in those occupations in which males dominate over female workers and which men perceive are becoming "feminized."

Court rulings and federal policy, such as the EEOC guidelines, have moved slowly but steadily toward establishing sexual harassment in its many forms as a violation of Title VII of the 1964 Civil Rights Act. Current law places principal liability on the company, not the harassing supervisor. The courts can order a company to prepare and disseminate a policy against sexual harassment. Other laws at the state level have been used to rectify damages done by sexual harassment. One area has been in employment law, where many states have guaranteed that women can receive unemployment if they are fired or have to quit due to sexual harassment. Criminal statutes, workers' compensation, and torts (civil lawsuits for acts such as infliction of emotional distress) have been pursued successfully by some women. But these only have value on a case-by-case basis and many cases are turned down. Sex discrimination laws hold the most promise for long-term social change of a class of persons. Feminist legal theorists have suggested that class-action suits are an even more effective vehicle that should be utilized in future years and are especially well suited to legal claims of a hostile work environment. (30)

Since the Supreme Court nomination of Judge Clarence Thomas has placed sexual harassment in the workplace on the front page, many more employers who previously had been indifferent to the treatment of women in their firms are putting a sexual harassment policy on their agenda. Efforts to eliminate sexual harassment need the support of top management and incorporation into the reward structure of that organization. (31) Policy statements against sexual harassment can be a mixed blessing. While they serve to alert potential harassers to the illegality of their acts and provide women with official channels to complain, they may also be used as a smokescreen by companies to make it appear that they have good intentions. For example, in the case of *Miller v. the Bank of America*, the company unsuccessfully argued that it could not be held responsible for the individual acts of a harasser since there was an official policy against such behavior.

Women, to ensure their best interests, have to engage in the struggle to eliminate harassment. Some of the progressive unions and women's committees already have dealt with sexual harassment as a health and safety issue. Both inside and outside of established unions, women have been active in organizing efforts against harassment. Many of the established women's and workers' rights groups, professional associations, and unions, along with sexual harassment service/advocacy groups, undertook educational programs based on research among their constituencies. The Modern Language Association, Women in Criminal Justice, the Coal Employment Project, and the American Federation of State, County, and Municipal Employees (AFSCME) have published handbooks explaining the issue and protections. (32)

One of the most effective local actions was taken by the Women's Committee of IUE Local 201. (33) After the entrance of a large number of women into the previously all-male professions, sexual harassment became a chronic and critical problem for women in General Electric plants. When a secretary at one plant was harassed by two managers, the members walked off their jobs in a wildcat strike. Eventually, the mounting publicity and unrest in the plant forced the company to remove both managers. "A major result of this mobilization was an enormous amount of discussion and education in the plant. Their program not only was helpful to victims, it probably prevented some new instances because it provided a new level of awareness of the problem." The women organizers concluded that, even in the cases that were successfully brought against managers and foremen, "constant vigilance and pressure" are required. (34)

The United Auto Workers (UAW) and AFSCME have issued policy statements condemning sexual harassment. They also have gone a step further by including a sexual harassment clause in contract provisions covering worker health and safety. The UAW has such clauses in its contracts with Chrysler and Ford. AFSCME, Local 3650, the Harvard Union of Clerical and Technical Workers, in conjunction with Harvard University, has five pages in its first contract outlining the university's responsibilities and an informal and formal process for resolution of grievances. (35)

Activist advocate groups, such as the Coalition of Labor Union Women (CLUW), WWI, the Women's Committees of unions, MassCOSH, and sex discrimination committees of the American Civil Liberties Union have given high priority to the issues involved in sexual harassment for more than a decade. These groups have put on workshops, put out fact sheets and resource guides, and taken dramatic actions against employers when the situation has called for them. CLUW has been engaged in a national poster campaign against sexual harassment in the workplace.

I have argued in this chapter that sexual harassment is more than a type of sexual discrimination. The growing recognition that sexual harassment creates stress on women, impairing their health, should give it a central place on the federal and state agendas for health and safety in the workplace. Since women are the main victims of sexual harassment, they have the largest investment in eliminating it. Their work against it is likely to be the most persistent and the most far-reaching. None of the actions discussed, however, is sufficient to prevent or eliminate sexual harassment. They must be combined with efforts to eliminate the basic cause of the problem—the inequality between women and men in the workforce, in the marketplace, and in the society.

NOTES

1. Conley, Frances, Lecturer at Schlesinger Library, October 23, 1991.

2. Telephone conversation with Ronald Ratney, Northeast Regional Administrator for OSHA, October 23, 1991, Boston, MA.

3. Emery, Margaret. "The Price of Saying No," *People*, October 1991, pp. 44–50.

4. Crull, Peggy, "The Impact of Sexual Harassment on the Job: A Profile of the Experiences of 92 Women," in *Sexuality in Organizations: Romantic and Coercive Behavior at Work*, eds. Dail Neugarten and Jay Shafritz. Oak Dale, IL, Moore Pub. Co., 1980, pp. 67–71. Also see Farley, Lin, *Sexual Shakedown: The Sexual Harassment of Women on the Job*. New York, McGraw-Hill, 1978. Gutek, Barabara A., *Sex and the Workplace: The Impact of Sexual Behavior and Harassment on Women, Men and Organizations*, San Francisco, Jossey-Bass, 1985; and Hearn, J., and Parkin, W., *'Sex' at 'Work': The Power and Paradox of Organization Sexuality*, Brighton, England, Wheatsheaf Books, 1987.

5. Goldsmith Kasinsky, R., "Rape, A Normal Act?" *Canadian Forum*, 40, September 1975; and Glass, Becky, "Workplace Harassment and the Victimization of Women," *Women Studies Int. Forum*, 11, 1, 1988:55–67.

6. Gutek, B., and Morasch, "Sex Ratios, Sex-Role Spillover, and Sexual Harassment of Women at Work," *Journal of Social Issues* 38, No. 4, 55–74.

7. Crull, Peggy, "Searching for the Causes of Sexual Harassment: An Examination of Two Prototypes," pp. 225–244 in eds. Bose, Felberg, and Sokoloff, *Hidden Aspects of Women's Work*, New York, Praeger, 1987.

8. Safran, Claire, "What Men Do to Women at Work," *Redbook*, November 1976.

9. Merit Systems Protection Board, *Sexual Harassment in the Federal Workplace: Is It a Problem?* 1981, pp. 2-3, and An Update, 1988, p. 39.

10. Collins, E.G.C., and Blodgett, T., "Sexual Harassment—Some See It—Some Won't," *Harvard Business Review*, 59, 2, 1981, 76-95.

11. Frankel, Paul, "Bared Buttocks and Federal Cases," *Society*, May/June 1991, 4–7. These figures may suggest that the most sexualized organizations may

be found in institutions that officially have sexual prohibitions. An interesting theme to pursue in future research.

12. *People*, October 28, 1991, p. 46.

13. *People*, Talk at Schlesinger Library, October 23, 1991. As a result of actions taken by the Dean of Stanford School of Medicine, Dr. Conley returned to her job.

14. Schroedel, Jean Reith, *Alone in a Crowd: Women in the Trades Tell Their Stories*, Philadelphia: Temple University Press, 1985, pp. 64–75, 70–71.

15. It is no accident that the court cases began to surface in the mid-'70s when women began to join the untraditional workforce in larger numbers. See Gutek, "Sex and the Workplace," and Hearn/Parkin, *'Sex' at 'Work.'*

16. Feminist theorists Liz Kelly and Elizabeth Stanko also present the behavior of sexual harassment as part of a continuum of sexual violence. They argue that nonroutine assaults such as rape are extensions of the more commonplace intrusions of sexual harassment. Stanko, Elizabeth, *Intimate Intrusions: Women's Experience of Male Violence*, London, Routledge & Kegan Paul, 1985, pp. 59–69, 70–82; and Kelly, Liz, *Surviving Sexual Violence*, Minneapolis, U. of Minnesota Press, 1988, pp. 97–137.

17. Strebeigh, Fred, *New York Times Magazine*, October 6, 1991. See also MacKinnon, Catharine, *Sexual Harassment of Working Women*, New Haven, Yale University Press, 1979. In *Meritor Savings Bank v. Vinson*, 1986, the Supreme Court ruled that hostile environment as well as quid pro quo was sex discrimination.

18. Crull, Peggy, "Sexual Harassment and Women's Health," in *Double Exposure: Women's Health Hazards on the Job and at Home*, ed. Chavkin, Wendy, New York, Monthly Review Press, 1984, p. 111. There is little empirical systematic evidence on the question of causality establishing a link between sexual harassment and stress symptoms. It suggests an important area of future research.

19. Crull, Peggy, "Sexual Harassment and Women's Health," p. 107.

20. Reith, Schroedel, *Alone in a Crowd*, pp. 59–60.

21. Crull, Peggy, "Searching for the Causes of Sexual Harassment," p. 236. (*Lipsett v. University of Puerto Rico*) 1983; Sexual Harassment Brief Bank, 1985, p. 279; *Guyette v. Stauffer Chemical Co.*)

22. Brewer, John, "Policewomen in the RUC," *British Journal of Sociology*, 42, No. 2, June 1991, 231–248, 239.

23. Gruber, James and Bjorn, Lars, "Blue Collar Blues: The Sexual Harassment of Women Autoworkers," *Work and Occupations* 9, No. 3, August 1982: 271–298.

24. Schroedel, *Alone in a Crowd*, pp. 132–139, 134–135, 138. I was not able to cover the effects of race and sexual harassment in this paper. See Ellis, Judy, "Sexual Harrassment and Race," *J. of Legislation* 8 (1), 1981, 30–45; and Gruber, J., and Bjorn, Lars, "Blue Collar Blues."

25. Wescott, Gail, *People*, 46.

26. Balamaci, Marilyn, *People*, 44–45.

27. Merit Protection Board, *Sexual Harassment in the Federal Workplace*, pp. 75–84.

28. Working Women (1988), cited in Schafran, Lynn, "The Harsh Lessons of Professor Hill," *The New York Times*, Forum, Sunday, October 13, 1991, F13.

29. Dr. Frances Conley, Schlesinger Library lecture, Oct. 23, 1991.

30. Estrich, Susan "Sex At Work," *Stanford Law Review* 43 (April 1991), 813-861, p. 860.

31. Gutek, Barbara, *Sex & The Workplace*, "The Impact of Sexual Behavior and Harassment on Women, Men and Organization," p. 174.

32. Coal Employment Project, "Sexual Harassment in the Mines: Legal Rights, Legal Remedies," AFSCME, "On the Job Sexual Harassment: What the Union Can Do;" and Modern Language Association of America, "Sexual and Gender Harassment in the Academy: A Guide for Faculty, Students and Administrators," New York, 1981.

33. Brown, Alex, and Sheridan, Laurie, "Pioneering Women's Committee Struggles With Hard Times," *Labor Research Review*, Spring 1988, 63-77, pp. 68-69. It is one of the oldest union women's committees in the country, established in 1987. It took root in one of the oldest, most democratic and progressive union locals in the labor movement.

34. Ibid., p. 69.

35. Harvard University and Harvard Union of Clerical and Technical Workers (AFSCME, AFL-CIO), *Personnel Manual*, July 1, 1989, pp. 38-43.

SECTION C

THE USE OF SCIENCE IN POLICYMAKING

12

Why Focus on Risk Assessment?

ROBERT GINSBURG

In the past 15 years, risk assessment, in particular quantitative risk assessment (QRA), has gone from an academic exercise to the apparently unassailable political and scientific basis of policy at the U.S. Environmental Protection Agency (EPA), the Occupational Safety and Health Administration (OSHA), and the Food and Drug Administration (FDA).

The extent of its involvement can be seen from just a few examples. FDA now uses QRA to determine compliance with the Delaney amendment regarding the threshold for the use of carcinogens as food additives. The 1990 Clean Air Act directed the National Academy of Sciences (NAS) to review the risk assessment procedures used by EPA. Various state air regulatory officials have stated that the NAS evaluation will be one of the predominating factors in determining the shape of their air pollution programs in the future. In the early 1990s, EPA proposed changes in the definition of hazardous waste that would result in the contamination of the drinking water for 13,000 more people. According to the agency's risk assessment, the contamination would not be significant when compared with the supposed savings to industry. EPA administrator William Reilly and many state regulatory programs have pushed to use QRA to set agency priorities. OSHA is beginning to use QRA to set workplace standards, presumably concluding it was effective in protecting public health from environmental pollution.

The Office of Management and Budget (OMB) recently requested OSHA to use "risk-risk" analysis, which ***Business Week*** called "a novel twist on cost-benefit analysis," to reevaluate proposed chemical-exposure limits. In 1993 the White House issued an Executive Order directing the Council on Competitiveness to develop guidelines to standardize risk assessment in federal agencies. The political concept and consequences of the increased use of risk assessment when it is evaluated with OMB's push for using cost-benefit analysis as a basis for public health decisions and the parallel attack on the use of animal data in evaluating human toxicity by Bruce Ames and Phil Abelson (editor of *Science*) and others—all are antienvironmental at heart.

In response to this thinly veiled political attack on environmental and occupational programs, there have been too few coherent and systematic challenges to both the politics and the science underlying use of QRA. In addition, too little attention has been devoted to developing the details of an alternative approach. This chapter and the four chapters that follow originally were the basis of presentations at the November 1991 meeting of the American Public Health Association. Hopefully, they represent a more concrete step in crystallizing the problems with risk assessment and will encourage or incite others to begin the hard work of developing more specific alternatives.

13

The Risks of Risk Assessment

ELLEN SILBERGELD

No one is very happy with risk assessment. Everyone suspects that it is a tool used by the other side to twist available data, or to fabricate data from very little, to achieve particular ends. The health advocacy community and the chemical industry have expressed similar criticisms of the subjectivity, unreliability, and politicization of the regulatory process since the introduction of quantitative risk assessment in the late 1970s. The results of risk assessment, in OSHA and EPA regulation, hazardous waste clean-up, and priority setting, are more often than not unacceptable to all parties in a dispute. Only two groups seem to flourish in the risk assessment era: an increasing horde of well-paid risk assessors, who hire themselves out to produce the large documents that usually obscure public debate, and the new discipline of risk communication experts, who have given propaganda a new life in the postideological age. (1)

Nevertheless, despite more than a decade of criticism, risk assessment seems invulnerable to attack, even on scientific grounds. Scientific criticisms have been raised from all perspectives, from Epstein to Ames, but they are dismissed on the grounds that we can't do better (as in a series of analyses by the National Research Council's Committee on Risk Assessment Methodology, an advisory group funded by the regulatory agencies). (2) The disquiet of citizens' groups and their advocates is often disparaged as an expression of technical illiteracy, ignorance, or an irrational fear of any risk, however infinitesimal. (3)

Many valid and searching criticisms are expressed in this and succeeding chapters of this section. The authors of these chapters rightly point out that risk assessment is restricted in its focus to one endpoint, carcinogenicity; it is unable to accommodate the real-world situation of multiple chemical exposures of varying doses and duration; it is malleable to the value-laden assumptions and politics of its practitioners and interpreters; it insulates and separates policymakers from citizens. More fundamentally, the practice of risk assessment at EPA and OSHA ducks the ethical problem of imposing risks on those who do not receive a commensurate share of the benefits. (4) The linking of cost–benefit analysis to risk assessment, by OMB and others (5), compounds the technical problems of risk assessment with those of simplistic economic analysis, as exemplified by the extraordinary "risk:risk" formulation by OMB, under which the "risks" of economic burdens on regulated industry are proposed to be directly comparable to the risks of death and disease on exposed workers.

If risk assessment is so unacceptable, why is its influence so persistent? As outlined by an Office of Technology Assessment (OTA) report (1987), the rise of quantitative risk assessment in regulatory policy in the United States was supported by at least two forces: the concern of occupational and environmental health advocates to identify and prevent exposure to chemical carcinogens, and the demands by courts, responding to industry litigation of federal agencies, for a rationalized basis of occupational and environmental regulation. (6) In the honeymoon period from 1976 through 1981, it seemed that both these forces could be accommodated through the use of quantitative risk assessment: chemical carcinogens could be expeditiously identified (through industry testing under the Toxic Substances Control Act and federally funded testing programs such as the National Toxicology Program) and objective, science-based methods could then quickly develop numerical standards, resistant to legal challenge. (7)

The honeymoon ended in disaster under the Todhunter years at EPA. The noble principles of the first risk assessment guidelines were never published in final form after their proposal in 1979, and EPA's new regime went on to discard most of the science, and many of the scientists, involved in the early years of risk assessment development. When Ruckelshaus returned in 1983, he attempted to salvage the agency's reputation in this area by proclaiming a strict division between the science and policy of risk assessment. (8) While the intent of insulating science from politics was laudable, it is not possible—nor desirable—to fully separate these human activities. In practice, the division of risk assessment from risk management encouraged a cult of the technological elite and the development of methodologies that did not have to be explained. Although these papers are from the perspective of health advocates, industry has also found risk assessment increasingly unacceptable. Some of the industry attack has been against the successes of risk assessment, such as the suit by the chemical industry to shut down the National Toxicology Program reporting system. But industry is also threatened by the instability of risk assessments and their apparent imperviousness to real data.

If not risk assessment, then what? Critics, including these authors, have been short on alternatives to the dominant methodology. It would be relatively easy for EPA and OSHA to develop standard methods for assessing risks of toxic chemicals in addition to carcinogens (as California has done for reproductive and developmental toxicants), or to acquire a more flexible and complex portfolio of risk assessment models (as EPA appears to be doing for the dioxins). But this might only multiply the problems of current practice. The philosophy of the Delaney clause (see Chapter 16, for a description) could be expanded into occupational and environmental regulation to make the qualitative finding of certain toxic properties the signal for absolute controls or for controls limited solely by technology. Since Delaney is currently more honored in the breach than in the observance at FDA, this is not an encouraging model. Public access to the risk assessment process could be increased by reforms in statute and regulation. But this would not solve the fundamental problems of risk assessment, only involve the public more directly in them. Toxics use reduction and product substitution could become more dominant in regulatory thinking, as in EPA's 33/50 program, but these approaches do not by themselves solve the implicit issue of how to select less toxic substances in production and consumption.

This chapter and whole section reflect the breadth and depth of concern over risk assessment in theory and practice in the 1990s. That the acceptability of this method of decision making is so low should be cause to challenge its continued use. What is hopeful is that these concerns are so widely shared. On this basis, it should be our goal to develop a consensus for rational change among the many parties at interest, including community-based environmental groups, labor unions, health advocacy organizations, scientific research institutions, and industry. If we do not seize this opportunity, we are guilty of the charge, like the proverbial conversationalist, of always complaining about the weather but never doing anything about it.

NOTES

1. For instance, Covello, V. T. (1989) "Communicating right-to-know information on chemical risks." *Environ. Sci. Technol.* 23: 1444–1449.

2. Committee on Risk Assessment Methods, National Academy of Sciences, documents on exposure assessment, maximally tolerated dose in bioassays, and two-stage models of chemical carcinogenesis, 1991–1992. Also, personal communication with Bernard Goldstein, MD, chair of CRAM, July 1992.

3. Zeckhauser, R., and Viscusi, K. (1991) "Risk within reason." *Science* 248: 559-564.

4. Sagoff, M. (1988) *The Economy of the Earth*. Cambridge: Cambridge University Press; Mayo, D., and Hollander, R. (eds.) (1991) *Acceptable Evidence: Science and Values in Risk Management*, New York: Oxford University Press.

5. Office of Management and Budget. (1991) *The Regulatory Program of the United States Government*. Washington: The White House.

6. Office of Technology Assessment. (1987). *Identifying and Regulating Chemical Carcinogens in the Federal Government.* Washington: GPO.

7. National Academy of Sciences. (1983) *Risk Assessment in the Federal Government: Managing the Process*. Washington: NAS Press.

8. See Silbergeld, E. K. (1991) "Risk assessment and risk management: The uneasy divorce." In Mayo, D., and Hollander, R., op cit., pp. 99-114.

14

Quantitative Risk Assessment and the Illusion of Safety

ROBERT GINSBURG

The U.S. Environmental Protection Agency (EPA) quietly announced in 1991 that it, rather than the contractors or the responsible parties, would be responsible for the preparation of Superfund site risk assessments used to select cleanup remedies. EPA noted that this change was made to ensure that the remedies *"protect human health and the environment"* (emphasis added). While this was designed, in part, to address some of the criticisms (1) raised about cleanups performed under Superfund, the announcement is far more revealing about EPA's understanding of risk assessment. EPA's decision is an explicit acknowledgment that the results of risk assessment evaluations/calculations substantively differ depending on who does the assessment. EPA's decision implicitly recognized the fundamental weaknesses of Quantitative Risk Assessment (QRA) and its dependence on highly uncertain and subjective assumptions. Unfortunately, EPA did not choose to emphasize that conclusion in its announcement.

THE POLITICAL CONTEXT

Despite such obvious reservations or perhaps because of them, QRA and its sister, cost-benefit analysis, have become enshrined as the basis for virtually all agency regulatory activity. Federal agencies, especially EPA, have begun to rely exclusively on QRA (which is what EPA actually means when it says "risk assessment") for all regulatory decisions. The most recent revisions of the Superfund RI/FS Guidance Manual, the 1990 "Hazardous Waste Land Ban" regulations, the 1990 benzene regulations, the 1991 Incinerator and Boiler and Industrial Furnace (BIF) regulations, the report prioritizing EPA programs based on the supposed level of "risk" to the population, and the proposed redefinition of Hazardous Waste under the RCRA regulations (2), all explicitly incorporate QRA as the basic requirement or justification for any EPA activity, permit, or cleanup limits. On the surface, given the consistent chorus from government and industry, it would appear that QRA was the greatest scientific tool for making "objective and sound" health policy decisions since the first canary was taken into a mine or the first rat used for toxicity testing.

The rush by the Reagan–Bush administrations (and to a lesser extent the Carter administration) to incorporate risk assessment has been prompted by the perceived "political" benefits of its use in pushing industrial deregulation. Its use is being explicitly advanced by The Council on Competitiveness headed by Vice President Dan Quayle (3), the President's Office of Science and Technology Policy (4), conservative environmental organizations such as the Conservation Foundation and Resources for the Future (5), which created its "Center for Risk Management" with EPA and industry funds, and the chemical industry (6), with its ceaseless cries against any restriction on its operations unless there is a number or calculation that irrefutably indicates what it considers a significant, biologically plausible, risk.

The dominance of risk assessment in EPA site-specific decision making is clearly at odds with the academic and scientific literature on the use of risk assessment. Most of those discussions argue for the need to provide comparisons between environmental risks and to provide an objective component for establishing funding and regulatory priorities. They argue that risk assessment is not accurate enough to provide an absolute basis for site-specific decisions due to the subjective nature of the process and the large uncertainty in the calculation. That is the essence of the National Research Council (7) Risk Assessment/Risk Management paradigm, which was published in 1983.

Acknowledgments of the limits of QRA are not hard to find. In testimony before the House Science and Technology subcommittee in May 1991, the president of the American Chemical Society (ACS) (8) noted the highly subjective and unscientific nature of QRA in stating that ". . . risk assessment requires *inferences* drawn from limited scientific data." A recent symposium (9) on QRA sponsored by the ACS contained numerous references to the uncertainties in risk assessments and various proposals to address the limitations imposed by such uncertainties. However, there are relatively few published discussions on the impact this uncertainty and variability of risk assessments has on the validity of the methodology. A 1986 study (10) of various estimates of TCE carcinogenic risk published by a group of EPA staff found the risk assessments legitimately could vary by seven to eight orders of magnitude when exposure estimates and transport modeling errors are included. They compared the choice provided by the risk estimates to trying to decide whether you had enough money to buy a cup of coffee or pay off the national debt. While this may be acceptable to some for establishing priorities, it is clearly a dubious basis for issuing permits, setting clean-up levels, and setting standards.

The question becomes why a methodology based on "inferences" and subject to dramatic variations in results has come to dominate EPA and industry policy and practice? Curiously, the public rhetoric from EPA and industry downplays the problems and paints QRA as just "good science" that gives "objective" (as opposed to the hysteria of citizen groups) and "scientific" evaluations of contamination problems. Furthermore, they focus on the supposed "excessively conservative assumptions" used in QRA calculations and suggest that spending money to save maybe a few lives from cancer over the next century is a waste. Despite carefully worded denials, most EPA and industry officials want people to take the QRA-derived numbers as the only measure of health impact to be considered. (11) Therefore, in that scenario, any objection to QRA-based decisions is due to ignorance and mindless opposition to science and technology. Thus, Union Carbide can refuse (12) to allow a neighborhood group to bring its experts on a plant inspection by stating that they would only offer a tour to residents who live within 4.5 miles of the plant to allay their fears. The bottom line is that a questionable scientific methodology is being pushed as "good science" without acknowledgment of its political underpinnings. It should be obvious that QRA has been aggressively pushed as the conservative political solution to environmental regulation. Any coherent attempt to challenge that methodology must begin to clearly articulate the technical problems with QRA and the political/social consequences of its application.

WHAT IS MEANT BY RISK ASSESSMENT?— THE FALSE PROMISE

Health Risk Assessments, in general, are an offspring of the Environmental Impact Statement process, which provided a public forum for discussion of health and environmental impacts of projects. In that original context, they provided a qualitative description of the array of effects that could be produced by a project or by exposure to pollution. QRA is, by definition, a more narrow process that attempts to provide a quantitative or numerical measure of human health impacts from a specific source of pollution or exposure to some identified pollutant(s) for direct comparison to other factors, in particular the cost of additional controls or clean-up options. The results are generally expressed as "so many extra cases of cancer beyond background levels when one million people are exposed to a certain concentration of a single pollutant." When EPA and industry discuss "risk assessment" in public forums or hearings, they describe the broad process and content of "health risk assessments," while in reality it is the more limited QRA solely based on cancer effects that is the basis for decisions. This deliberate confusion is designed to give scientific credibility to the QRA-based decisions and avoid discussions about the validity and propriety of QRA in setting actual permit limits and clean-up levels.

QRA calculations are composed of four steps:

a. Deciding whether a specific substance or substances can increase the incidence of a disease;
b. Estimating the types and amounts of pollutants that might be emitted;
c. Estimating what concentration of pollutants "might" be transported to a point of exposure; and
d. What extra risk exposure to that concentration might exist (for example, "one extra case of cancer in a million people so exposed").

EPA and the National Academy of Sciences call the first step "Hazard Identification"; the second and third steps, "Exposure Assessment"; and the fourth step "Dose-Response Assessment and Risk Characterization." Each of those steps needs to be examined carefully to understand the decisions that went into the calculation and the effect they have on the final result.

TECHNICAL LIMITATIONS OF QRA— LIMITED SCOPE OF ASSESSMENT

First, QRA calculations are almost entirely limited to the risk of cancer based on studies of cancer incidence in animals and occasionally in workers. The original basis for that restriction was the need for standardized test data, which had only been developed for cancer and not other effects such as reproductive, nervous system, or immune system effects. In addition, cancer was a chronic health effect that was reported on death certificates. Thus, cancer became the surrogate measure for any and all chronic health effects. In the last 15 years, research has shown that estimates of cancer effects may not be the most sensitive indicator of risk. For example, comparison of the reproductive and carcinogenic potential of 2,3,7,8-

tetrachlorodibenzodioxin ("dioxin") and benzene (13) showed reproductive effects to be 9.5 and 2.5 times greater, respectively. Other researchers (14) have suggested that effects on the nervous system may be greater than cancer at the same level of exposure. This concern becomes even greater when consideration is given to real-life exposures where numerous substances are present and can interact.

The resulting risk level from QRA calculations (for example, one in a million) is restricted to cancer effects. While it is valid to argue whether or not the calculation overstates the risk of cancer, it says almost nothing about other health effects or the likelihood of any *meaningful* adverse health effect (including rashes, headaches and dizziness, breathing disorders, allergies, liver and kidney effects, reproductive effects, and so forth) showing up in the exposed population. EPA concluded that cancer is the most fatal effect and translated that into an assumption that the risk assessment calculation for cancer is sufficient to protect people from all other effects. There is no scientific basis for such an assumption even if one assumes that the original conclusion is valid.

Any meaningful analysis of the interaction of these chemicals with each other and with existing community health problems is almost impossible when all noncancer effects are either ignored or merged into a single number. (15) For example, there are a large number of substances emitted that are lung irritants. Recent research (16) indicates that the severity of health effects associated with inhalation of irritants (ranging in severity of irritant properties from solvents to acids, ammonia, and chlorine) is more dependent on a person's pre-existing conditions and (by inference) other factors related to his or her health status than the irritant potential of the pollutant. Another study (17) of different communities in Los Angeles County shows that residents in the South Bay area (as indicated by pollution levels in Long Beach) have been exposed to a mix of pollutants that has resulted in greater lung damage than people living in Lancaster, which is subject only to high ozone levels.

Any health assessment for toxic substances, especially in urban areas, must take into account related emissions. For example, the carbon monoxide levels (18) present in Lynwood, in the southeastern area of Los Angeles, are among the highest in the United States and present a major health threat to that community and the surrounding communities. However, that area is also subject to very high levels of dichloromethane emissions. (19) Some effects of dichloromethane on the human cardiovascular system are virtually identical to carbon monoxide. The health effects in that community from exposure to both will be magnified. Similar analyses can be made for asthma (which has recently been shown to be a major health problem in low-income and African-American communities) (20) and hypertension.

BROAD RANGE OF POSSIBLY VALID RESULTS

The final results (that is, risk level) from QRA calculations have a large variability and uncertainty. The uncertainty in any QRA calculation refers to errors or uncertainty in the input data and method of calculation such as emission estimate errors, modeling errors, limited or inappropriate toxicological data, and so forth. Variability refers to different assumptions that can legitimately be made in doing the calculation such as use of scaling factors, food and water consumption estimates, use of nonpositive epidemiological data, estimates of duration of exposure, and so forth. When a report states that the risk from exposure to pollutants from a specific source, such as an incinerator, is one-in-a-million, the risk calculation could have produced results somewhere between one-in-100 to one-in-10 billion. Some sources of the variability and uncertainty are:

a. *Emissions estimates*. The prediction of cancer risk is directly proportional to the estimate of emissions. While it is possible from an engineering and analytical perspective to accurately determine emissions, this does not happen in practice for several reasons. First, the emissions do not include the contribution from accidents, nonrecurring emissions such as equipment start-up emissions, or emissions from temporary equipment failure such as flares. Second, the emission calculations are based on the use of average emission factors that may not be representative of actual plant operations, which can cause variations due to slightly different processes, variations in throughput and emissions. Third, fugitive emissions are a major component of the total, and they are generally recognized to be underreported. This underreporting was acknowledged recently (21) by the refinery operators in Los Angeles in their offer to "allow" better monitoring and sampling of fugitive emissions as an incentive for marketable permits. These can result in variations in the emission estimates by a factor of 10 to 50 (22) (that is, depending on how one estimates emissions, the highest value for the emissions used in the QRA calculation will be 10 to 50 times larger than the lowest possible value).

b. *Exposure estimates.* Generally, EPA uses mathematical models to calculate what someone might be exposed to at some specified point such as a fence line or a well. The error involved in using models to calculate the point-of-exposure concentration (used in the risk assessment calculation) based on the emission/release source data (either a stack or a landfill) can vary by a factor of 10 depending on the models used and the various input factors. Models used to estimate transport in soil and water can have substantially larger errors. (23) The results from air dispersion models are generally made consistent with actual monitoring results, when available, in order to improve predictions in other areas without monitoring data. However, in some cases monitoring data have been ignored, if the data indicate higher exposure levels. For example, estimates of the risk of cancer from air pollution in Southeast Chicago by EPA ignored sampling results in the area that showed higher levels than their models predicted. At the hearing, EPA staff indicated that the higher sampling numbers were not used because the "risk assessment calculation was too conservative and higher exposure levels would show too high a risk." (24)

c. *Health effects or risk estimates.* This is perhaps the greatest source of uncertainty in the entire process. A survey of QRA calculations and regulatory levels from several countries applicable to 2,3,7,8-dioxin found results that varied by a factor of nearly 2,000. (25) Using some of the assumptions from chemical and incinerator industries (such as whether and how to use inconclusive epidemiology and animal studies, whether to include in each calculation a factor related to duration of exposure, and so forth) can push that to a factor of 10,000. A recent evaluation of the carcinogenic potential of perchloroethylene (26) demonstrated that the calculation of the unit risk factor could vary by a factor of nearly 2,000 depending on the animal studies chosen and the assumptions on human metabolism used in the calculation.

Furthermore, these risk estimates do not consider a wide variety of related factors that affect the assessment of risk. The calculation does not usually take into account simultaneous exposure to the same or similar substances from adjacent sources or through various routes of exposure (such as contamination of food, air, and water). The cancer risk calculation does not take into account a variety of interpersonal factors including variability between people, especially between adults and children. Children are not simply small adults but have generally higher rates of exposure per pound than adults. In addition, children can absorb more contaminants than adults. For example (27), upon exposure, lead is nearly four times as biologically available in children as compared with adults. The net result is that children are more sensitive to exposure to chemicals than adults.

d. *Finally, QRA does not consider the likelihood of failure*. For example, in nuclear power plants the calculated risk of an accident may be small but it becomes more likely the longer the plants run and especially when compounded by operator error. This has been demonstrated at both Three Mile Island and Chernobyl. NASA (28) had calculated that the risk of the space shuttle exploding was one in 100,000 prior to the Challenger explosion. After that accident, they "recalculated" and determined that the risk was closer to one-in-78 during the first five years of operation.

Thus a "one-in-a-million" risk of cancer can reasonably vary by a factor ranging from 10,000 to 100,000 and that does not even include the risk from other effects. EPA and industry also acknowledge that even with their own calculations they cannot tell you what the most likely risk is going to be. What is unstated is that a small "risk" does not mean that no one will be affected or that an exposure is safe. Rather it is an indication of the *relative* potential for a specific effect to be produced from a specific hazard *and not an absolute measure* or even an absolute comparison (as between hang gliding and cancer).

WHY HAVE EPA AND INDUSTRY ADOPTED QRA?

This approach has been so aggressively promoted by EPA and industry because of a fundamental view of EPA's responsibility. EPA and industry believe the legal requirement of agency decision making is to manage exposures to toxic substances as compared to the citizen view that the function of the agency should be to prevent exposure. With management of pollutants and polluting facilities as its goal, EPA has adopted methods of controlling releases rather than protecting public health by controlling production and generation of toxic substances.

In that context, by 1984 EPA needed to find a way to continue the substance of the Gorsuch–Hernandez deregulation efforts but with the appearance of scientific "objectivity." William Ruckelshaus, then in his second turn as EPA administrator, took the risk assessment efforts developed by EPA, FDA, and NRC over the previous 15 years, added cost-benefit analysis (under the rubric of risk management), and proclaimed his agency's commitment to protecting public health and the environment in a "cost-effective way." The result was an "objective" methodology, supported by the National Academy of Sciences, that enabled the EPA to set permit levels and clean-up levels for toxic substances. QRA theoretically gives them a method of determining what levels of pollution or pollutants are "acceptably safe" without having to evaluate available technology, alternative processes, alternative substances, or community concerns.

QRA serves three regulatory and political functions:

a. *QRA allows EPA to set clean-up levels or emission levels by a predetermined calculation.* The calculation will set a concentration goal at "a point of exposure" (which can vary) that, according to the theory, will not result in any "significant risk" at that "point of exposure." In reality this is a sophisticated form of the *dilution* solution. This concentration is by definition an average at the site chosen (the averaging period is determined separately and is limited by models and monitoring capabilities), and in almost all cases there will be some people exposed to higher levels and those exposed to lower levels. However, this is quite consistent with the goal of managing exposures.

In addition, QRA provides EPA with a much simpler administrative decision-making process. Once an emission level or clean-up level has been determined by QRA, then regulators don't have to worry about changing regulations or permits for a long time since, supposedly, a safe level has been scientifically determined and emissions at that level won't harm anyone. This was explicitly raised in a paper (29) by researchers at Rohm and Haas and Dow Chemical presented at the 1991 American Chemical Society symposium on the uses of risk assessment. While noting that *either* decreasing actual exposure or increasing acceptable exposure (that is, toxicological data) will lower the assigned level of risk, they argued for increasing refinements in determining acceptable exposure as the most efficient and effective approach. Thus for industry there is a clear incentive to use toxicological research and QRA as a substitute for process changes. This could result in ever-increasing levels of exposure allowed for the same risk level without any changes in the manufacturing process.

b. *EPA can now calculate a minimum level of exposure which is independent of any particular site and which is independent of any ability to achieve lower emission or clean-up levels or even eliminate the use, discharge or exposure to the contaminant.* EPA also has effectively cut out public participation by reducing the risk discussion to a technical calculation that requires technical expertise to do the calculations and argue the numerous choices and assumptions. The procedure also allows an increase in exposure levels through recalculation and reassessment of the unit risk factor by "experts" that can avoid public discussion or be easily manipulated. This is what the Chlorine Institute did with the data on dioxin. (30) Furthermore, the general public is put on the defensive by appearing to oppose "good, state-of-the-art" science. The effort to put opponents on the defensive is supported by a variety of irrelevant and misleading comparisons such as comparing the risk of hang gliding with drinking contaminated well water, as has been proposed by some industry apologists. Such comparisons confuse voluntary and controllable risks with involuntary and individually uncontrollable risks. Other misleading comparisons include the comparison of naturally occurring carcinogens in food to exposure to carcinogens in the environment as the result of industrial practices or products as has been proposed by Bruce Ames.

c. *The results are put in terms that are already familiar to the public.* We all are too conversant with gambling odds so that the QRA results appear familiar and understandable. However, the government handles this "gambling" differently to suit its purpose. People are strongly encouraged by government to play various lotteries where the odds of winning are one in 13 million (no matter how many people buy tickets). The government then wants the public to think that its calculation of "one-in-a-million" means that there is virtually no chance of "winning their prize" at Superfund sites, and so forth.

CONSEQUENCES OF QRA

QRA poses a significant obstacle to active citizen and worker participation in protecting their health. As Tom Webster and Paul Connett recently wrote (31):

> At its worst, risk assessment can become a method for disenfranchising the public, appropriating essentially political decisions about choices of technology (with all of their consequences). When the range of questions are artificially restricted and cloaked in scientific jargon, experts become the only competent decision-makers. Some people might desire a world ruled by scientists and engineers; we don't.

The use of QRA as a basis for EPA decision making, in general, and toxic substance control, specifically, is fundamentally incompatible with Pollution Prevention and Toxics Use Reduction. QRA will apply to a limited number of substances and will set "allowable discharge levels." Industry can then use these calculations to avoid changing processes or adopting even reasonably available technology (bans or best technology won't have to be considered since control is only required up to the calculated limit). Between 1988–1991, the chemical industry challenged air toxics regulations in Wisconsin using the argument that no limitations on processes or emissions of hazardous air pollutants can be required unless a site-specific assessment shows a significant risk. The lawsuit several years ago against the National Toxicology Program (NTP) for listing Dichlorobenzene in the Annual Report on Carcinogens contained the same basic argument. Finally, the argument on dioxin regulation is the same. If such an argument is even partially successful, pollution prevention efforts and toxics use reduction plans will be made toothless.

ALTERNATIVE APPROACHES

Industry and EPA have argued for the last several years that there is no alternative to the use of QRA to set national priorities, issue permits, or in any way regulate toxic substances. That issue has not been clearly and effectively rebutted, and the use of QRA has taken on the aura of inevitability. While there is no simple answer to the question of how to address toxic pollution, several approaches can be identified that are more consistent with the application of long-standing public health principles *and that do not rely on QRA*.

A public health-based approach would focus on what could be done to prevent exposure. Such an approach would recognize the severe limitations in our ability to quantify the effects from exposure to toxic substances. This does not mean that research to better quantify the effects of exposure to toxic substances is inappropriate or that efforts to quantify the risk from exposure have no role in an agency's decisions. However, the basis of regulatory programs needs to be firmly rooted in public health principles established over the last 200 years and focus on preventing exposure through reducing the use and release of toxic substances.

There should be clear recognition that quantities generated, used, or released and, to a lesser extent, either modeled or measured exposure data are the only sufficiently accurate and measurable bases for identifying hot spots even if there is not the ability to definitively document their impact. The sheer volume of emissions and releases, the level of exposure, and the variety of serious health effects that those emissions and releases could contribute to or cause justifies moving to find ways to reduce or eliminate the emissions. By focusing on emissions, the endless bickering over health effects could be circumvented and replaced with more productive discussions on how to make industrial practice and manufacturing more environmentally and economically sustainable.

This was the basis of centralized water treatment and sewage collection programs just after the turn of the twentieth century (without, I might add, proper human epidemiological data proving that sewage dumped in drinking water caused typhoid and cholera). This approach is analogous to the work of John Snow in nineteenth-century England. During a cholera epidemic he noted that the cholera cases were clustered on the basis of which water company people got their water from. Rather than waiting for several weeks or months or years to try to prove that the water from the well used by the water company caused the cholera (which he couldn't have done without germ theory), he stopped the flow of toxins into the community by removing the handle of the well. The best public health policy to adopt is to start removing the handle on these sources "pumping" toxic chemicals into the environment.

ACKNOWLEDGMENTS

The author gratefully acknowledges the Joyce Foundation for partial financial support of some of the research underlying this chapter. Parts of this chapter were developed in the context of testimony on behalf of the Labor/Community Strategy Center regarding strategies to protect residents of Los Angeles from the harmful effects of toxic air pollution. Helpful discussions and comments on various sections by Eric Mann, Chris Mathis, Dean Toji, Ross Vincent, Gerry Poje, and Stephen Lester are gratefully acknowledged. However, the opinions and conclusions expressed in this chapter represent solely those of the author.

NOTES

1. U.S. Congress, Office of Technology Assessment, "Coming Clean: Superfund's Problems Can Be Solved . . . ," OTA-ITE-433 (Washington, DC: U.S. Government Printing Office, October 1989).

2. *Federal Register*, May 20, 1992, p. 21450.

3. Federal Focus, Inc. (Thorne Auchter, Exec. Director), Proposal to Council on Competitiveness. *Toward Common Measures: Recommendations for a Presidential Executive Order on Environmental Risk Assessment and Risk Management Policy*. Washington, DC, June 1991.

4. Letter dated March 6, 1992, from D. Allen Bromley, Director OSTP, to Henry Habicht, Deputy Administrator, USEPA, indicating "unanimous approval by the Federal Coordinating Council for Science, Engineering and Technology (FCCSET) for the Ad Hoc Working Group on Risk Assessment (WGRA) to undertake an inventory of Federal government program activities in the area of Risk Assessment."

5. Resources for the Future, press release, February 1992.

6. *Chemical and Engineering News*, May 4, 1992. "Dow Chief Views Issues Management As Vital To Expanding Leadership Role."

7. National Research Council, *Risk Assessment in the Federal Government: Managing the Process* (Washington, DC: National Academy Press, 1984).

8. *Chemical and Engineering News*, June 3, 1991, p. 34. "ACS views on environmental risk assessment, NIH funding."

9. Symposium on "The Potential for Using Quantitative Risk Assessment in Setting Environmental Priorities," American Chemical Society National Meeting, New York, NY, August 25–30, 1991.

10. Cothern C., Coniglio W., Marcus W., "Estimating Risk to Human Health: Trichloroethylene in Drinking Water as an Example." *Environ. Sci Technology* 20(1986): 111–116.

11. Statement by Dr. Paul Pappinek, South Coast Air Quality Management District (SCAQMD) hearing, Wilmington, CA, April 4, 1992.

12. Petition of Grievance to the Chemical Manufacturers Association—March 16, 1992, Regarding Union Carbide Corporation—Seadrift, TX Plant.

13. Gaylor, D.W. "Comparison of Teratogenic and Carcinogenic Risks," *Regulatory Toxicology and Pharmacology* 10(1989)138–143.

14. Kilburn, K.H. "Is The Human Nervous System Most Sensitive to Environmental Toxins?" *Arch Env Health* 44(1989)343–344.

15. California Air Pollution Control Officers Association (CAPCOA), Risk Assessment *Guidelines for the Air Toxics "Hot Spots" Program*. SCAQMD, Diamond Bar, CA, January 1991.

16. Blanc et al., *Journal of the American Medical Association* 1991, 266:664-669.

17. Detels R, et. al., "The UCLA Population Studies of CORD: X. A Cohort Study of Changes in Respiratory Function Associated with Chronic Exposure to SOx, NOx and Hydrocarbons." *American Journal of Public Health* 1991;81:350-359.

18. *Los Angeles Times*, April 3, 1992, p. A1.

19. Emission Inventories reported to SCAQMD under AB2588, March 1992.

20. See for example Weiss, K.B. and Wagener, D.K.. "Changing Patterns of Asthma Mortality: Identifying Target Populations at High Risk." *Journal of the American Medical Association* 1990;264:1683.

21. South Coast Air Quality Management District (SCAQMD) hearing, Wilmington, CA, April 4, 1992.

22. Summerhays, J., and Croke, H., "Air Toxics Emission Inventory for the Southeast Chicago Area," USEPA, Region V, Chicago, IL, July 1987; Shikiya D., Liu, C., Nelson, E., and Rapoport, R. "The Magnitude of Ambient Air Toxics Impacts From Existing Sources in the South Coast Air Basin." SCAQMD, Planning Division. Diamond Bar, CA, June 1987.

23. McKone, T.E., and Bogen, K.T. "Predicting the Uncertainties in Risk Assessment." *Environ. Sci. & Technology* 25(1991)1674-1681; Grogan, P.J., Heinold, D.W., Anderson, P.D. "Uncertainty in Multipathway Health Risk Assessments." Paper presented at the Air Pollution Control Association Meeting, Dallas, TX, June 1988.

24. Summerhays, J., "Estimation and Evaluation of Cancer Risk Attributed to Air Pollution in Southeast Chicago," USEPA—Region V, Chicago, IL, January 1989; personal communication by John Summerhays, USEPA—Region V, May 1989.

25. Tollefson, L., *Regulatory Toxicology and Pharmacology*; 13(1991): 150-169.

26. CA Department of Health Services, "Health Effects of Perchloroethylene." August 1991.

27. Plunket, L.M., Turnbull, D., Rodricks, J.V. "Differences between Adults and Children Affecting Exposure Assessment." *Proceedings of the Conference on Similarities and Differences between Children and Adults: Implications for Risk Assessment* (1991). Available from ENVIRON International, Arlington, VA.

28. Freudenburg, W., "Perceived Risk, Real Risk: Social Science and the Art of Probabilistic Risk Assessment." *Science*, 242 (1988):44-49.

29. Jaycock, M.A., and Hawkins, N.C. Abstract of paper presented at American Chemical Society/Division of Environmental Chemistry, August 25-30, 1991, in New York, NY, p. 110 of Abstracts for Division of Environmental Chemistry.

30. Bailey, J. "Dueling Studies: How Two Industries Created A Fresh Spin on the Dioxin Debate." *Wall Street Journal*. February 20, 1992.

31. Webster, T., and Connett, P., "Risk Assessment: A Public Health Hazard?" *J. Pesticide Reform*, 10(1990):26-31. Spring 1990.

15

Risk Assessment and Occupational Health

OVERVIEW AND RECOMMENDATIONS

C. MARK SMITH
KARL T. KELSEY
DAVID C. CHRISTIANI

The modern working environment imposes a variety of risks on worker health and safety. Many of these risks involve toxic chemicals, which may have serious acute, or chronic, impacts on the health of workers and their families. (1-3) Occupational exposures to such chemicals are widespread and are often at levels considerably greater than those attributable to environmental pollution. (4) Significant numbers of workers are exposed to toxic chemicals on the job each year, and thousands of these individuals suffer significant adverse effects as a result. (1)

Toxic chemicals can lead to both reversible and irreversible health effects, some of which may not be clinically detectable for years following exposure. (3, 5) Respiratory diseases (6), neurological effects (7, 8), reproductive impairment and birth defects (9-12), cardiovascular effects (13, 14), as well as a variety of cancers (15-18) have all been associated with occupational exposures to toxics. Depending on the chemical(s), the dose, and the susceptibility of those involved, the risk of experiencing adverse outcomes following such exposures may range from minimal to severe. Fortunately, once identified, most of these risks are controllable. (1) Yet exposures to many toxic agents in the workplace are allowed under current regulations, often at levels that may pose highly significant risks to worker health and safety. Risk assessment methods, when used appropriately, can provide useful tools to help identify, evaluate, and justify appropriate regulation of such occupational risks and are being increasingly relied upon by regulatory agencies. (19–21)

The following chapter discusses the use of risk assessment in the occupational health arena, providing a general overview of this developing field and concluding with a number of recommendations that we believe will improve the use of risk assessment in occupational health. We have attempted to highlight the rationale for, and strengths and weaknesses associated with, the increasing use of risk assessment to guide the regulatory agenda concerning occupational chemicals. Some of the key areas of current controversy in risk assessment methods and their application to controlling occupational risks are discussed, drawing on the authors' perspectives as public health scientists engaged in occupational health and toxicological research.

RISK ASSESSMENT AND RISK MANAGEMENT

From the standpoint of regulatory decision making, a distinction is usually made between the activities and responsibilities of risk assessors and risk managers. (21, 22) The term "risk" refers to the possibility of an adverse outcome and often involves some estimate of the magnitude of the outcome, its nature, and the probability that such an event will occur. The field of risk assessment has developed to help both identify and quantify risks, including those posed by exposures to toxic chemicals in the workplace. (22) Alternative approaches to the evaluation and regulation of occupational chemical hazards exist that do not explicitly consider risk levels (for example, relying on hazard identification schemes to identify potential dangers or structuring regulations to minimize significant chemical exposures regardless of risk). Most of these alternative approaches do, however, implicitly consider the magnitudes of the risks involved especially when applied to actual regulatory decisions. Thus, the distinctions between these non-risk-based approaches and more formal risk assessments are often less than clear-cut .

Risk management, within the context of this chapter, refers to the process whereby decisions are made about what is to be done (or not done) to reduce a given health risk. In the occupational arena, such risk management efforts most obviously include the promulgation and enforcement of occupational standards by the Occupational Safety and Health Administration (OSHA). Important decisions that significantly impact the management of occupational health risks are, however, also made at a variety of other levels including the courts (2), the Office of Management and Budget (OMB) (23, 24), and other elements of the executive branch, to name but a few. In a broader sense, significant risk management decisions in the workplace also may be made by a variety of other groups, including occupational health professionals (for example, industrial hygienists and occupational physicians) and representatives of management and labor. Such risk management decisions are often, but not always, guided by an explicit evaluation of the risks in question. Other criteria and information, however, are also usually taken into account by risk managers. These may involve considerations of technological feasibility, economic impacts, ethics, and politics. Clearly, these latter factors often play a more significant role in risk management decisions than the nature and magnitudes of the risks themselves.

In contrast to risk management, risk assessment is a somewhat more scientific undertaking. Scientists and others involved in risk assessment strive to use the available scientific data to gain insight into the nature and degrees of risk posed by various activities and chemicals. The goal of risk assessment is to provide the various risk managers with objective information upon which to base their decisions. (22) Ideally, risk assessment should be based on solid science. In reality, the endeavor is by no means a fully objective scientific undertaking, devoid of policy choices. Due to the inherent uncertainties in the process, scientific judgments and assumptions regarding the quality of various data, analyses, and models are always required (the science involved is rarely definitive and is often open to differing interpretations). Considerable debate has occurred in recent years over the appropriateness of many of the assumptions commonly made in risk assessment and over approaches for making these assumptions. Several groups have been reviewing risk assessment methodologies, including the National Academy of Science's Committees on Risk Assessment, the Office of Science and Technology Policy's Federal Coordinating Council on Science, Engineering, and Technology, and the Committee on Risk Assessment for Hazardous Air Pollutants, among others. These groups are considering a wide range of methodological issues pertaining to issues such as high-dose to low-dose extrapolation models, pharmacokinetics, mechanistic models of disease causation, methods for assessing and reporting uncertainty, and markers of susceptibility, among others. Attempts to coordinate and, perhaps, standardize risk assessment methods between various federal agencies also are being made. The Office of Management and Budget (OMB) also has begun to play a more active role in this process, expressing concern that chemical risk assessment as practiced by many government agencies (notably the Environmental Protection Agency [EPA], National Institute for Occupational Safety and Health [NIOSH], and Occupational Safety and Health Administration [OSHA]) has become too "conservative" (health-protective) in its assumptions. (24) OMB has argued that this may lead to a misdirection of regulatory attention toward areas of little real significance to public health and at great governmental and private expense. These ongoing debates, particularly those surrounding the potential conservativeness of the current risk assessment process, may lead to significant changes in federal risk assessment efforts in the future.

Risk assessment relies on a variety of different methods and may consider many types of scientific information including data from short-term tests (for example, the Ames test) (25), mechanistic studies (for example, pharmacokinetics and pharmacodynamics) (27), animal bioassays (26), and human epidemiological studies. (15) Assessing the risks posed by toxic chemicals requires that several factors be considered. Paramount among these are: (a) the inherent toxicity of the chemical, including the nature of its effects and potency; (b) the expected range of exposures likely to be experienced by the target population; and (c) the susceptibility of those exposed. It should be noted that these factors can be considered both *quantitatively* and *qualitatively*, and both of these approaches have been used to guide occupational and environmental health efforts. In all of these cases, the purpose of the risk evaluation process is fundamentally the same—to evaluate the possible adverse health effects that a given activity or chemical exposure may exert upon a population, an individual, or the environment. Because of this commonality of goals and for the sake of simplicity, these various risk evaluation approaches will all be referred to as risk assessments in this paper. It should be noted, however, that the field of risk assessment has, unfortunately, come to be associated to a large degree with quantitative assessment approaches, particularly the quantitative extrapolation methods that are used to generate numerical estimates of human cancer risk from animal bioassay data. (28, 29) Although these quantitative methods have been extensively used and are often appropriate, other approaches also are available (see below) that may often be preferable given the inherent uncertainties in the underlying science. As has been noted many times in the risk assessment literature, quantitative risk assessments require numerous assumptions (both in potency and exposure estimations), yielding numerical estimates of risk that are highly uncertain. (31–33) In the face of these fundamental limitations, we believe that risk managers should view computationally and structurally complex models that have not been adequately validated with skepticism and suggest that qualitative methods receive renewed attention by the risk assessment and risk management communities.

Some of the less quantitative approaches to risk assessment include (a) qualitative characterizations of risks, where potential health risks are identified but not quantified (for example, hazard evaluations [22], carcinogen classification schemes [34, 35], and so forth); (b) qualitative risk estimation approaches, where chemicals are ranked or classified by broad categories of risk (for example, potency classification schemes [36, 37]); (c) semiquantitative approaches, where effect levels (for example, no observable adverse effect levels, benchmark doses [38], and so forth) are used in combination with uncertainty factors to establish "safe" exposure levels. Each of these approaches can and has been used to assess occupational health risks, and, in a broad sense, each constitutes a form of risk assessment. Thus, the field of risk assessment should be viewed as encompassing both quantitative and qualitative methods. For many chemicals, it can be argued that the qualitative approaches provide a more realistic picture of the ability of scientists and risk assessors to estimate human risks.

RISK ASSESSMENT AND OCCUPATIONAL HEALTH

Efforts to regulate occupational exposures to toxic agents are at their best when adequate pre-existing epidemiologic data are available from which to directly estimate the human toxicity of a chemical. Such epidemiological studies provide a very valuable source of information for risk assessment and, furthermore, offer perhaps the best chance to validate extrapolative risk assessment models. Waiting for these data to emerge, however, is not an ideal approach to public health since disease, often irreversible, will have occurred in at least some segment of the population. For a variety of reasons, epidemiological data are often not available, or are of limited usefulness, to occupational health specialists and regulators. (3) For new chemicals, human exposures may not yet have occurred. For others, exposures may have been so low or restricted to such a small population that effects of the chemical are nearly impossible to detect. Classical epidemiological studies also have well-known design and analytical limitations. (3) Many health effects of concern, including diverse outcomes such as respiratory impairment, neurological effects, and cancers, may not arise or be detectable until years after the causative exposures have ceased or may be the result of progressive accumulation of subclinical damage that is not immediately identifiable. (1-3) Such delayed effects often are not easy to associate with specific occupational exposures and may be difficult to detect epidemiologically. Other limitations of epidemiology, including issues of statistical power, exposure measurement uncertainty, the healthy worker effect, and misclassification also may make it difficult to detect potentially significant health risks. (3)

Due to these limitations, much effort in risk assessment has gone into the development of methodologies to address less well documented risks, using models of disease and injury that attempt to predict human health outcomes. Predictive models like these are advantageous to public health protection since they allow potentially significant threats to public health to be considered before disease or injury occurs or is detectable. Unfortunately, all of these predictive approaches are fraught with many uncertainties and limitations that make precise estimates of risk levels impossible. For example, occupational exposures to toxins often involve multiple agents. (1, 3) In this situation, complex dose-response relationships and potential interactions between toxic compounds make it difficult to predict the human health impacts of even well-characterized toxins. For chemicals where human data are lacking or limited, the use of information from animal or other model systems may be required. (22, 26, 39) In these cases, risk assessment must rely on high-dose to low-dose and interspecies extrapolations that introduce considerable uncertainty into potency calculations. Uncertainties with respect to intraspecies variability in response and in exposure estimates further limit the accuracy of many risk assessments.

In spite of all these uncertainties, the application of scientific principles to assess risks is widely, although by no means universally, accepted. What is the rationale for this? The simplest case for risk assessment is that, without some procedure to identify toxic chemicals and consider their potencies, regulators must either assume the default positions that all chemicals are deadly and should be equally and stringently controlled or that all chemicals are safe and we need not worry about any of them. Clearly, these extreme options do not make much sense. Yet, without some approach to assessing risks, regulatory agencies face just this dilemma.

Indeed, without some evidence of harm or the potential for significant harm, regulatory agencies and society as a whole often choose not to regulate environmental or occupational exposures to chemicals. This situation continues today where many chemicals with weak epidemiological data, and most toxicologically untested chemicals are, by default, essentially regulated as though they are safe. (23, 40) Risk assessment can play an important role in identifying such potential risks and mobilizing regulatory action.

Most efforts at chemical risk assessment have been driven by three factors. One is the diverse nature of the toxic responses and disease endpoints involved; some toxic endpoints are clearly worse than others. The second is the wide range of potencies that toxic agents exhibit; some toxins are clearly "stronger" than others. The third is that toxicity depends on dose, which varies considerably between exposure situations. Toxicologists have demonstrated that, at a given level of exposure, the toxicity of different chemicals can vary by up to several million fold. These differences in potency have been observed in animal model systems, and for a smaller number of chemicals in humans as well, for both acute and delayed effects like cancer. Human exposures to chemical agents also can vary considerably. These variations in potency and exposure levels imply that the human risks posed by different chemicals will vary by many orders of magnitude. From a public health standpoint, occupational health regulatory efforts should be directed toward the worst chemicals (a function of the probability of harm and the potential severity of the outcomes). Ideally, the potent toxins that cause severe, irreversible effects should be more stringently regulated compared to those of lower potency. Greater efforts should be made to minimize occupational exposures to these agents, perhaps by reducing their use in the workplace, and they also should be the focus of strict enforcement and monitoring efforts to ensure compliance with regulatory standards. Risk assessment techniques have largely been developed to help regulators take these wide variations in risk into account in their risk management efforts.

OCCUPATIONAL RISK ASSESSMENT AND RISK MANAGEMENT STRATEGIES

Occupational health decision making and standard setting have always been guided to some degree, either explictly or implicitly, by evaluations of risks to worker health and safety. Both quantitative and qualitative approaches have been used in such assessments. In the occupational setting, chemical risk reduction efforts usually include one or more of the strategies outlined in Table 15.1.

Risk assessments, including both qualitative and quantitative approaches, can play a role in directing such efforts, providing guidance to risk management decisions that might otherwise be driven solely by one-sided analyses of control costs, technological feasibility, or by wholly sub-

TABLE 15.1. Summary of Occupational Risk Reduction Strategies

1. *Regulation of exposures, through*
   a. Engineering controls (for example, ventilation, closed-loop production technologies, and so forth)
   b. Process changes, including toxic use reduction and chemical substitution (that is, eliminating or reducing the use of toxins or substituting toxins with non- or less toxic agents)
   c. Worker training
   d. Monitoring of exposure levels via area or personal sampling
   e. Use of personal protection devices (for example, gloves, respirators, and so forth)
2. *Minimization or treatment of effects*
   a. Screening for toxic effects with medical removal and/or treatment
   b. Monitoring of dose (for example, blood lead levels) with removal and/or treatment

TABLE 15.2. Potential Uses of Risk Assessment in Occupational Health

1. Identification of risks
2. Targeting of regulatory attention
3. Targeting of enforcement and compliance efforts
4. Evaluation of risk tradeoffs
5. Identification of *de minimus* risks
6. Investigation of risk/benefit distributions (even if hypothetical)
7. Identification of data gaps
8. Guidance in the design of worker training programs and risk communication materials
9. Guidance in the design of medical monitoring and surveillance efforts
10. Establishing need for technology-forcing standards

jective political considerations. Table 15.2 briefly summarizes some of the potential uses of risk assessment in the occupational setting. In addition to these specific uses, the procedural aspects of risk assessment are themselves important, as they provide a framework that can ensure that occupational health risks are at least considered. These methods can provide information useful in setting exposure limits and in establishing other components of occupational health standards such as hazard communication requirements, engineering controls, and medical removal provisions, among others. Careful assessments of risks also can help regulatory agencies to better target their limited resources toward the most significant risks. Provided *appropriately conservative* assumptions are used, quantitative approaches also can identify *de minimus* risks, ones too small to warrant regulatory attention (although, as will be discussed below, determining appropriate analytical approaches and acceptable risk levels is problematic).

Targeting and choosing appropriate engineering controls also can be facilitated by judicious use of risk assessment, which can identify potentially significant health risks that should be investigated further or regulated. They also may help in choosing between competing process, design, or control alternatives. For example, some engineering controls, such as ventilation, may shift risks from the workplace to the more general environment. If this results in significant risks to other populations, then alternative management strategies such as toxic use reduction (TUR) should be considered. Risk assessment provides a mechanism that allows these and related questions of risk distribution to be addressed.

With respect to TUR, which may involve substitution of potently toxic chemicals with ones less dangerous, risk assessment allows for the initial identification of the toxic substances to be reduced, replaced, or eliminated. Although one can argue that all chemical use in the workplace should be reduced, given the demands of a modern society, such reduction efforts are likely to occur only if a reasonable case can be made that the chemicals in question pose some significant risk. Qualitative and quantitative risk assessment provide tools that occupational health decision makers can use to more effectively make such a case and can provide a framework within which comparisons of risks between potential chemical or process substitutes can be made. Simplistic evaluations of such chemical risks can, and have in the past, led to the substitution in the workplace of one set of chemical hazards for another.

Risk assessment also may be useful in considering other risk tradeoffs as well, including the potential creation of new workplace risks in the process of dealing with the old. The targeting and design of worker training programs, workplace monitoring efforts, medical surveillance, and epidemiological studies can all benefit from both quantitative and qualitative risk assessment, which can focus attention on the key risk endpoints. In some situations risk assessments also may be useful in establishing probabilities of causality in occupational and environmental compensation disputes.

Finally, quantitative or semiquantitative risk assessments also may play a role in establishing and justifying technology-forcing standards or goals. Depending on factors such as control costs and technological feasibility, currently available control methods may not be able to eliminate a given chemical exposure. Thus, substantial, and in some cases unacceptably high, residual risks may remain. Risk assessment provides a tool to determine whether the feasible control strategies are sufficiently protective of worker health. In some situations, additional reductions in exposure, which will require the development of new technologies, may be required. In other situations, an outright ban on the use of the chemical may be justified.

Although of great importance, risk assessments do not, however, provide a panacea for managing occupational health risks. Some of these assessments may be so uncertain as to provide little practical guidance to the decision makers, and their ability to quantitatively estimate human health risks should not be oversold lest excessive reliance on such assessments lead regulators to ignore other important elements in setting occupational standards. These limitations should not, however, cause us to abandon their wise use.

ROLES OF OSHA AND NIOSH

At the federal level, risk management efforts in the occupational arena, including the promulgation and enforcement of occupational standards, are largely the responsibility of OSHA. OSHA is required to consider recommendations from other agencies including the National Institute for

Occupational Safety and Health (NIOSH). (2) NIOSH plays an important role in developing and recommending occupational safety and health standards and, among other duties, provides OSHA with scientifically based assessments of worker health and safety issues. Quantitative risk assessment methods are being increasingly relied upon by NIOSH in these evaluations. (5, 19, 20) OSHA is not, however, required to base its decisions on these NIOSH assessments and often relies on the recommendations of other groups including the American Conference of Governmental and Industrial Hygienists (ACGIH). (41) Although NIOSH, ACGIH, and other groups often evaluate similar risks, an overall lack of coordination and consistency in the application of risk assessment methodologies between these various groups exists.

The OSH Act established both OSHA and NIOSH and provides the statutory basis for their activities. (2) Among other requirements, the OSH Act states that OSHA must establish standards

> dealing with toxic materials [which] most adequately assures, to the extent feasible, on the basis of the best available evidence, that no employee will suffer material impairment of health or functional capacity even if such employee has regular exposure to the hazard. In addition to the attainment of the highest degree of health and safety protection for the employee, other considerations shall be the latest scientific data in the field, [and] the feasibility of the standards . . . (OSH Act Sec. 6(b)5)

The direct reference to toxic materials in this section implicitly requires OSHA to make an initial determination as to the toxic potential of a chemical. Such a determination requires, at the very least, a qualitative assessment of the potential hazard or risk posed by the agent in question. Furthermore, the language embodied in the act, which reads, "*no employee will suffer material impairment of health*," "*to the extent feasible*," suggests that the agency must consider the nature, the likelihood of occurrence, and the feasibility of reducing risks to even one employee. Subsequent case law has further delineated OSHA's responsibilities in this regard. Largely as the result of two Supreme Court cases (on the agency's proposed benzene and cotton dust standards), OSHA has been required to demonstrate that its proposed standards reduce "*significant risks to worker health*" (42 a, b). "*Significant*" was suggested to be an increase in lifetime risk of developing cancer of greater than one in 1,000 and that it was more likely than not that harm would occur without regulation. In essence, these rulings have required OSHA to support its standards with some type of risk assessment demonstrating that "significant risks" to worker health are being addressed. Although the court did not explicitly require the use of quantitative risk assessments, these approaches do provide one of the few tools able to generate reasonable inferences or assumptions about the magnitudes of the risks involved. In practice, OSHA and NIOSH are now using quantitative techniques whenever possible (19, 20) to determine whether proposed regulations pass the threshold test of "significant risk" using both epidemiology and animal bioassay data.

The significant risk threshold effectively established by the Supreme Court can be criticized on several grounds. Although the court did not require that OSHA risk estimates be scientifically precise, the requirement that harm be more likely than not to occur if no regulations were to be enacted suggests a higher burden of proof than is actually achievable in most risk assessments. Furthermore, given the uncertainties inherent in the process, risks higher than those calculated using current risk assessment procedures may actually be experienced by susceptible subgroups of the population (see below for additional comments). A very significant limitation of this requirement is that a numerical significant risk threshold cannot be met for a large number of chemicals because insufficient toxicological data exist to allow for any quantitative estimate of risk to be made. Thus, chemicals that exhibit potent toxicity based on qualitative data (for example, structure activity relationships, short-term tests indicative of genotoxicity, and so forth) may not be regulated. Lastly, aggregate risks resulting from multiple chemical exposures are not explicitly considered. Thus, exposures that result in worker risks considerably above the risk threshold may completely escape regulatory attention.

BRIEF DISCUSSIONS OF SEVERAL IMPORTANT ISSUES

Acceptable Risk

In general, higher risks have been deemed to be acceptable in the management of occupational risks compared to those experienced in environmental settings. (61) For examp1e, excess lifetime cancer risks of less than one in 1,000 have usually been deemed acceptable for workers (see discussion above), while risks as low as one in a million have often been deemed unacceptable in the regulation of environmental carcinogens. The differences in acceptable risk levels may in fact be greater than this because of differing assumptions used to determine the risk estimates—more conservative assumptions are often used in estimating risks due to environmental exposures to carcinogens (for example, NIOSH used a body weight species scaling factor to extrapolate butadiene cancer risks from rodents to humans [19] vs. the more conservative surface area scaling factor currently used by EPA [26]).

The notion of what constitutes an acceptable level of risk in regulatory standard setting is controversial and is clearly an issue of ethics and policy rather than one of science. (62) Extensive discussion of this issue is beyond the scope of this paper, but there are several points worth mentioning. Regulatory decisions regarding acceptable risk levels generally consider two aspects of the risk matrix: (a) the public health implications of risk level, that is, the total number of expected disease cases or deaths that may be attributable to the exposure in question; and (b) issues of ethics and equity pertaining to the imposition of risks on limited numbers of people where the population burden of disease would be expected to be small. (43, 44) Clearly the imposition of small risks on large populations may result in a highly significant exposure-induced disease burden and should be prevented. Just as clearly, the issue goes beyond statistical deaths or cases of disease; a one in 100 excess risk of death applied to a population of five would be expected to yield no actual deaths but would be considered by most to be an unacceptably high risk level. Furthermore, an extensive body of work (as well as a good dose of common sense) indicates that all risks of equal magnitude are not of equal concern to most people. (63, 64) Society cares about other issues as well as the risk level. These issues include whether the risk is voluntary, the individual's perceived and actual control over the risk, the benefits (if any) associated with the risk, and so forth.

Several additional points should also be kept in mind when comparing a risk estimate with an acceptable risk level. First of all, risk estimates in themselves are only that—estimates. Although conservative assumptions are generally used in estimating these risks, our ability to judge the degree of conservativeness for any individual chemical is usually limited (see below). A comprehensive reporting of the uncertainties involved in most numerical estimates of risk would reveal wide probability distributions,which are also likely to be of unknown shape. Actual risks for any given chemical may be higher as well as lower than the estimates calculated using standard quantitative risk estimation procedures, and assignment of a meaningful percentile (for example, fraction of the probability distribution of risk that falls below any given risk level) to these risk estimates is clearly difficult. Secondly, some risk estimates are better than others, and the weight given a risk estimate in the risk management process should be related in some explicit way to our confidence in it. Current risk assessments, which usually do not adequately acknowledge their inherent uncertainties, make such evaluations impossible. Thirdly, risk estimates often fail to consider issues such as susceptible individuals and aggregate risks, which may well distort regulatory efforts toward less health protective avenues.

Aggregate Risks and Sensitive Populations

Aggregate risks are rarely considered in either occupational or environmental risk management decisions. Workers are often exposed to multiple agents on the job, may be exposed to the same agents via multiple pathways (including exposures that occur outside the job), and may experience additional, nonchemical, job-related risks as well. Many regulations, however, are targeted toward single compounds or single exposure pathways. If, in theory, one accepts a "one in 1,000 excess cancer risk" as "significant" and as a threshold risk level for regulatory action (as implied in the U.S. Supreme Court cases discussed previously (41, 42), then we believe that aggregate risks must be considered when determining whether an occupational regulation is warranted.

In addition to aggregate risks, the possibility that some members of the population may be particularly susceptible to toxic insults must also be considered. Progress in detecting and identifying genetic traits associated with increased probabilities of contracting disease and increased susceptibility to toxins suggests that wide variations in susceptibility to these agents may occur. (45–48) Failure to consider such sensitive subgroups may lead to underestimates of both individual and overall population risks. For regulations that are supposed to protect all workers this is a critical issue and provides a strong argument for retaining "conservative assumptions" in occupational health risk assessment (since at this time we have no way of knowing whether sensitive individuals in the human population are more or less sensitive than the animal bioassay systems from which we often extrapolate human risks).

Poorly Characterized Chemicals

Untested or poorly characterized chemicals present another major problem to risk assessment. Of the more than 70,000 chemicals synthesized by humans, less than 10 percent have been toxicologically characterized to any appreciable degree. (26) To date, most untested chemicals have been, by default, assigned toxic potencies of zero in both the occupational and environmental regulatory arenas. (23, 40) Of course establishing an experimentally based quantitative risk estimate for these agents is clearly not feasible (statistical approaches to assigning default potencies to untested chemicals are being considered but to our knowledge have not been used in a regulatory setting). Thus, a blind reliance on threshold risk determinations, such as the one-in-1,000 excess cancer risk discussed earlier, would preclude regulating many chemicals of unknown toxicity. Clearly, alternative approaches need to be considered in these situations, including the application of technology-based standards where risk levels cannot be estimated.

Conservativeness

There is currently considerable debate over conservativeness in risk assessment. (23, 40, 49–53) Although most risk assessors and toxicologists try their best to use the most defensible scientific evidence and methods available to guide their decisions, such judgments may, nonetheless, vary considerably from scientist to scientist. These differences can lead to divergent risk estimates and usually arise from honest scientific disagreements. In the face of these uncertainties, regulators, in consultation with scientists, have attempted to choose "reasonably" (consistent with scientific theory or data) "conservative" (ones judged *unlikely* to lead to an underestimate of true risk) assumptions when faced with scientific uncertainty.

Some argue that the procedures currently in use, notably those associated with extrapolations of human cancer risks from animal bioassays, are too conservative, that is to say, that they consistently exaggerate actual risks. If true, such a situation could lead to an inefficient allocation of public health resources toward controlling exposures that actually present minimal risks. (49) Others argue that several of the assumptions in risk assessment are either clearly not conservative or are of questionable conservativeness. A full discussion of this issue is again beyond the scope of this paper, and the reader is urged to refer to the additional references as noted above for a fuller discussion. A few points, however, warrant discussion. First, the empirical evidence pertaining to the issue of conservativeness in carcinogen risk assessment appears to us to be weak and inconclusive. This is largely because of limitations in the human epidemiology available; potency estimates based upon epidemiology are available for fewer than 30 agents, and these estimates are themselves highly uncertain. (4, 15, 54) Although the accuracy of cancer potency factors derived from animal bioassays is open to question, significant evidence suggests that they do not consistently exaggerate human risk levels. For example, cancer potency factors based on animal bioassays were not inconsistent with human epidemiological data for 18 out of 20 chemicals investigated. (4) For two chemicals, the animal bioassay results appeared to overestimate human risk. (Note that the term *not inconsistent* is used because the results do not really allow for comparisons of the low-dose potencies of the chemicals in question). Furthermore, a recent epidemiological study by NIOSH suggests that the "conservative" EPA cancer potency factor for dioxin cannot be ruled out by the human cancer experience of the best studied cohort of workers occupationally exposed to this agent. (55, 56) Although this study cannot address the possibility that the human low *dose* potency of TCDD is less than that predicted by the extrapolative approach currently used by the EPA, these results clearly do not support this contention. Also note that the epidemiological studies considered above did not, in general, explicitly consider potentially susceptible subgroups of the population. Such subgroups are likely to exist and would be expected to experience risks greater than those of the cohorts studied.

In any case, many of the specific issues involved in this debate are not resolvable over the short term, and others, such as dose extrapolation models, may never be. Overall, the best *judgment* of many scientists is that the current approach to estimating human cancer risk based on animal bioassays is likely to be "conservative" for most chemicals. The degree of conservativeness, however, is unknown, and for some individual chemicals the methods now used may actually underestimate risks. (40) A brief discussion of some of these issues follows.

The MTD Debate

Animal carcinogenicity bioassays are usually performed at the maximum tolerated dose (MTD) and/or some large fraction thereof. (26, 57, 58) MTD dosing is used to enhance the power of these bioassays to detect a carcinogenic effect (power is an issue because technical and economic considerations limit the number of animals that can be tested to relatively small groups). MTD dosing has been criticized for several reasons. Most significantly, critics argue that these high doses (which are usually much higher than expected human doses) may lead to organ toxicity, inflammation (which may lead to the production of endogenously generated mutagens), and an increase in cell proliferation in target organs of the test species (for example, regenerative proliferation in the liver). (36, 57) Many scientists and a substantial body of research suggests that cell proliferation may enhance carcinogenesis by increasing the rate at which mutations occur in the target organ and/or allowing altered cells to clonally expand their populations. (59, 60) Although this hypothesis is plausible, we should keep in mind that organ toxicity can, in theory, act in the opposite direction. For example, most cancer chemotherapeutic agents in use today are cytotoxins (many of them DNA-damaging agents that are thought to themselves be carcinogenic) that preferentially kill rapidly dividing cells including many that are cancerous and precancerous. MTD-related cytotoxicity may in theory act in a similar fashion, preferentially killing such cells, thus leading to a potential decrease in the ultimate number of tumors observed. This may be a particular problem with animal strains that exhibit high background rates of cancers. For most carcinogens, insufficient data are currently available on the relationships between MTD dosing and the kinetics of cytotoxicity and cell proliferation in target organs to evaluate the significance of these arguments. A more complete understanding of these issues may well allow for better low-dose estimates of risk in the future. Finally, the significance of the MTD issue with respect to many occupational exposures must also be questioned because of the high exposure levels that often occur in the workplace. In the occupational setting, exposures to carcinogens often occur at a significant fraction of the animal MTDs. (59) These higher doses suggest that occupational cancer risks for many chemicals may more nearly mimic those observed in high-dose animal experiments (that is, the uncertainty attributable to the high-dose to low-dose extrapolation should be less).

Susceptibility

The extrapolated carcinogenic potencies of chemicals are usually based on the most sensitive species tested. Some of the species (strains) commonly tested have high background rates of cancer. Taken together these two observations have been used to argue that many of the cancer potencies used by regulatory agencies overestimate human risks. However, one must keep in mind that: (a) only two species are usually tested (mice and rats), which does not provide a very broad-based look at interspecies differences in potency; (b) the human population exhibits a relatively high rate of cancer (although the organ distribution is quite different from that observed in the animal bioassays); and (c) the animal strains tested are highy inbred and will not exhibit the genetic variability associated with disease susceptibility in the human population. Sensitive individuals (for example, those bearing mutations in the tumor suppressor gene p53 [48]) may be at even greater risk than the animals tested. In conclusion, we cannot tell where the species tested fall compared to either the average human or susceptible subgroups in the population. Continued reliance on the most sensitive species appears to constitute a reasonable default assumption unless compelling evidence to the contrary is available.

CONCLUSIONS AND RECOMMENDATIONS

In conclusion, we believe that risk assessment, in some form, has always been and should continue to be used to guide occupational health decision making. If used appropriately, these approaches can provide useful tools, but are clearly not a panacea, for directing occupational health efforts. Risk assessment methods can provide important procedural guidance to the evaluation of risks and provide a framework that allows risks to be considered before actual impacts to human health have occurred. It is, however, crucial that risk assessment be performed and applied in a fashion that realistically reflects its strengths and weaknesses. Furthermore, the use of risk assessment in the occupational arena must be consistent with occupational health principles. Table 15.3 lists some of the impor-

TABLE 15.3. Recommendations to Improve the Use of Risk Assessment in Occupational Health

1. Occupational health risk assessment methods should be standardized to the extent feasible, while allowing flexibility to consider the best science available.
2. Additional attention should be paid to the use of risk assessment data by risk managers and regulators. For example, risk managers and regulators should be required to justify decisions that appear to discount risk assessment data.
3. Risk assessment should rely on careful use of the best science available. To maintain integrity and public trust, the uncertainties inherent in the process, including those that may lead to underestimates as well as overestimates of risk, must be better identified. A fair, balanced approach must be used to evaluate the available scientific information.
4. The present attempt to be reasonably conservative in the choice of assumptions in risk assessment should remain. Conservative assumptions are the appropriate starting point for more detailed analysis. They do, however, need to better consider the available science, including mechanisms of toxicity. Mechanistic models also should be considered conservatively.
5. New scientific evidence regarding the toxicity of chemicals should be validated before it is used to supplant conservative assumptions. Guidelines need to be developed for determining what constitutes appropriate validation.
6. Aggregate risks, untested chemicals, and sensitive populations are issues that need critical attention and are not treated conservatively in current approaches to risk assessment. The significance of these issues to worker health risks should receive increased attention.
7. Risk managers should keep in mind that complex analyses and models are not necessarily better; they often just obfuscate the process, making it more difficult for diverse participation in the regulatory process itself. Applying the principle of Occam's razor, we urge that computationally and structurally complicated models that have not been demonstrated to do a better job of predicting risks be viewed with skepticism.
8. Qualitative representations of risk should receive additional attention since numerical estimates often imply more precision than our current scientific understanding warrants.
9. *De minimus* or threshold risk findings, because of their inherent uncertainty and the inability to even generate such estimates for many chemicals, should not themselves close the door on regulatory action.
10. Risk assessment should not be the sole tool used to determine occupational risk management decisions. Other values that should be explicitly considered include: the distribution of hypothetical risks and benefits in the population, the essentiality or need of the process or product, the longevity of the chemical(s) in the environment or in people, feasible alternatives that reduce exposures regardless of risk.
11. Precautionary principles should receive more attention in regulating occupational risks, especially when dealing with poorly characterized chemicals or complex exposure scenarios. For chemicals that have a poor database, technology-based regulations should be required.
12. Since risk assessment cannot provide precise, reliable estimates of risk, the best approach for regulating many chemical exposures will require a combination of technology- and risk-based methods.

(*cont.*)

TABLE 15.3. (*cont.*)

13. Risk assessment and risk management decisions should be clearly elaborated and explained via an open process with opportunity for scientific, labor, community, and management participation. Because of the unequal distribution of expertise in this field, we urge that a fund be established to enable worker groups to hire their own experts on toxicological and risk assessment matters.
14. Worker exposure information should be made more accessible to independent researchers. This will facilitate epidemiological investigations in the work environment, leading to better protection of worker health and ultimately allowing for more robust validations of our risk assessment methodologies against actual human data.

tant issues that we believe need to be considered when using risk assessment to evaluate and to guide efforts to reduce risks to worker health. We believe the recommendations summarized in this table will improve the use of risk assessment in the occupational health arena and urge that the occupational health community consider them carefully.

ACKNOWLEDGMENTS

C.M.S. was supported in part by grants ES 00002, from the National Institute of Environmental Health Science, and CA 09078, from the National Cancer Institute.

NOTES

1. Levy, B. S., Wegman, D. H., "Occupational Health in the United States: An Overview." In: Levy, B. S. and Wegman, D. H. (eds.) *Occupational Health: Recognizing and Preventing Work-Related Disease*. Boston, MA: Little, Brown, 1988.

2. Ashford, N. A., Caldart, C. C. (eds.) *Technology, Law and the Working Environment*. New York, NY: Van Nostrand Reinhold, 1991.

3. Monson, R. R. Occupational Epidemiology. Boca Raton, FL: CRC Press, 1990.

4. Goodman, G., Wilson, R. "Quantitative Prediction of Human Cancer Risk From Rodent Carcinogenic Potencies: A Closer Look at the Epidemiological Evidence for Some Chemicals Not Definitively Carcinogenic in Humans." *Regul. Toxicol. Pharmacol.* 14(2):118–146, 1991.

5. Millar, J. D. "Summary of Proposed National Strategies for the Prevention of Leading Work-Related Diseases and Injuries, Part I," *Am. J. Ind. Med.* 13:223, 1988.

6. Wegman, D. H., Christiani, D. C. "Respiratory Disorders." In: Levy, B. S., Wegman, D. H. (eds.) *Occupational Health: Recognizing and Preventing Work-Related Disease*. Boston, MA: Little, Brown, 1988.

7. Silbergeld, E. K., Landrigan, P. J., Froines, J. R., Pfeffer, R. M. "The Occupational Lead Standard: A Goal Unachieved, A Process in Need of Repair." *New Solutions* 1:20–30, 1991.

8. Baker, E. L. "Neurologic and Behavioral Disorders." In: Levy, B. S., Wegman, D. H. (eds.) *Occupational Health: Recognizing and Preventing Work-Related Disease*. Boston, MA: Little, Brown, 1988.

9. Hatch, M. C., Stein, Z. A., "Reproductive Disorders." In: Levy, B. S., Wegman, D. H. (eds.) *Occupational Health: Recognizing and Preventing Work-Related Disease*. Boston, MA: Little, Brown, 1988.

10. Hass, J. F., Schottenfeld R. "Risks to the Offspring from Parental Occupational Exposures." *J. Occup. Med.* 21:607–615, 1979.

11. Rosenberg, M. J., Feldblum, P. J., Marshall, E. G. "Occupational Influences on Reproduction: A Review of the Recent Literature." *J. Occup. Med.* 29: 584–598, 1987.

12. Curran, W. J. "Dangers for pregnant women in the workplace." *New Engl. J. Med.* 312:164–165, 1985.

13. Theriault, G. P. "Cardiovascular Disorders." In Levy, B. S., Wegman, D. H. (eds.) *Occupational Health: Recognizing and Preventing Work-Related Disease*, Boston, MA: Little, Brown, 1988, pp. 431–441.

14. Pirkle, J. L., Schwartz, J., Landis, R., and Harlan, W. R. "The relationship between blood lead levels and blood pressure and its cardiovascular risk implications." *Amer. J. Epidemiol.* 121:246–258, 1985.

15. Monson, R. R. "Occupation and Cancer" In *Occupational Epidemiology*, Boca Raton, FL: CRC Press, 1990.

16. Doll, R., Peto, R. "The Causes of Cancer: Quantitative Estimates of Avoidable Risks of Cancer in the United States Today." *J. Natl. Cancer Institute*, 66: 1191–1308, 1982.

17. Cole, P. "Cancer and Occupation: Status and Needs of Epidemiologic Research." *Cancer* 39: 1788–1791, 1977.

18. Brandt-Rauf, P. W., preface. In: Brandt-Rauf, P. W. (ed.), Occupational Cancer and Carcinogenesis. *Occupational Medicine, State of the Art Review*. Philadelphia, PA; Hanley and Belfus, Inc., 1987.

19. Occupational Safety and Health Administration, "Occupational Exposure to 1,3-Butadiene: Proposed Rule and Notice of Hearing," *Federal Register*, vol. 55, no. 155, August 10, 1990.

20. Occupational Safety and Health Administration, "Occupational Exposure to 4,4'-Methylenedianilline (MDA): Proposed Rule," *Federal Register*, vol. 54 no. 91, 20672–20744, May 12, 1989.

21. Millar, J. D. "Risk Assessment. A Tool to be Used Responsibly," presented to the American Mining Congress Session on Occupational Safety and Health, New Orleans, LA, 1990.

22. Committee on the Institutional Means for Assessment of Risks to Public Health, Commission on Life Sciences, National Research Council, *Risk Assessment in the Federal Government: Managing the Process*. Washington, DC: National Academy Press, 1983.

23. Evans, J. S., Graham, J. D., Gray, G. M., Hollis, A., Ryan, B., Smith, A., Smith, M., Taylor, A. "Summary of Workshop to Review an OMB Report on Regu-

latory Risk Assessment and Management." *Risk: Issues in Health and Safety*, 3:77-83, 1992.

24. U.S. Office of Management and Budget. "Current Regulatory Issues in Risk Assessment and Risk Management," Regulatory Program of the United States, April l, 1990-March 31, 1991.

25. Ames, B. "The Detection of Chemical Mutagens with Enteric Bacteria." In: A. Hollaender (ed.), *Chemical Mutagens*. New York: Plenum Press, 1971.

26. Huff, J., Haseman, J., Rall, D. "Scientific Concepts, Value, and Significance of Chemical Carcinogenesis Studies." *Annu. Rev. Pharmacol. Toxicol.* 31:621-652, 1991.

27. Travis, C. C. (ed). *Carcinogen Risk Assessment*. New York: Plenum Press, 1988.

28. Crump, K. S. "An improved procedure for low-dose carcinogenic risk assessment from animal data." *J Envir. Pathol. Toxicol.* 5:339-346, 1981.

29. U.S. Environmental Protection Agency. "Guidelines for Carcinogenic Risk Assessment." *Federal Register* 51(185), CFR 2984, No. 185 (Sept. 24) pp. 33,992-34,003. 1986.

30. Portier, C. J., Kaplan, N. L. "Variability of Safe Dose Estimates When Using Complicated Models of the Carcinogenic Process. A Case Study: Methylene Chloride." *Fundamental and Applied Toxicology* 13:533-544, 1989.

31. Finkel, A.M. "Confronting Uncertainty in Risk Management: A Guide for Decision Makers," Center for Risk Management, Resources for the Future, Washington, DC, 1990.

32. Hoel, D. G., Kaplan, N. C., Andersen, M. W. "Implications of Non-linear Kinetics on Risk Estimation in Carcinogenesis." *Science* 219:1032-1037, 1983.

33. Paustenbach, D. J., Jernigan, J. D., Finley, B. L., Ripple, S. R., Keenan, R. E. "The Current Practice of Risk Assessment: Potential Impact on Standards for Toxic Air Contaminants." *J. Air Waste Mgm. Assoc.* 40:1620-1630, 1990.

34. Matula, T. I., Somers, E. "The Classification of Chemical Carcinogens." *Regul. Tox. Pharmacol.* 10:174-182, 1989.

35. Ashby, J., Johannsen, F. R., Raabe, G. K., et al. "A Scheme for Classifying Carcinogens." *Regul. Tox. Pharmac.* 12:270-295, 1990.

36. Ames, B. N., Magaw, R., Gold, L. S. "Ranking Possible Carcinogenic Hazards." *Science*. 236:271-280, 1987.

37. Doull, J., Bruce, M. C. "Origin and Scope of Toxicology." In: Klaassen, C. D., Amdur, M. O., Doull, J. (eds.) *Toxicology. The Basic Science of Poisons*. New York: Macmillan Publishing Co., 1986.

38. Dourson, M. L., and Stara, J. F. "Regulatory History and Experimental Support of Uncertainty (Safety) Factors." *Regul. Tox. Pharmacol.* 3:224-238, 1983.

39. Paustenbach. D. J. "Important Recent Advances in the Practice of Health Risk Assessment: Implications for the 90's." *Regul. Tox. Pharmocol.* 10:204-243, 1989.31

40. Finkel, A. M. "Is Risk Assessment Really Too Conservative? Revising the Revisionists." *Colombia Journal of Environmental Law* 14:427-467, 1989.

41. Paxman, D. G., Robinson, J. C. "Regulation of Occupational Carcinogens Under OSHA's Air Contaminants Standard." *Regul. Tox. Pharmacol.* 12:296-308, 1990.

42. *Industrial Union Department AFL-CIO v. American Petroleum Institute*, 448 U.S. 607, 1980. *American Textile Manufacturers Association vs. Donovan*, 452 U.S. 490, 495, 1980.

43. Travis, C. C., Richter, S. A., Crouch, E. A. C., Wilson, R., Klema, E. D. "Cancer Risk Management." *Environ. Sci. Technol.* 21:415-420, 1987.

44. Travis, C. C., Hattemer-Frey, H. A. "Determining an Acceptable Level of Risk." *Environ. Sci. Technol.* 22:873-876, 1988.

45. Hattis, D., Erdreich L., Ballew, M. "Human Variability in Susceptibility to Toxic Chemicals—A Preliminary Analysis of Pharmocokinetic Data from Normal Volunteers." *Risk Analysis* 7: 415-426, 1987.

46. Brain, J. D., Beck, B., Warren, A. J., Shaikh, R. A. (eds.) *Variations in Susceptibility to Inhaled Pollutants. Identification, Mechanisms and Policy Implications*. Baltimore, MD: Johns Hopkins University Press, 1988.

47. Finkel, A. M. "Uncertainty, Variability and the Value of Information in Cancer Risk Assessment." Doctoral Thesis, Harvard University, Boston, MA, 1987.

48. Malkin, D. F., Li, P., Strong, L. C., Fraumeni, J. F., Nelson, C. E., Kim, D. H., Kassel, J., Gryka, M. A., Bischoff, F. Z., Tamsky, M. A., Friend, S. H. "Germ-line p53 Mutations in a Familial Syndrome of Breast Cancer, Sarcomas and Other Neoplasms." *Science* 250:1233-1250, 1990.

49. Nichols, A. L., and Zeckhauser, R. "The Perils of Prudence: How Conservative Risk Assessments Distort Regulation." *Regul. Tox. Pharmacol.* 8:61-75, 1988.

50. Bailar III, J. C., Crouch, E.A.C., Shaikh, R., Spiegelman, D., "One-hit Models of Carcinogenesis: Conservative Or Not?" *Risk Analysis*, 8:485-497, 1988.

51. Sielken, R. L., "Driving Cancer Dose-Response Modeling With Data, Not Assumptions." *Risk Analysis* 10:207-208, 1990.

52. Park, C. N. "Underestimation of Linear Models." *Risk Analysis* 10: 209-210, 1990.

53. Bailar III, J. C., Crouch, E.A.C., Spiegelman, D., Shaikh, R. "Response to Park." *Risk Analysis* 10:211-212, 1990.

54. Tomatis, L., Aitio, A., Wilbourn, J., Shuker, L. "Human Carcinogens So Far Identified." *Jpn. J. Cancer Res.* 80:795-807, 1989.

55. Fingerhut, M. A., Halperin, W. E., Marlow, D. A., et al. "Cancer Mortality in Workers Exposed to 2,3,7,8-terachlorodibenzo-p-dioxin." *New Engl. J. Med.* 324:212-218, 1991.

56. Webster, T. "Estimation of the Cancer Risk Posed to Humans by 2,3,7,8,-TCDD." Presented at the 11th International Symposium on Chlorinated Dioxins and Related Compounds at R.T.P., No. Carolina, Sept. 1991.

57. Apostolou, A. "Relevance of Maximum Tolerated Dose to Human Carcinogenic Risk." *Regul. Tox. Pharmacol.* 11:68-80, 1990.

58. Haseman, J. K. "Issues in Carcinogenicity Testing: Dose Selection." *Fund. Appl. Tox.* 5:66-78, 1985.

59. Gold, L. S., Backman, G. M., Hooper, N. K., Peto, R. "Ranking the Potential Carcinogenic Hazards to Workers from Exposures to Chemicals That Are Tumorogenic in Rodents." *Envir. Hlth. Perspectives* 76:211-219, 1987.

59. Cohen, S. M., Ellwein, L. B. "Genetic Errors, Cell Proliferation and Carcinogenesis." *Cancer Res.* 15(24):6493-6505, 1991.

60. Pitot, H. C., Dragan, Y. P., Neveu, M. J., et al. "Chemicals, Cell Proliferation, Risk Estimation and Multistage Carcinogenesis." *Prog. Clin. Biol. Res.* 369: 517-532, 1991.

61. Hashimoto, D. H., Brennan, T., Christiani, D. C. "Should Exposure to Asbestos in Buildings be Regulated on an Occupational or Environmental Basis." *Ann. of NY Acad. Sci.*, 1992.

62. Almeder, R. F., Humber, J. M. "Quantitative Risk Assessment and the Notion of Acceptable Risk." In Almeder, R. E., Humber, J. M. (eds.) *Quantitative Risk Assessment, Biomedical Ethic Reviews.* Clifton, NJ: Humana Press, 1986.

63. Slovic, P. "Perceptions of Risk." *Science* 236:280-285, 1987.

64. Drake, K., Wildavsky, A. "Theories of Risk Perception: Who Fears What and Why," *Daedalus* 119:41-60, 1990.

16

The Risk Wars

ASSESSING RISK ASSESSMENT

DANIEL WARTENBERG
CARON CHESS

Environmental and occupational policies—from Superfund clean-ups to occupational standards—increasingly are based on Quantitative Risk Assessment (QRA), a technique designed to forecast risks that cannot be measured directly. While officials of the Environmental Protectional Agency (EPA) and other agencies laud QRA as a means to make policy decisions more objective, environmentalists lambast QRA and the decisions based on it as far too sensitive to subjective judgments masquerading as science: "By changing an assumption here, tweaking an equation there, risk assessors can justify virtually any policy they desire." (1) Regardless, QRA is the methodology that often dominates the battles about environmental health issues. Understanding QRA is fundamental to understanding most skirmishes. The ongoing controversy about dioxin's potency is no exception.

Early in 1983, television news broadcast images of EPA scientists in moon suits assessing the impact of dioxin contamination in Times Beach, Missouri. Dioxin, a trace contaminant in oil sprayed on local roads, was believed to be extremely dangerous—so much so that the federal Centers for Disease Control (CDC) had cautioned against "long-term contact" with the substance even at the barely measurable concentration of one part per billion. Soil samples collected at Times Beach showed dioxin concentrations of from 50 to 100 times that level, prompting Vernon Houk, director of the CDC's National Center for Environmental Health and Injury Control, to call for a buyout of homes and evacuation of all 2,240 residents. In calling for this residential buyout, Houk argued that "we are worried about the overall long-term health effects." (2) By no means were such measures considered extreme. Indeed, some toxicologists believed that even the seemingly stringent CDC guidelines were too soft on dioxin. One EPA official testified in court: "There is presently no safe level of dioxin in the environment." (3)

Within five years, as new biological data became available on the effects of dioxin, some environmental scientists dramatically changed their views about the dangers of dioxin. In 1988, for instance, CDC proposed a 20-fold relaxation of its dioxin standard and EPA proposed easing its standard by 16-fold. In August 1991, Houk went so far as to recant his decision on Times Beach, maintaining that new information suggested "the evacuation was unnecessary." (4) Houk's statement reflected a widely held belief in the research community that the reevaluation of dioxin was state-of-the-art science.

Many environmentalists saw the revisionism as a sham, orchestrated to make it appear as if science were driving policy decisions that, in fact, they believe were motivated by pressure from industry. Environmentalist Barry Commoner criticized EPA's proposal to revise the standard as based on "ludicrously bad science." (5) Even some academic scientists, such as Columbia University's cancer researcher I. Bernard Weinstein, were perplexed. Contradicting the federal agencies, Weinstein declared that he knew "of no new scientific data or reason for the EPA to change its guidelines." (6)

KNOTTY QUESTIONS

The dioxin story raises a number of knotty questions: How do scientists reach judgments on the existence of a health hazard or, given its existence, on the severity of that hazard? How can purported experts on such matters wildly disagree about the margin of safety those governmental standards provide? How can public officials make major policy reversals and still argue that they are making judgments consistent with data?

All those questions have the same answer: on the basis of QRA. As the case of dioxin illustrates, using QRA to set government policy may raise as many questions as it answers. The basic tenet of QRA is that data on health effects detected in small populations exposed to high concentrations of a suspect chemical can be extrapolated to predict health effects in large populations exposed to lower concentrations of the same chemical. Thus, if workers in a factory are exposed to a high level of a suspect chemical and, after a sufficient number of years of follow-up, no one develops cancer, it is reasonable to infer that the substance is less dangerous than one from which workers have been known to get cancer.

Furthermore, one infers that if community residents are exposed to a low level of the same substance, they will not incur substantial increased risk of cancer either.

Like weather forecasts, QRAs are based on data, mathematical equations, and intuition based on years of experience. And, as we know all too well, weather predictions are sometimes far off the mark. But because these predictions are expressed numerically ("there is an 80 percent chance of rain tomorrow"), they take on the appearance of accuracy. It is easy enough to check the accuracy of a weather report by walking outside. But for QRA, whose predictions are also presented numerically, the feedback is not nearly so straightforward. It is generally difficult, if not impossible, to evaluate their accuracy.

In concept, QRA is meant to be an objective approach to risk evaluation for making informed public policy. Although components of the methodology allow for some flexibility, permitting risk assessors to accommodate a range of circumstances, many federal agencies have explicit guidelines for carrying out QRA. (7) Thus, federal officials argue that there is little room for subjective judgments.

In practice, however, even some EPA officials who are staunch supporters of QRA acknowledge its imprecision, noting that it entails a measure of subjective judgment. William Ruckelshaus, the EPA Administrator who championed basing agency decisions on QRA, acknowledged, "Risk assessment data can be like the tortured spy. If you torture it long enough, it will tell you anything you want to know." (8)

Although some policymakers see the numerical format of QRA as improvement over the even more ambiguous characterizations previously used to evaluate risk, others see these numerical gradations as deceptive, shrouding a highly subjective and uncertain process in the aura of scientific and numerical certainty.

There are merits to both sides of that argument. What seems clear is that society needs not only to develop more precise assessments of risk but also to develop better ways of grappling with disagreements about them.

A BRIEF HISTORY OF QRA

Until the late nineteenth century, predictions of the likelihood of disease or death were applied mainly to the major killers of the day—pneumonia, influenza, and tuberculosis. But by the twentieth century another, newer focus was coming to the fore: the health hazards caused by exposure to chemicals. Scientists began to explore ways to assess the risks presented by various substances.

Amid debate over the regulation of food and drugs, scientists and policymakers were discussing the ethical and practical limitations of using human volunteers as test subjects. (9) They argued that more detailed information would be available if exposed animals could be sacrificed and their pathology studied in the laboratory. This notion that animal response to substances can serve as a predictor of human response is another central tenet of QRA.

The rise of the use of pesticides on crops in the 1920s brought a new concern about the effects of chronic exposure to those chemicals. Scientists increased testing of the effects of chemicals on laboratory animals and developed special inbred strains of rodents to increase comparability among replicate tests. Typically, studies determined the highest concentration laboratory rodents could tolerate without becoming ill—the so-called no-observable-effects level, or NOEL.

In the early 1940s, on the basis of NOEL, regulators at the U.S. Food and Drug Administration (FDA) introduced the idea of safety factors. In the absence of data on human health effects, the regulators proposed that the standard maximum exposure for humans allow for a 10-fold greater sensitivity in people than in test animals, and a 10-fold variability in sensitivity to a chemical from person to person. Thus, regulators proposed that a standard 100 times below the NOEL would adequately protect humans from adverse effects of chemical exposures. (10) This rule of thumb is still widely applied to noncarcinogenic toxins.

By the mid-1940s, when these studies were looked to for policy guidance, some scientists began to challenge the notion that a NOEL could apply to carcinogens. They argued that no one could prove the existence of a threshold below which no risk would be incurred, and that any exposure—no matter how small—increased a person's risk of cancer. By 1950, the FDA concluded that no safety factor could be justified for a carcinogen.

In response, Congressman James Delaney proposed in 1958 to amend the Food, Drug and Cosmetic Act of 1938 to outlaw the addition to the food supply for any purpose of any chemical carcinogen, regardless of the strength of the carcinogen. The stringency of this amendment, the so-called zero tolerance rule, has been under fire ever since. Some scientists argue that chemicals which are carcinogenic, but only slightly, constitute "acceptable" risks in foods. But the no-threshold school of thought has since colored much of the debate over carcinogens and risk assessment.

In the 1960s and 1970s, public concern about environmental hazards and the consequences of pollution increased dramatically. Concern over exposure to radiation was heightened with controversy over the atomic bomb tests. Rachel Carson's prophetic book *Silent Spring* (11) sensitized the nation to the dangers of environmental contamination. In 1970, the EPA and the Occupational Safety and Health Administration

(OSHA) were founded to address increasing environmental and workplace concerns.

In this climate of growing awareness of chemical risks, research efforts intensified to model mathematically the risk of cancer imparted by exposure to specific amounts of various substances—the beginnings of what evolved into modern QRA. Rather than accepting an all-or-nothing approach, as embodied by the no-threshold philosophy, such quantitative estimates of carcinogenic strength made it possible—in theory, if not in practice—to balance the health risks of, say, a proposed new chemical plant against the benefits: employment opportunities, the development of useful products, and so on. By comparing risk of lives lost with the potential for dollars gained, many regulators thought they had derived a fundamental basis for decision making on public policy. Critics, including labor unions and environmentalists, assailed the comparisons, arguing that there was more to human life than simple dollar values and pointing out that the people who stood to profit were rarely the ones at risk. (12)

Among federal regulatory agencies, the FDA and EPA took the lead in codifying and advancing the use of QRA during the 1970s and 1980s. In the early 1970s, the FDA proposed guidelines for extrapolating from high-dose exposure of laboratory animals to low-dose exposure in people. The agency also defined a regulatory application for QRA, setting a maximum acceptable lifetime risk as one excess cancer in 100 million people exposed. Later, this guideline was relaxed to a risk of one excess cancer in a million people exposed. This guidance has become a rule of thumb currently applied in a wide range of environmental and occupational settings.

In 1976, six years after EPA's creation, the agency followed FDA's lead and published guidelines for cancer evaluation that incorporated high-dose to low-dose extrapolation. Unlike FDA, however, it did not set a specific level of unacceptable risk, but generally used values in the range of one excess cancer in a million to one excess cancer in 100,000. In the wake of scandals at EPA in the early 1980s, reappointed Administrator William Ruckelshaus reaffirmed and strengthened the use of QRA in EPA decision making. More recently, the method has become a principal tool by which regulators at all levels of government make decisions regarding the acceptability of a variety of risks.

THE QRA PROCESS

To understand how QRA works, consider how the technique would be applied to the evaluation of cancer risks from the air emissions of a proposed municipal solid waste incinerator. Routine household trash is collected from a community, delivered to the incinerator and burned under controlled conditions. Exhaust gases and particulates are passed through various filters and then emitted into the outside air. QRA can be used to evaluate the potential carcinogenic effects of breathing air that contains these emissions. (Although other aspects of incineration merit evaluation, including cancer risks from ash disposal and acute risks from other emissions, we explore here only risks from air emissions of carcinogens.)

Most federal agencies define QRA as a four-stage process. Although each stage is defined scientifically through guidance documents, risk assessors must also make some subjective judgments based on scientific interpretation and personal values. While some argue that these subjective decisions should reflect community values, in practice scientists conducting the QRA usually make these calls, imposing their own best judgment.

Hazard Identification

The goal of this first stage is to identify all situations or substances that can, under any circumstances, pose a risk to human health, and to predict all possible adverse health effects. To assess incineration, a risk assessor compiles a list of materials that waft out of the smokestack: heavy metals such as cadmium, chromium, and nickel, and organics like dioxins and polyaromatic hydrocarbons (PAHs). A list is made of all potential adverse health effects of each of these substances: cancers, nerve disorders, reproductive effects, and immune system dysfunctions.

At a later stage, substances or health effects can be excluded if they turn out to be unimportant. However, omission of compounds or specific health effects from consideration at this stage can undermine the validity of a QRA.

Exposure Assessment

In this stage, risk assessors estimate for each of the substances the amount a typical person is likely to encounter. To do so, the risk assessor must first determine the source and amount of the material at the facility in question—data that can be obtained from direct measurement, from historical records, or from information gathered at other similar locations. For proposed incinerators, for example, emissions data for extant facilities often are applied, after adjustment for the size and technical characteristics of the proposal under evaluation.

Next, with the help of complex computer models, the risk assessor tries to determine how contaminants move through the environment and ultimately come into contact with people. This is usually done using complex material flow models, such as air dispersion models, that have been calibrated with some data specific to the site under evaluation, such as daily wind speeds and directions. For example, a risk assessor might model the settling of particulates onto agricultural fields, the uptake by plants and vegetables, the subsequent "bioconcentration" of these materials through farm animals that eat crops grown on these fields, and their path into human food. Translating this complex process necessarily involves some judgment calls. For example, early risk assessments of many incinerators failed to consider the settling of dioxin onto agricultural fields, even though this pathway can contribute substantially to total risk for the population.

But exposure assessment should not stop with modeling how people might come in contact with a contaminant; the risk assessor must estimate the actual exposure of a typical individual living nearby or working on site. For example, people living near an incinerator may breathe carcinogenic particles from plant emissions. While assessment of hazard based on these air concentrations may be relatively high, the individual's risk from breathing the air may be mediated by the physics of inhalation. The largest particles in the air never reach the lungs because they could be blocked by nasal hairs and tracheal mucosa. And many smaller particles enter and leave the lungs without depositing their toxic load.

Similarly, scientists have demonstrated that even if a suspect toxic is present in the environment, it may not be in a chemical form that will lead to exposure. For example, studies of dioxin-contaminated soil in New Jersey and Times Beach have shown that upon human contact with Times Beach soil, nearly all the dioxin was released, while at the New Jersey site as much as 90 percent was chemically bound in the soil and would not affect people. (13)

While this "bioavailability" is an important consideration in QRA, calculations must usually be made based on models rather than specific measurements for the facilities under consideration.

How does one define a "typical" person's exposure to a contaminant? To be on the safe side, risk assessors generally imagine what would happen under the most dire circumstances possible—the so-called worst-case scenario. The thinking goes that if the risk in this unlikely situation is sufficiently small, then under normal operating conditions the object of study must be safe.

But risk assessors often disagree about what kinds of worst-case scenario should be considered; the criteria are subjective. In California, for example, a company applying to build an incinerator must detail in its proposal the potential consequences of a total failure of the facility's air pollution control equipment: which materials would exit the stack, how long it would take to shut down the incinerator for repair, and what would be the likely effects on public health. (14) Other states, however, dismiss such a technical failure as too unlikely to account for.

Risk assessors also disagree about how bad the worst-case scenario should be. Some apply what they refer to as a realistic worst case—for example, a partial, rather than total, breakdown of equipment. The distinctions are important. The definition of worst case, acknowledged by risk assessors to be subjective, to a large degree determines the severity of the final risk estimate.

Worst-case scenarios may also vary by definition of populations at risk. For example, when considering the risk of drinking milk from cows that graze on grass contaminated with small amounts of particulates deposited from incinerator emissions, some risk assessors use data on the average milk consumption for the entire U.S. population. Others may consider, instead, American adults who consume the greatest proportion of their diet as milk, a decision that on the face of it would seem protective of public health. However, some risk assessors have shown that nursing infants may be at substantially greater risk than even adults who consume substantial amounts of milk products. They argue that worst case scenarios must be modeled with infants, rather than adults. Similarly, using average figures for vegetable consumption may fail to account for the effect of contaminants on strict vegetarians. Some risk assessors argue that worst-case scenarios must take into account these sensitive sub-populations.

Dose–Response Modeling

The third stage of QRA is dose-response modeling—determining how much exposure to a given hazardous substance is harmful to public health. This varies according to the chemical's "cancer potency."

In this stage the use of laboratory animals as models for people comes into play. But particularly for the assessment of carcinogens, the merits of animal testing are controversial. There are solid data for only a few carcinogens comparing their effects on animals with their effects on people. Most often the effects are similar; but in certain cases, differences between human and laboratory animal sensitivity to the same chemical can be substantial (for example, benzene), and use of animal data can lead to a great deal of inaccuracy when it comes to estimating risk. When the sensitivity to a substance varies among species of laboratory animal, the implications are even more confusing. Another problem is that animal data are typically based on tests in which only a few animals are exposed to extremely high concentrations of the chemical under study. This may preclude detection of chemicals that are weakly or moderately carcinogenic, but which may be extremely important if human exposure is widespread.

However, critics of QRA who believe that the methodology is overly protective of public health argue that tests based on laboratory animal responses at high dose exposures can overestimate the potency of some carcinogens and even make carcinogens out of materials that would be harmless at the concentrations to which people are usually subjected.

There also is no consensus on an appropriate mathematical formulation for the dose-response model. Much dispute centers on the no-threshold model, which maintains that exposure to even one molecule of a carcinogen increases the risk of getting cancer. Some investigators assert that the cellular damage caused by very low levels of exposure to certain chemicals can be repaired by other processes within the body. It has also been suggested that certain compounds mediate the effects of other substances so that people must be exposed to chemicals in a prescribed order for cancer to develop.

Conversely, other critics of QRA argue that substances can exacerbate one another's effects. If true, exposure to more than one substance could result in substantially greater cancer risk than the sum of effects from exposure to each compound individually, so-called synergistic activity. The choice of model, again a necessarily subjective decision, will determine the results of QRA.

For dioxins (emitted from incinerators or otherwise), the choice of the correct model has stirred a major controversy. When its potency was estimated mathematically from animal data on the basis of the no-threshold model, dioxin seemed to be the most potent carcinogen known. From that line of thinking came the EPA estimate that exposures of one part per billion increase the risk of cancer. Later on, however, when its potency was estimated from mathematical models of biological processes—taking into account, for instance, that dioxin is carcinogenic only if it occupies certain receptor sites in the body—dioxin appears markedly less dangerous. It was the later estimate that led to Vernon Houk's conclusion that the Times Beach evacuation was unnecessary.

But the science seems to be more complex than even these models suggest: recent, far more sophisticated studies suggest that the effects of dioxin have no threshold and that relaxation of stringent exposure guidelines is not yet warranted. (15) (A thorough review of these results is under way.) At a forum on dioxin toxicity sponsored by the Banbury Center, dioxin researchers debated the validity of the models, with the University of Maryland's Ellen Silbergeld—a prominent environmentalist—arguing that "replacing one stupid model with another" won't necessarily improve the risk estimate. (16)

In view of weak and contradictory human data, many environmentalists advocate retention of stringent standards as the most prudent strategy for protecting public health. They also worry publicly about the precedent of relaxing the standards for what was once touted as the most potent carcinogen known to man. Other scientists argue just as vehemently that existing regulation must reflect new data.

Risk Characterization

In this final stage, the information from the three other stages is combined into an overall estimate of risk. For each chemical listed in the first stage, the assessor calculates, in stage two, a predicted cumulative worst-case exposure for a person over an entire average lifetime, and multiplies this total exposure by the potency estimated in stage three to derive a predicted risk of cancer for an individual exposed to that amount of each chemical over a lifetime. The risk assessor then adds up these risks for each chemical to get the total risk for the incinerator—or whatever activity is being evaluated. For an incinerator, most assessors add the risks of getting cancer from exposure to dioxins to the cancer risk of other organic compounds and of heavy metals to get an overall risk for the incinerator—say, one cancer death per million people exposed.

That approach has its pitfalls. For one thing, it assumes that when risks are combined chemicals never exacerbate, reduce, or otherwise modify the effects of others. However, we know from studies of animals exposed to pesticides two at a time that some chemicals increase the effects of each other, while other chemicals cancel each other out. Since there are little data about these types of interactions, EPA recommends that risk assessors simply add the risks. However, if that approach is wrong—and an assessor fails to account for even one large chemical interaction—results of the QRA would greatly miscalculate the risk.

IMPROVING RISK ASSESSMENT

These differing assumptions throughout all the stages of the risk assessment process make for QRAs that vary dramatically. Some critics have illustrated the subjectivity of QRA by showing it is possible to use government guidelines for QRA and come up with numbers quite different than government risk assessors'. For example, in 1989 the Natural Resources Defense Council reassessed the risk of Alar, a crop treatment applied to apple trees. They calculated a risk of more than two cancers per 10,000 children, a value 25 times higher than EPA's most extreme calculation for the same situation. The differences in the figures result from differing assumptions regarding exposure and potency. According to economist Paul Portnoy, Acting Director of the Center for Risk Management of Resources for the Future, "It sounds like a case where no one is right or wrong . . .

like there are plausible reasons for both sides' estimates." He further underscored the uncertainty fundamental in the whole QRA process, noting that he suspects that "the actual risk is several orders of magnitude below EPA's number . . . but there is also a chance that it could be higher." (17)

Because QRA is so sensitive to subjective assumptions, many environmentalists argue against using the method at all. Others advocate improving rather than abandoning QRA. For example, Bernard Goldstein, a former assistant administrator at EPA who has played a role in refining the use of QRA, acknowledges that the subjectivity of the process has led to some problems about credibility. One way of making the tool useful, he argues, is to codify QRA "under guidelines that are logical, explicit, consistent, and openly derived"—but that could be allowed to evolve as new data and scientific methodologies become available. (18)

However, because scientists may view fundamental issues in risk assessment very differently, this may not be easy. A recent survey of members of the Society of Toxicology about toxicological concepts, assumptions, and interpretations found profound differences of opinions, particularly about the ability to predict human health effects from animal data. (19)

Given the controversy swirling around QRA, a substantial effort has been made to improve its scientific underpinnings. In assessing exposures, a major focus has been on making more exhaustive, detailed measurements—sampling each person's exposure to a toxic substance, for instance, instead of making one estimate for an entire neighborhood. Biomarkers, complex biological compounds resulting from exposure, can indicate how much of a substance makes its way into the body. To improve dose-response modeling, researchers are exploring the complex biochemistry of how the human body responds to specific compounds. Just as such studies have led to debate about the potency of dioxin, the consideration of the ways other chemicals interact with human metabolic processes will result in reevaluation of the potency of other chemicals.

Policymakers, long preoccupied with cancer, are beginning to broaden the scope of risk assessments to consider other health effects from exposure to hazardous substances: birth defects and neurological and immunological problems.

Some analysts are also advocating making the uncertainty and subjectivity of the risk assessment process explicit from the start. (20) The limitations of the data employed in the assessment—say, the statistical parameters for measuring the exposure, or the estimates of carcinogenic potency—should be quantified and made public, argue some researchers. So should the assumptions and judgment calls a risk assessor makes throughout the risk assessment process. Just as critical, risk assessors should divulge the entire range of risk estimates that could have been obtained from the data if different assumptions had been made. And risk assessors should give their determination of the most likely estimate of the risk.

Changes are also being made in how the results of QRA are presented. It is becoming clear that presenting a single risk number—such as one cancer death per million people exposed—implies a degree of certainty unwarranted by the methodology. Instead, investigators are beginning to take into account the uncertainty of the numbers that go into their calculations, and they are presenting the QRA as a range of estimates. For example, in considering the risks of cancer from an incinerator, some assessors now offer separate QRAs for men, women, and children, because each group has different physical characteristics, metabolisms, and exposures, and hence different risks. Furthermore, risk assessors are evaluating eating habits more carefully, allowing for the diets of a range of groups: vegetarians, people who fish and eat their own catches, and people who raise and eat their own vegetables. Each group of diners has a different exposure, but all deserve to be protected from unnecessary risk.

TRANSLATING QRA INTO POLICY

Even more controversial than the figures that come out of QRA is how they are used. Often these estimates determine a government agency's position on the acceptability of an activity, such as siting a solid waste incinerator. In some situations, such as consideration of hazardous waste sites, EPA has set a risk guideline—a lifetime cancer risk of less than one in 100,000. If a QRA estimates a lifetime cancer risk of less than this guideline, the activity generally is considered acceptable.

In other situations, such as the regulation of pesticides, EPA tries to balance the estimated health risk with the potential benefits of the activity, such as increased crop yield and decreased pest damage. Similarly, OSHA is mandated to consider factors other than the potential health effects of workplace exposure. Thus, factors other than the risk may influence the regulatory process.

Public opinion—and the views of the regulated community—also play a role. An EPA report suggests that "the remaining and emerging environmental risks considered most serious by the general public today are different from those considered most serious by the technical professionals charged with reducing environmental risk." (21) For example, people find involuntary risks, such as cancer risk due to pollution from a factory, much more abhorrent than voluntary risks, such as cancer risk from smoking cigarettes. Moreover, people tend to find processes controlled by others—such as the siting of an incinerator—more risky than actions under their own control, such as driving a car. Conversely, risks that are familiar—such as ones a worker encounters daily—seem less risky than unfamiliar ones, such as the introduction of a new process. (22) There are many other considerations that QRAs can't measure. For example, the impact of an environmental hazard on property values may be as important to some homeowners as public health concerns. (23)

There are also significant problems translating public health assessments into terms that have relevance to individuals' lives. (24) William Ruckelshaus explains the problem by telling of an individual who is told that the risk of cancer from a 70-year exposure to a carcinogen in the water supply is between one in 100,000 and one in 10,000,000. The individual responds, "Yeah, but will I get cancer if I drink the water?" (25)

To resolve this confusion, risk managers must not only find more accessible language to express risk but also to address the fundamental concerns of workers and community residents. Risk managers also may need to acknowledge that their views of the scientific underpinnings of their decisions are not necessarily the only "correct" views. The examination of the opinions of members of the Society of Toxicology indicated that differences in opinions among toxicologists were correlated with whether they worked in industry, academia, or government. (26) This study suggests that values may play a role not only in disagreements between scientists and the public but also among scientists.

DEFINING COMMON GROUND

The disparity between the concerns of the public and those of scientists in regulatory agencies does not necessarily invalidate either. Some regulators argue that they both should be taken into account when the time comes to establish policy. Indeed, situations could be ranked on the basis of QRA, but decisions could also require explicit consideration of public priorities and offer alternative policy options. But in far too many instances, the public is left out of the loop. Edicts from regulators routinely infuriate affected citizens who have not taken part in the evaluation. (Declarations of scientific "truths" may also irritate scientists who do not interpret the data the same way as the scientists who took part in the evaluation.)

The sense of disenfranchisement must be addressed head on. Local residents and workers should be considered vital sources of scientific information. Their daily experiences confronting a hazard often provide more information than the best equipment science has to offer. For example, residents and workers often know the history and variability of contamination that scientists cannot easily reconstruct or measure. Residents also understand, far better than any outsiders, the patterns of living that expose people in a community to a hazardous substance. One California official tells of an epidemiological study that was redesigned after residents explained a pattern of exposure that state investigators had mischaracterized. (27) For this guidance alone, local residents should be part of the QRA process from inception. Similarly, workers can provide pieces vital to solving scientific puzzles by providing researchers information about specific work practices and on-the-job exposures.

However, people's values vary from community to community and even within communities. Within specific workplaces, people may have quite different responses to risk. Failure to take account of this variation may invalidate a risk management decision in the eyes of those affected, irrespective of the quality of science employed in the QRA. Residents and workers must be brought into the risk evaluation process early on so that they can articulate their concerns, fears, and desires. Government agencies can no longer comfortably rely on pro-forma public hearings that perpetuate a "decide-announce-defend" strategy for dealing with environmental controversies.

Despite many government scientists and policymakers who bristle at the notion of asking for input from the lay public, some government officials have gradually realized that imposing decisions on a resistant community is often fated to end in environmental gridlock (28)—if not overt conflict. There is grudging acceptance that just as weapons experts at the Pentagon are not the sole determinants of defense policy, neither should scientific experts be the sole determinants of environmental policy. Some agency scientists are growing weary of the risk wars and are looking for ways to forge truces. Although agencies have been slow to translate this realization into practice (29), there have been a few high-profile and many quiet experiments at peacemaking.

One of the most well-known examples took place in Tacoma, Washington, where EPA tried to balance local and scientific concerns when regulating air emissions from a smelter. Thus, QRA's predictions of adverse health effects were discussed with the affected communities, which also explored the economic losses that would result if the facility were closed. Rather than dictating what they believed to be the best policy, EPA officials presented the information to the citizens of Tacoma in an attempt to develop a plan to minimize the effects on health without closing the plant. Although EPA reserved the final authority to regulate the plant, Tacomans explored the potential trade-offs between economic realities and elevated cancer risks. (Ultimately the issue was rendered moot: the plant had to close for reasons not directly related to the pollution question.) (30)

A National Research Council report suggested that the most productive interactions concerning such issues are those that "treat outside parties as fully legitimate participants so that two-way exchange occurs." (31) This means involving "outside" groups early and routinely. Agencies are beginning to use citizen advisory committees to provide input before decisions are made. For example, the state of California has developed advisory committees to provide input into the development of epidemiological studies. Technical assistance grants, which provide funding for community groups at Superfund sites to hire their own experts to review agency documents, such as risk assessments, are now mandated by legislation. One study suggested that not only environmental groups but also agency officials may find such grants useful to resolving conflicts at sites. (32)

However, the science of risk assessment is commanding far more attention than investigation of how to use QRA to forge truces rather than to escalate battles. Uncertainties cannot be removed from the process of making a QRA, but they can be defined clearly. Subjective choices in QRA methodology cannot be eliminated, but they can be made consistent. Societal values, equity, and acceptability of risks cannot be viewed as uniform, but such variations can be acknowledged. True progress in managing risks, however, can be made only if the people affected by the problem are made part of the solution. This will take at least as much dedication as the scientific research.

EDITORS' NOTE

This chapter is an expanded version of a commentary published in the March/April 1992 issue of *The Sciences*, pp. 16–21.

NOTES

1. Joe Thornton, "Risking Democracy," *Greenpeace*. March/April 1991, pp. 14–15.
2. Robert Reinhold, "U.S. Offers to Buy All Homes in Town Tainted by Dioxin," *The New York Times*, February 23, 1983.
3. E. R. Shipp, "U.S. Wins Suit on Clean-up of Dioxin," *The New York Times*, February 1, 1984.
4. Keith Schneider, "U.S. Backing Away from Saying Dioxin Is a Deadly Peril," *The New York Times*. August 15, 1991.
5. Laurie Hays, "Proposals from Federal Agencies to Ease Dioxin Standards Renew Debate on Risk," *Wall Street Journal*, June 27, 1988.
6. *Ibid*.
7. Environmental Protection Agency, "Proposed guidelines for carcinogen risk assessment; request for comments," *Federal Register* 49(227), 46293–46301,1984.
8. William D. Ruckelshaus, cited in *Wall Street Journal*, January 3, 1985.
9. Peter Barton Hutt, "Use of quantitative risk assessment in regulatory

decision-making under federal health and safety statutes," in D. G. Hoel, Richard A. Merrill, Frederica P. Perera (eds.), *Risk Quantitation and Regulatory Policy*, Cold Spring Harbor, NY: Cold Spring Harbor Laboratory, 1985.

10. *Ibid.*

11. Rachel Carson. *Silent Spring*, New York: Fawcett Crest Books, 1962.

12. John D. Graham, Laura C. Green, and Marc J. Roberts, *In Search of Safety: Chemicals and Cancer Risk*, Cambridge, MA: Harvard University Press, 1988.

13. Thomas H. Umbreit, Elizabeth J. Hesse, and Michael A. Gallo, "Bioavailability of dioxin in soil for a 2,4,5-T manufacturing site," *Science* 232, 497-499,1986.

14. Health Risk Assessment Guidelines for Non-Hazardous Waste Incinerators. August, 1990. State of California Air Resources Board.

15. Leslie Roberts, "More pieces in the dioxin puzzle," *Science* 254, 377,1991.

16. Leslie Roberts, "Dioxin risk revisited," *Science* 251, 624-626, 1991.

17. Leslie Roberts, "Alar: The numbers game," *Science* 243, 1430, 1989.

18. Bernard D. Goldstein, "Risk assessment/risk management is a three-step process: In defense of EPA's risk assessment guidelines," *J. Amer. Coll. Toxicology*, 7: 543-549, 1988.

19. Nancy Kraus, Torbjorn Malmfors, and Paul Slovic, "Intuitive toxicology: Expert and lay judgments of chemical risks," *Risk Analysis* 12, 215-231,1992.

20. Adam M. Finkel, *Confronting Uncertainty in Risk Management: A Guide for Decision-Makers*, Washington DC: Center for Risk Management, Resources for the Future, 1990.

21. *Reducing Risk: Setting Priorities and Strategies for Environmental Protection*. Report of The Science Advisory Board: Relative Risk Reduction Strategies Committee to William K. Reilly. September 1990.

22. See, for example, Paul Slovic, Baruch Fischhoff, and Sara Lichtenstein,"Facts and Fears: Understanding Perceived Risk," in R. C. Schwing and W. A. Albers, Jr. (eds.), *Societal Risk Assessment: How Safe Is Safe Enough?*, New York: Plenum Press, 1980.

23. See, for example, Gary H. McClelland, William D. Schulze, and Brian Hurd, "The effect of risk beliefs on property values: A case study of hazardous waste sites," *Risk Analysis* 10, 485-497, 1990.

24. See, for example, Harold Isadore Sharlin, "EDB: A case study in the communication of health risk," report for Office of Policy Analysis, USEPA, 1985.

25. William Ruckelshaus, "Risk in a Free Society," *Risk Analysis* 4, 157-162, 1984.

26. Nancy Kraus, Torbjorn Malmfors, and Paul Slovic, "Intuitive toxicology: Expert and lay judgments of chemical risks," *Risk Analysis* 12, 215-231,1992.

27. Billie Jo Hance, Caron Chess, Peter M. Sandman, *Improving Dialogue with Communities: A Manual for Government*, New Jersey Department of Environmental Protection, Trenton, 1988.

28. The term "environmental gridlock" was coined by Christopher Daggett, former Commissioner of the New Jersey Department of Environmental Protection, see Carl E. Van Horne, *Breaking the Environmental Gridlock*, New Brunswick, NJ: The Eagleton Institute of Politics, Rutgers University, 1988.

29. Caron Chess and Kandice Salomone, "Rhetoric and reality: Risk communication in government agencies," *Journal of Environmental Education* 23, 28–33, 1992.

30. Sheldon Krimsky and Alonzo Plough, *Environmental Hazards: Communicating Risks as a Social Process*. Dover, MA: Auburn House, 1988.

31. National Research Council, *Improving Risk Communication*, Washington, DC: National Academy Press, 1989.

32. Caron Chess, Stephen K. Long, and Peter M. Sandman, *Making Technical Assistance Grants Work*, New Brunswick, NJ: Environmental Communication Research Program, 1990.

17

Alternatives to Risk Assessment

THE EXAMPLE OF DIOXIN

MARY H. O'BRIEN

The current reliance on risk assessment by U.S. public health officials and environmental regulators is extreme and worth examining. The U.S. public interest environmental community is increasingly rejecting the management of toxic materials on the basis of Quantitative Risk Assessment.

Many of us who have spent years wrestling with the scientific inadequacies and political manipulation of risk assessments are now urging citizens to reject the government and industry process of using risk assessment to determine safe or acceptable levels of toxics. We reject the process because it is scientifically indefensible, ethically repugnant, and practically inefficient.

While it is standard on a theoretical basis to separate risk assessment (a supposedly scientific process) from risk management (the public policy and decision-making process), such a separation would be real only if risk assessments started with the world: with assessment of toxics or other hazardous exposure, cumulative impacts, effects on wildlife development, and so forth.

The vast majority of toxic risk assessments, however, start with risk management. Some corporation wants a permit to dump toxins into the environment or a registration for a pesticide formulation, or there is a spill or leakage at a site, or there are complaints to the health department by parents living next to a municipal incinerator. So the risk assessment is born in a political context of existing toxic discharges and existing industrial, agricultural, municipal, and domestic practices.

My experience with such risk assessments is that they are not, in reality, separate from risk management. Economic power and political influence prevent the separation. So when I speak in this presentation of "risk assessment," I am speaking of the unified risk assessment/risk management process.

The entire risk assessment process is based on an assimilative capacity approach to toxics. It asks, "How much of each toxic chemical can humans and Earth take without buckling at the knees?"

The alternative process, the one I join others in urging, is based on the precautionary principle. It asks different questions. It asks, "What is the *least* tinkering we can do with earth's life systems? What opportunities do we have to eliminate discharges of toxic chemicals into the environment?"

THE EXAMPLE OF DIOXIN

For examples in this presentation, I am going to use dioxin, or, more precisely, the most toxic dioxin, called 2,3,7,8-TCDD. The risk assessment approach to the discharge and presence of dioxin is a classic example of why toxics risk assessment is unacceptable. Risk assessment of pesticides, hazardous waste incinerators, nuclear waste, and others would serve as equally compelling examples.

Further, I am not going to examine the scientific indefensibility of toxics risk assessment except to say that it ultimately arises from the attempt to show something that is scientifically unknowable in the real world: namely, a "safe" or "acceptable" level of toxics exposure.

The scientific failure of toxics risk assessment involves in part inaccurate and inadequate data; it also involves indefensible narrowness of the scientific questions that are asked. A risk assessment of 2,3,7,8-TCDD, for instance, does not address the question of the effects of 2,3,7,8-TCDD in real-world concert with other 2,3,7,8-substituted dioxins and furans, coplanar PCBs, endocrine disrupters, cancer initiators, and immune suppressants. Real-world cumulative impacts are not now and will not be addressed by real-world risk assessments.

Instead of focusing on such scientific indefensibilities of toxics risk assessment, I wish to briefly discuss the ethical failure of and practical alternatives to toxics risk assessment, using dioxin examples.

TOXICS RISK ASSESSMENT: ETHICALLY REPUGNANT

The management of toxic discharges by risk assessment is unethical on the basis of a number of values.

a. *The decision to expose people to toxic chemicals without their consent and, in the case of many informed people, against their will, is an attack on the bodily integrity of those people.* As the California Supreme Court noted in a case regarding a patent's right to refuse medical treatment, this ethical principle is a part of our Constitution: "The constitutional right of privacy guarantees to the individual the freedom to choose to reject, or refuse to consent to, intrusions of his [or her] bodily integrity." (1)

If people have a constitutional right to refuse a treatment that is supposedly medically beneficial, they surely have the constitutional right to refuse invasion of their bodies by toxic chemicals.

b. *The decision to manage toxic chemicals on the basis of some hypothetical "average" person facilitates the victimization of particular subpopulations in our society.* The U.S. Environmental Protection Agency (EPA) estimates that many Native Americans, Asian Americans, and poor people living along the Columbia River consume large quantities of fish (100 grams or 3.5 ounces of fish a day) contaminated with approximately 6.5 ppt 2,3,7,8-TCDD equivalents (considering only 2,3,7,8-dioxin and 2,3,7,8-furan). The EPA estimates that such people ingest nine picograms/kilogram body weight of 2,3,7,8-TCDD per day (2), which is nine times over the EPA's so-called Acceptable Daily Intake for 2,3,7,8-TCDD (one pg/kg/day) for reproductive and developmental effects and which the EPA associates with 1,500 cancers per million Native Americans. (3)

Children, chemically sensitive people, embryos, infants, immune-compromised people—all these people are routinely put at potentially high risk by decisions based on risk assessments designed around average adults.

c. *The decision to issue permits for toxic pollution that will cause some people to contract cancer is premeditated murder.* Suppose the practice of risk assessment advanced to the point where the EPA could not only *determine* that, say, *exactly* one person out of every 1,000 Native Americans will get cancer from eating dioxin-contaminated fish, but the EPA knew the names of those individuals. Then, when the agency published the permit for pulp mills to discharge dioxin into the river upstream of those people, it published the names of those who would get cancer. Those people would be able to go to court to stop the U.S. government from depriving them of their constitutional right to life.

But when government agencies permit the intentional, unnecessary discharge of toxics that they estimate will cause some individuals to die of cancer, they are getting away with precisely this premeditated murder *because it is anonymous.* (4)

As Alabama State Attorney General Jimmy Evans stated recently:

> The risk assessment technologies . . . say people will die as a result of dioxin emissions. Then they say that is perfectly acceptable.

"That is really, really—to me—outrageous and bizarre. It reflects an elitism, a plantation mentality. I think it amounts to a confession. It is very, very simple to me. It is a moral issue. They have said people will die, and we are supposed to accept that. As attorney general of this state, I can't. (5)

d. *The decision to permit discharges of toxic chemicals via quantitative risk assessment is antidemocratic*, because understanding and critiquing the formulas, assumptions, and models involved are technically beyond the time and financial resources of the average citizen. The numbers look objective; the scientific terms look impressive; but the average person does not know that the assumptions are loaded with values, and that critical gaps in knowledge render the models and formulas almost infinitely vulnerable to political manipulation.

Those with the money to hire specialists who will argue numbers from particular models, assumptions, and data will control the outcome of the risk assessment.

The decision to invade people's bodies with toxic chemicals must be considered in terms which the proposed victims can judge. If people cannot understand the terms of the debate about the fate of their bodies, they are not being included in decision making regarding their own bodies.

e. *The decision to permit discharges of human-made chemicals via risk assessment assumes that chemicals are innocent until proven guilty*. If nothing is known about the neurotoxicity or immunotoxicity of a particular chemical, for instance, that neurotoxocity or immunotoxicity is treated as zero, as nonexistent, in the risk assessment model.

There are those of us who believe that living beings, not synthetic chemicals, should be given the benefit of the doubt. The burden of proof should be on the polluter to show that the pollution of air or water or people's food supply will not harm humans and other living beings. The burden of proof should not be on the potential victims to come up with a body count before they can halt the stream of a toxin into their water, food chain, and bodies.

Adherence to the precautionary principle would eliminate and prevent emissions if there is reason to believe damage or harmful effects are likely to be caused. Use of this precautionary principle is a more ethical approach to life than the assimilative capacity approach whereby toxic pollution is permitted up to some amount estimated to be tolerable to living beings and ecosystems. (6)

f. *These decisions to invade people's bodies with toxic chemicals, these decisions to cause high cancer rates among particular subpopulations, are ethically repugnant* because the invasions are not being done as a last resort.

Do we need to use chlorine compounds to bleach pulp white or cream-colored? No.

Do any functions of paper depend on the paper being bleached with chlorine compounds? No.

Do we need, therefore, to discharge dioxin from pulp mills at a rate that is predicted to cause some people to die of cancer? No.

Once again referring to a legal decision regarding forcible medication with antipsychotic drugs, the 10th Circuit Court wrote:

> Our constitutional jurisprudence long has held that where a state interest conflicts with fundamental personal liberties, the means by which that interest is promoted must be carefully selected so as to result in the minimum possible infringement of protected rights . . . Thus, *less restrictive alternatives . . . should be ruled out* before resorting to antipsychotic drugs. (7, emphasis added)

By the same principle, alternatives to the discharge of organochlorine compounds by industry should be ruled out before the state permits chemical trespass of people with these toxic compounds. If this process were followed in the United States, every one of the 104 chlorine-using pulp mills would be on a time line to phase out its use of chlorine compounds. Economically feasible alternatives to the use of chlorine compounds exist.

RISK ASSESSMENT: PRACTICALLY INEFFICIENT

Let us now look at the practicality of risk assessment as a tool for protecting public health and the environment. If we want to find less toxic ways of managing and using our resources, which of the following two questions is going to get us there faster?

a. The risk assessment question, which is, "How much of each toxic chemical can living systems and ecosystems assimilate?"
b. The clean production question, which is, "What are the options for discharging the least amount of toxins?"

I would contend that asking and answering the clean production question is more likely to lead to protection of public and environmental health, but the risk assessment question is the one almost exclusively asked by the Environmental Protection Agency, state and local health departments, other government agencies, and industry.

Again, the case of dioxin is a compelling one. A moderate-sized, modern, chlorine dioxide-using pulp mill will discharge 16.5 tons of organochlorine chemicals every day into the river or lake from which it takes water. (The figure of 16.5 tons is derived assuming 1,000 tons of pulp production a day; water discharges of 1.5 kg AOX/ton pulp with AOX being a measure of the chlorine bound to organochlorine molecules; and multiplication of the 1.5 kg by a factor of 10 to account for the nonchlorine atoms in the organochlorine molecules.)

On the other hand, the production of white and cream-colored chlorine-free paper in Sweden, Germany, Austria, and the United States allows mills to eliminate completely their discharge of organochlorine chemicals.

We can either continue to ask risk assessment questions about the precise toxicity of the 2,3,7,8-TCDD discharged by chlorine-using pulp mills (along with studying the toxicity of each of the approximately 2,000 other organochlorine compounds discharged by such mills into water, air, landfilled waste, and paper products, as well as the toxicity of mixtures of these organochlorine compounds and their metabolites), or we can ask what options we have for alternatives to organochlorine discharges.

When asking questions about "clean production" of pulp and paper or of any other product, we should ask about the options we have that avoid or eliminate hazardous waste and hazardous products, and use a minimal amount of raw materials, water, and energy.

A commitment to actively search for, analyze, and publicly consider clean production methodology in our country would involve the systematic, ongoing use of environmental audits of all stages of production: resource extraction, manufacturing, transport, product use, and waste disposal. The environmental audits would examine all reasonable, environmentally sound alternative practices.

The regulations governing the preparation of environmental impact statements under the National Environmental Policy Act (NEPA) offer a beginning model as to the types of questions to ask and the public participation processes to use. The entire process of environmental audits would need to be enforceable by citizens in court, as are environmental impact statements under NEPA. A bottom-line measure of the legal adequacy of an environmental audit might be its explicit, public consideration of relevant, clean production technology that is being successfully employed somewhere in the world. A company undertaking an environmental audit might learn of such clean production technology through its own search or through informed public comment on draft environmental audits.

Questions about the availability of alternative technologies can be answered more scientifically than can questions about how much of a toxic chemical the earth can assimilate without damage. While we cannot know what all the cumulative impacts of dioxin, dioxin-like compounds, and other organochlorine and toxic compounds are, we can know whether pulp mills are able to produce white paper without using any chlorine compounds.

CONCLUSION

From the point of view of industry and the government, the Quantitative Risk Assessment approach to management of toxic chemicals is a good deal.

Industry can use risk assessment to greatly delay requirements to change their practices until each toxic chemical in question can be shown to be causing significant damage. Industry can *actively* delay changes in their practices by throwing money, experts, time, and sophisticated-looking formulas and factors into increasingly complex risk assessment models.

Risk assessment is a good deal for government, as well. Public agencies can use risk assessment to provide hiding places for bureaucrats. If there is a risk assessment a bureaucrat can hide behind while permitting business-as-usual and toxic discharges, that bureaucrat can try to remain aloof from the ethics involved in shuttling toxic chemicals into people's and other animals' bodies without their consent, against their will, and in the absence of a determination of necessity.

Risk assessment is not such a good deal for the real and potential victims of exposure to toxic chemicals, however. These victims will find that risk assessment does not offer avenues of relief from the chemicals. Risk assessment is generally used to estimate how much of each toxic chemical the victims can take.

The assimilative capacity approach to toxic chemicals fails to ask the question that is scientifically answerable, ethically precautionary, and practical. This question is, "What are our options for eliminating discharge of these toxic and potentially toxic chemicals?"

NOTES

1. *Bartling v. Supreme Court*, 163 Cal. App. 3d 186, 209 Cal. Rptr. 220 (1984), cited in Sher, Vic. 1987. Poisons and consent: The constitutional right to be free from exposures to pesticides in state eradication projects. *Journal of Pesticide Reform* 7(2):16–19.

2. McCormack, Craig (U.S. Environmental Protection Agency, Office of Policy Planning and Evaluation), and David Cleverly (U.S. Environmental Protection Agency, Office of Research and Development). April 23, 1990. Analysis of the potential populations at risk from the consumption of freshwater fish caught near paper mills. Draft.

3. *Ibid.*

4. See Merrell, Paul, and Carol Van Strum. 1990. Negligible risk: Premeditated murder? *Journal of Pesticide Reform* 10(1):20-22.

5. Kipp, Stephen. "Evans will ask EPA for tougher dioxin standards." *Birmingham Post Herald* (Alabama), October 28, 1991.

6. See Greenpeace International. July 1990. Protection of the environment through the "precautionary action" approach. Unpublished paper.

7. *Bee v. Greaves*, 744 F.2d 1387, 1396 l0th Cir. (1984), cert denied, 105 S.Ct. 1187 (1986).

18

Playing Industrial Hygiene to Win

EILEEN SENN TARLAU

ITEM—Numerous workers become sensitized to toluene diisocyanate (TDI) at a plant manufacturing foam automobile seats. Personal air sampling conducted by corporate industrial hygienists consistently shows levels of TDI to be within all legal and recommended standards.

ITEM—Dozens of workers in a new office building suffer eye, nose and throat irritation. Vendors who supplied the furniture, partitions and carpeting all reveal that they used formaldehyde in their products. Air samples collected by an indoor air quality consultant, however, show formaldehyde levels in compliance with the new Occupational Safety and Health Administration (OSHA) standard.

ITEM—Workers at a construction site become ill, and bulk samples of the soil reveal high levels of phenols and many other chemicals. Industrial hygienists from OSHA collect personal air samples but can find no violations of OSHA's Permissible Exposure Limits (PELs).

ITEM—Machinery noise levels at a carburetor rebuilding factory create stressful working conditions and damage workers' hearing. An OSHA industrial hygienist measures noise levels high enough for management to require workers to wear ear plugs but not high enough to require management to quiet the machinery.

Industrial hygienists can be extremely helpful to workers by identifying, evaluating and recommending controls for health hazards on the job. Experience has shown, however, that industrial hygienists' personal exposure monitoring and exposure limits have been used to "scientifically prove" that working conditions are "safe" when they were not, even when workers were getting sick.

The idea of measuring how much of a chemical, radiation, noise, or other hazard a worker is exposed to and comparing this to a level that has been proven to be safe is not a bad idea. But there are many problems with the way this has worked in practice.

Problem 1. Most chemicals and other hazards have not had adequate long-term tests conducted to determine whether they can cause cancer, damage brain and nervous system function, lung function, immune and hormone systems function, reproductive system function or many other vital bodily functions.

Problem 2. Legal or recommended limits are often thousands of times too high to protect health. Many limits were set using unscientific, irregular procedures, and corporations have strongly influenced or even dictated the outcome.

Problem 3. OSHA now claims to have "updated" its PELs. However, it did not use the latest scientific methods or information to do so, and the new PELs are only slightly more protective than the old ones.

Problem 4. Most measurements of worker exposure have been incomplete and inaccurate.

Problem 5. Industrial hygienists' preoccupation with the ritual of air sampling has given the impression that this is the best way to approach an occupational health problem and detracted from other, more useful activities that industrial hygienists can perform such as evaluating controls.

Because of these problems, industrial hygiene monitoring usually paints a rosy picture of conditions in the workplace by giving it a "Clean Bill of Health." Even though this is often false, it is hard to dispute because it appears to be quantitative and scientific.

BEWARE THE "CLEAN BILL OF HEALTH"

Workers and their unions should be aware of the potential use of industrial hygiene monitoring against them. Certain wording in an industrial hygiene report is often a tip-off that a "Clean Bill of Health" is in the making. Two examples are:

a. "No violations of OSHA standards were found." This statement seems designed to give peace of mind to those unfamiliar with the inadequate nature of most OSHA standards. It also implies that compliance with every standard was checked—a highly unlikely possibility.

b. "The concentrations of *x*, *y* and *z* were found to be well below any legal or recommended standard." This statement is easily misinterpreted to mean that no adverse health effects are expected at the levels found. The statement also implies that exposure to these particular chemicals is the only health concern at the workplace.

DEALING WITH THE INDUSTRIAL HYGIENIST

Who the industrial hygienist works for greatly influences how willing and able he or she will be to listen to and assist workers and unions. The majority of industrial hygienists work for corporations either directly or as hired consultants. Many want to do a good job but are hampered by their employer. Some are used as little more than technicians to measure air contaminants. Their professional judgment and experience rarely are called upon to make recommendations for improvement in the workplace. In some cases, a union may be powerful enough to have a corporate industrial hygienist do a proper evaluation and make useful recommendations. Their tendency, however, will be not to talk to the union unless it demands involvement. Contract language that mandates union involvement is particularly helpful.

Obviously, it is ideal if the industrial hygienist works for the union directly or is hired as a union consultant. The courts have ruled that, under the National Labor Relations Act, unions have the right to have their own health and safety expert make workplace surveys and recommend improvements. Ongoing relations between the union health and safety committee and the industrial hygienist are necessary to evaluate the workplace, make educated tactical choices, and educate members. These judgments require union committees to think about priority setting, both to get victories and to confront the most serious hazards.

Because of their dependence on air sampling, poor quality exposure limits, and other problems discussed later, government industrial hygienists working for OSHA really have their hands tied at this time unless the problem is one of the few where OSHA has a useful standard (see the accompanying list): in most cases involving chemical exposure, they will be unable to issue citations. However, they should be asked to write a report on the hazards they observed but could not cite.

Hygienists working for the National Institute for Occupational Safety and Health (NIOSH) or state governments usually have the freedom to write reports containing strong recommendations and should be asked to do so. It will be totally up to the union to negotiate with management to implement the recommendations, however, since these agencies have no enforcement power.

MISINTERPRETATION OF OSHA HEALTH INSPECTIONS

OSHA health inspections that do not result in citations are especially subject to misinterpretation as a "Clean Bill of Health." When OSHA cannot find violations of its exposure limits for chemicals, radiation, or noise, it is not permitted to issue a citation ordering the employer to correct the problems. The employer is then likely to claim that there are no health hazards in the workplace even when workers are experiencing health problems. Even if a citation is issued, people mistakenly may believe that the citation lists everything wrong in the workplace.

In reality, there may be many hazards in the workplace that OSHA did not observe or OSHA's industrial hygiene sampling did not pick up. Some of the reasons for this could be:

- The process used by OSHA is backwards. Instead of focusing on health complaints, the OSHA industrial hygienist looks at exposure numbers. For workers, this adds insult to injury.
- OSHA sampling only evaluates how much of a chemical enters the body by being breathed in. But many chemicals also are absorbed through the skin or accidentally eaten due to contaminated lunchrooms.
- OSHA most often measures only for one or two chemicals, rather than for all those to which workers are exposed.
- OSHA does not consider the combined effects of chemicals and other hazards such as heat and noise.
- OSHA usually collects air samples for only one day and may easily miss peak exposures occurring during maintenance, leaks, and emergencies, especially if they occur after first shift.
- Variations due to season or production schedules also may be missed.
- The employer may slow down production or change the chemicals or procedures used to try to hide bad conditions.

CHOOSING WHEN TO FILE OSHA HEALTH COMPLAINTS

Before filing an OSHA health complaint you should be fairly sure that there are violations of OSHA health standards in your workplace. Otherwise, there is a good chance that the problems you are concerned about will not be cited by OSHA, giving the impression that everything is fine.

There are several chemical-related OSHA health standards that apply to all workplaces and that are widely violated. These include 1910.1200/ Hazard Communication, 1910.20/ Access to Medical and Monitoring Data, and 1904/ Recording and Reporting Occupational Injuries and Illnesses.

Whenever you file a complaint, these should be mentioned if you have reason to believe the employer is out of compliance.

The OSHA noise standard also is widely violated and may be something you will have success with in an OSHA complaint.

When the problem you face is reducing exposure to chemicals, however, you come right up against the problem of PELs that are too high to protect workers from health effects. There are only a few chemicals for which OSHA has even a small chance of finding overexposures when it collects air samples. These are listed in Appendix A. There are no solvents on this list; OSHA has virtually no chance of finding solvent exposures above the solvent PELs, which are hundreds of times too high.

There is another group of chemicals that has comprehensive OSHA standards (see Appendix B). This means that the standard specifies a whole range of requirements for air monitoring, personal protective equipment, engineering controls and work practices, medical monitoring, and employee education and training. These requirements go into effect only when the PEL or half of the PEL (the "action level") is exceeded. The PELs in comprehensive standards are, on the whole, much more protective than PELs in the OSHA Z tables. So if you have a problem with one of the chemicals with a comprehensive standard, an OSHA inspection may be helpful.

If employees are required to wear respirators or eye or face protection, check to see if the provisions of 1910.133 and 1910.134 are being followed. Improper gloves or chemical protective clothing may be cited under 1910.132.

Some helpful provisions on the other chemical-related standards are given in Appendix C. These standards should be consulted to see if the employer is violating any provisions.

WRITING YOUR OWN RULES FOR THE CORPORATE NUMBERS GAME

Because corporate industrial hygiene monitoring is most often used against workers, it should be approached with extreme caution. The union must do its best to assure that monitoring will benefit workers and not undermine their demands to clean up the workplace. The union should take full advantage of its legal right to bargain over health and safety, which is provided by the National Labor Relations Act.

Certainly there is little need for corporate exposure monitoring with carcinogens, mutagens, or teratogens since the position of labor and many health professionals is that there is no "safe" level for these. Instead, insist that the emphasis be put on reducing exposures to such chemicals to the lowest possible level.

In deciding whether sampling may be justified, the union should determine whether the following circumstances exist:

- The employer plans to install engineering controls and wishes to compare exposures before and after controls. This is a legitimate purpose and usually can be accomplished using area samples or direct reading instruments, rather than personal samples worn by workers.
- Personal samples worn by workers to evaluate exposures will be used for one or more of the following:
  - To decide where controls are needed
  - To create a record of current exposures
  - To compile exposure data for future use in epidemiology studies or other research.

The union will want to seriously consider whether it will support sampling for the last two purposes. These are in the realm of research and may not ever actually benefit workers. Sampling to decide where controls are needed is much more legitimate and likely to result in immediate improvement of working conditions.

In all cases, it is important to assure that the following conditions are met:

- All of the chemical exposures in question are identified and will be sampled.
- The union will observe monitoring and can assure that sampling will take place during the worst exposures.
- The employer agrees that exposure data will be given in full to the union and applicable parts to all workers who have been sampled or have similar exposures, with a copy to these workers' permanent personnel files.
- The employer commits in writing to make the data available to public health professionals acceptable to the union who wish to use it for epidemiology or other research purposes.
- Exposure limits that are protective of health will be used to evaluate the samples. Unions and workers will get more protection if they use health guidelines based on an excellent EPA database rather than OSHA or other exposure limits. A list of these EPA-based guidelines appears in a booklet entitled "Health-Based Exposure Limits and Lowest National Exposure Limits" circulated for discussion by the Occupational Health Section of the American Public Health Association.

If the above outcomes and conditions cannot be assured, the union should devise a strategy to change the situation. The following may assist union attempts to negotiate an acceptable agreement with management:

- Advising management that the union will label all sampling results as fraudulent and will give no credibility to them.
- Advising management that the union will boycott the sampling by advising workers not to wear sampling devices.
- Other actions designed to motivate management to renegotiate.

Obviously, the union must be confident of its ability to protect workers from discipline if some of these actions are actually carried out.

The union should also bear in mind that some workers may decide on their own to contaminate, destroy, or otherwise sabotage samples. This is likely to happen especially where workers have had past experiences where sampling resulted in no improvements in the workplace. Such sabotage arises out of curiosity about how bad things have to be before management acts and frustration that poor working conditions are not recognized as needing improvement. Such workers may need union guidance and protection.

USEFUL ACTIONS FOR THE INDUSTRIAL HYGIENIST

Have the industrial hygienist spend time looking at the work as it is carried out, evaluating controls, and talking with workers to find out when they experience irritation, smell odors, see dust; in this way, the worst exposure periods can be pinpointed. Appendix D is a checklist for evaluating the potential for chemical exposure by routes of entry.

The industrial hygienist should note all potential exposures to chemicals, noise, radiation, heat, cold, vibration, repetitive trauma, bacteria, viruses, and other biological hazards. In order to document that exposure is taking place, it is helpful to have photographs or videotape showing skin contact or visible contaminants.

The hygienist can collect bulk or wipe samples for analysis when there is no other way (such as a Material Safety Data Sheet) to find out if a substance is present. These types of samples are useful if you can use a "yes" or "no" answer to act. Bulk samples often are taken of materials suspected to be asbestos. Wipe samples of surfaces such as lunch tables may be useful where contamination by chemicals that are toxic when ingested, such as lead, is suspected.

The hygienist should observe and investigate all of the following and note whether they are effective:

- Labeling, placarding, and communication of hazards to employees.
- Level of health and safety expertise and staffing among management.

- Worker training and education in their job duties as well as health and safety practices and controls.
- Confined space entry procedures and practices.

Have the industrial hygienist and/or the union interview workers in private concerning any health problems, symptoms, or complaints they may have. Results of such interviews must be confidential. Any reports should reveal only the employee's job title, not his or her name.

DEMAND A GOOD REPORT

Unions and workers should insist that the industrial hygienist write a report and that they get a copy of it. Unions and workers are entitled to all reports under OSHA regulation 1910.20. The report should state clearly the limitations of their investigations. For example such a statement could read as follows:

> The findings and exposure data reported here are accurate only for the workplace conditions existing at the time of the evaluation. Not all potential occupational health problems or exposures were evaluated. Inadequate information is available concerning what exposures to most hazards are safe for workers.

Unions and workers also should insist that the report fully describe the conditions that the industrial hygienist observed in the workplace. This in itself will go a long way toward combatting management's claims that all is well and good.

Most importantly, the report should list all of the possible improvements that could be made in the workplace. The union can decide which ones it wants the most. Changes in the work environment should be stressed over personal protective equipment. (Appendix C is a listing of common industrial hygiene recommendations.)

If NIOSH has published control technology recommendations for specific operations, these should be highlighted in the report. Some industries and operations for which control technology reports have been published are identified in Appendix E.

Nontechnical recommendations should be included, such as increased health and safety staffing and training among management personnel and better worker training. If the union thinks it will be beneficial, a recommendation for a joint union-management health and safety committee can be made.

DOING YOUR OWN INDUSTRIAL HYGIENE

As we have seen, industrial hygiene often has misplaced an emphasis on technical sampling methods rather than good investigations into health hazards and innovative problem solving. The more we realize that the best industrial hygiene is problem solving, the more accessible it is for union health and safety representatives and activists. Here is a review of the traditional concepts of industrial hygiene and how a labor union might approach them.

Hazard Recognition

The best way to recognize potential health hazards is to know work operations and the associated hazards in the particular workplace. Inspections that utilize visual observations of the workplace for those hazards and talking to the "experts" (workers) are the best ways to learn the hazards.

Hazard Evaluation

This currently is done by sampling. Instead, evaluation can be performed by observing visible contaminants, noting odors, predicting exposure from situations such as open containers or spraying operations, or from interviewing workers about their health symptoms and complaints.

Controlling Hazards

This requires problem-solving techniques, coupled with solutions. Some solutions, such as designing a ventilation system, require a certain amount of technical expertise, while others require job expertise to recognize ways to change work organization and work practices. All solutions need to be trial-tested and modified until they are workable.

ACKNOWLEDGMENTS

Peter Dooley contributed wonderful encouragement, many helpful reviews, and clarifying discussions. Thoughtful reviews of earlier drafts were done by Buck Cameron, Debbie Nagin, Thurman Wenzl, Richard Youngstrom, and Grace Ziem. Barry Castleman and Grace Ziem conducted the original exposé of exposure limits that laid the foundation for the ideas in this chapter.

APPENDIX A

Chemicals for Which OSHA Has Better Than a 1 in 10 Chance of Finding Exposures Greater Than the PEL

OSHA Personal TWA Samples Collected—January 1985—December 1989

	SUBSTANCE	PERCENT OVEREXPOSURES
1.	Silver metal and soluble compounds	37.2
2.	Coke oven emissions	31.1
3.	Respirable silica	28.5
4.	Lead—inorganic	27.5
5.	Wood dust	22.9
6.	Carbon monoxide	21.4
7.	Chromic acid and chromates	20.4
8.	Total dust	15.6
9.	Beryllium and compounds	14.5
10.	Coal tar pitch volatiles	13.9
11.	Copper dusts and mist	12.9
12.	Mercury	12.4
13.	Welding fume, total particulate	11.2
14.	Ethylene oxide	10.5
15.	Arsenic and organic compounds	10.1

APPENDIX B

Useful OSHA Health Standards*

CHEMICALS WITH COMPREHENSIVE STANDARDS

1910.1001	Asbestos
1910.1017	Vinyl chloride
1910.1018	Inorganic arsenic
1910.1025	Lead
1910.1028	Benzene
1910.1029	Coke oven emissions
1910.1043	Cotton dust
1910.1044	1,2-Dibromo-3-chloropropane
1910.1045	Acrylonitrile
1910.1047	Ethylene oxide
1910.1048	Formaldehyde

CHEMICAL-RELATED STANDARDS

1910.94	Ventilation
1910.107	Spray finishing using flammable and combustible materials
1910.108	Dip tanks containing flammable or combustible liquids
1910.120	Hazardous waste operations and emergency response
1910.252	Welding, cutting, and brazing
1910.1200	Hazard communication
1910.1450	Occupational exposure to hazardous chemicals in labs

PERSONAL PROTECTIVE EQUIPMENT STANDARDS

1910.132	General requirements
1910.133	Eye and face protection
1910.134	Respiratory protection

GENERAL

1910.141	Sanitation
1910.151	Medical services and first aid

NONCHEMICAL HAZARDS

1910.95	Occupational noise exposure
1910.96	Ionizing radiation
1910.97	Nonionizing radiation

RECORD KEEPING

1904	Recording and reporting occupational injuries and illnesses
1910.20	Access to medical and monitoring data

*Up to 1991.

APPENDIX C

Standard Industrial Hygiene Recommendations

1. [ ] Provide employees immediately with short-term protection against the toxic material(s) by providing the following properly selected, fitted and maintained personal protective and emergency equipment:
   - a [ ] Respirators—see OSHA Reg. 1910.134
   - b [ ] Gloves—see OSHA Reg. 1910.132
   - c [ ] Chemical-protective clothing—see OSHA Reg. 1910.132
   - d [ ] Chemical splash goggles—see OSHA Reg. 133
   - e [ ] Chemical-protective boots—see OSHA Reg. 1910.132
   - f [ ] Eye-wash fountain—see OSHA Reg. 1910.151
   - g [ ] Body-wash shower—see OSHA Reg. 1910.151
   - h [ ] Spill clean-up kits
   - i [ ] Other________________________________
2. [ ] Permanently reduce exposure to the toxic material(s) by instituting the following engineering and work practice controls:
   - a [ ] Substitute a less toxic material
   - b [ ] Isolate or enclose the operation
   - c [ ] Install local exhaust ventilation
   - d [ ] Provide dilution ventilation
   - e [ ] Eliminate skin contact
   - f [ ] Other________________________________
3. [ ] The following elements of an effective respirator program should be instituted—see OSHA
   Reg. 1910.134(b):
   - a [ ] Written standard operating procedures governing the selection and use of respirators
   - b [ ] Proper selection on the basis of the hazards to which workers are exposed
   - c [ ] Training of users in proper use and limitations of respirators
   - d [ ] Assignment of respirators to individual workers for their exclusive use
   - e [ ] Regular cleaning and disinfecting after each day's use
   - f [ ] Storage in a convenient, clean, and sanitary location
   - g [ ] Inspection during cleaning and replacement of worn or deteriorated parts

h [ ] Surveillance of work area conditions and degree of employee exposure or stress

i [ ] Regular evaluation to determine the continued effectiveness of the program

j [ ] Annual review of respirator user's medical status for physical ability to perform the work and use the equipment

4. [ ] Improve housekeeping as follows:

a [ ] Keep floors and work surfaces free of visible contaminants—see OSHA Reg. 1910.22(a)

b [ ] Eliminate dry sweeping

c [ ] Eliminate the use of compressed air for cleaning—see OSHA Reg. 190.242(b)

d [ ] Use a HEPA vacuum for cleaning

e [ ] Use wet wiping or mopping for cleaning

f [ ] Clean up spills promptly using properly trained and equipped employees—see OSHA Reg. 1910.120

g [ ] Eliminate vermin—see OSHA 1910.141(a)(5)

h [ ] Other______________________________

5. [ ] Improve lunchroom, locker and lavatory facilities as follows:

a [ ] Prohibit eating in work areas—see OSHA Reg. 1910.141(g)

b [ ] Require vacuuming of clothing before entering lunchroom

c [ ] Keep lunchroom clean—see OSHA Reg. 1910.141(g)

d [ ] HEPA vacuum lunchroom daily

e [ ] Provide separate locker facilities for work and street clothing—see OSHA Reg. 1910.141(e)

f [ ] Assure that employees wash hands and face prior to eating, drinking, or smoking

g [ ] Keep lavatories clean

h [ ] Provide soap, towels, and warm water in lavatories—see OSHA Reg. 1910.141(c)(2)

i [ ] Provide additional lavatories—see OSHA Reg. 1910.141(c)

j [ ] Provide additional hand-washing facilities

k [ ] Provide showers

l [ ] Assure that employees shower before going home

m [ ] Keep showers clean

n [ ] Provide soap, towels, and warm water in showers—see OSHA Reg. 1910.141(c)(3)

o [ ] Provide potable drinking water—see OSHA Reg. 1910.141(b)

p [ ] Other______________________________

6. [ ] Assure that employees receive comprehensive information and training concerning hazardous chemicals—see OSHA Reg. 1910.1200 or Right-to-Know Laws.

7. [ ] Maintain an OSHA Log of Injuries and Illnesses and post Summary every February—see OSHA Reg. 1904.

8. [ ] Notify employees of their rights to obtain copies of medical and monitoring data and provide such copies to employees upon their request—see OSHA Reg. 1910.20.
9. [ ] Form and hold regular meetings of a joint worker-management health and safety committee.
10. [ ] Obtain information on the following comprehensive OSHA standards:
    - a [ ] Asbestos—see OSHA Reg. 1910.1001
    - b [ ] Formaldehyde—see OSHA Reg. 1910.1048
    - c [ ] Ethylene Oxide—see OSHA Reg. 1910.1047
    - d [ ] Lead—see OSHA Reg. 1910.1025
    - e [ ] Arsenic—see OSHA Reg. 1910.1018
    - f [ ] Benzene—see OSHA Reg. 1910.1028
    - g [ ] Noise—see OSHA Reg. 1910.95
    - h [ ] Other____________________________

APPENDIX D

Checklist for Evaluating Chemical Exposure

A. EVALUATE THE POTENTIAL FOR AIRBORNE EXPOSURE
   1. Exposure Sources (rank high/medium/low)
      a. Types and amounts of chemicals in use or created by combustion or decomposition.
      b. Visible leaks, spills or emissions from process equipment, vents, stacks or from containers.
      c. Settled dust that may be resuspended into the air.
      d. Open containers from which liquids may evaporate.
      e. Heating or drying that may make a chemical more volatile or dusty.
      f. Odors. Consult an odor threshold table to get an estimate of concentration.
      g. Do air monitoring where the presence of a contaminant is suspected but cannot be verified by sight or smell.
      h. Visualize exposure by taking photographs or videotape.
   2. Job Functions (estimate hours/day)
      a. Manual handling in general.
      b. Active verb job tasks such as grinding, scraping, sawing, cutting, sanding, drilling, spraying, measuring, mixing, blending, dumping, sweeping, wiping, pouring, crushing, filtering, extracting, packaging.
   3. Control Failures
      a. Visible leaks from ventilation hoods, ductwork, collectors.
      b. Hoods that are located too far from the source or that are missing or broken.
      c. Ductwork that is clogged, dented, or has holes.
      d. Insufficient makeup air to replace exhausted air.
      e. Contamination inside respirators.
      f. Improperly selected, maintained, or used respirator.
      g. Lack of or inadequate housekeeping equipment.
      h. Lack of or inadequate doffing and laundering procedures for clothing contaminated by dust.

B. EVALUATE THE POTENTIAL FOR ACCIDENTAL INGESTION
   1. Exposure Sources (rank high/medium/low)
      a. Types and amounts of chemicals in use or created by combustion or decomposition. Solids are of primary concern.

b. Contamination of work surfaces that may spread to food, beverage, gum, cigarettes, hands or face.
c. Contamination of hands or face that may enter mouth.
d. Do wipe sampling to verify the presence of a contaminant on work surfaces, hands, face, and so forth.

2. Control Failures
a. Contamination of inside of respirator which may enter mouth.
b. Contamination of lunchroom surfaces which may spread to food, beverage, gum, cigarettes, hands or face.

C. EVALUATE THE POTENTIAL FOR SKIN CONTACT AND ABSORPTION

1. Exposure Sources
a. Types and amounts of chemicals in use or created by combustion or decomposition. Check dermal absorption potential. Do not rely upon OSHA SKIN notations. Assume most liquids will penetrate skin.
b. Consider whether one chemical can act as a "carrier" for other chemicals.
c. Visualize dermal exposure by taking photographs or videotape.

2. Job Functions
a. Dipping hands into material.
b. Handling of wet objects or rags.

3. Control Failures
a. Contamination of inside of gloves.
b. Improperly selected, maintained, or used gloves.
c. Improperly selected, maintained, or used chemical protective clothing.
d. Lack of or inadequate facilities for washing of hands and face close to work areas.
e. Lack of or inadequate shower facilities.

APPENDIX E

Some Industries and Operations for Which NIOSH Control Technology Reports Have Been Published

NIOSH Numbered Publications

74-100 Lead Exposure at an Indoor Firing Range

74-114 Cotton Dust Control in Yarn Manufacturing

75-108 Development of Design Criteria for Exhaust Systems for Open Surface Tanks

75-115 Engineering Control of Welding Fumes

75-165 Compendium of Materials for Noise Control (Revised: See 80-116)

76-130 Lead Exposure and Design Considerations for Indoor Firing Ranges

76-179 Abrasive Blasting Operations: Engineering Control and Work Practices Manual

76-180 Engineering Control Research Recommendations

76-186 Recirculation of Exhaust Air

78-109 An Evaluation of Cotton Dust Control Systems

78-141 The Recirculation of Exhaust Air . . . Symposium Proceedings

78-159 Engineering Control Technology Assessment for the Plastics and Resins Industry

78-165 Control of Exposure to Metalworking Fluids

79-114 An Evaluation of Occupational Health Hazard Control Technology for the Foundry Industry

79-125 Assessment of Selected Control Technology Techniques for Welding Fumes

79-143A Validation of a Recommended Approach to Recirculation of Industrial Exhaust Air—Vol. I (Spring Grinding, Chrome Plating, Dry Cleaning, Welding and Vapor Degreasing Operations)

79-143B Validation of a Recommended Approach to Recirculation of Industrial Exhaust Air—Vol. II (Lead Battery, Woodworking, Metal Grinding and Enamel Blending Operations)

80-107 CIB-33—Radiofrequency (RF) Sealers and Heaters: Potential Health Hazards and Their Prevention

80-112 Industrial Hygiene Characterization of the Photovoltaic Solar Cell Industry

80-114 Control Technology for Worker Exposure to Coke Oven Emissions

80-136 Engineering Control Technology Assessment of the Dry Cleaning Industry

80-143	Control Technology Assessment: The Secondary Nonferrous Smelting Industry
81-113	Evaluation of Air Cleaning and Monitoring Equipment Used in Recirculation Systems
81-118	Control of Emissions from Seals and Fittings in Chemical Process Industries
81-121	An Evaluation of Engineering Control Technology for Spray Painting
83-115	Occupational Health Control Technology for the Primary Aluminum Industry
83-127	Comprehensive Safety Recommendations for Land-Based Oil and Gas Well Drilling
84-102	Engineering Control of Occupational Safety and Health Hazards: Recommendations for Improving Engineering Practice, Education, and Research, Summary Report
84-106	NIOSH Alert—Request for Assistance in Controlling Carbon Monoxide Hazard in Aircraft Refueling Operations
84-110	Health Hazard Control Technology Assessment of the Silica Flour Milling Industry
84-111	Control of Air Contaminants in Tire Manufacturing
85-102	Control Technology Assessment: Metal Plating and Cleaning Operations
88-108	Safe Maintenance Guide for Robotic Workstations
88-119	Guidelines for Protecting the Safety and Health of Health Care Workers
89-107	Guidelines for Prevention of Transmission of HIV Virus and Hepatitis B Virus to Health-Care and Public-Safety Workers
89-115	Current Intelligence Bulletin 52—Ethylene Oxide Sterilizers in Health Care Facilities —Engineering Controls and Work Practices
89-120	Control Technology for Ethylene Oxide Sterilization in Hospitals
89-121	Control of Asbestos Exposure During Brake Drum Service

NIOSH documents are sometimes available at no cost from NIOSH Publications. Some OSHA and NIOSH offices, especially regional offices, have libraries that may have NIOSH documents available for use.

PART III

SOCIAL CONFLICT AND THE POLITICS OF HEALTH AND SAFETY

We have emphasized throughout this volume that a full understanding of the issues surrounding the work environment encompasses a complex set of political, economic, and social institutions. In Part I we provided material that examined the broad political economy of occupational health and safety. In Part II we examined the regulatory, legal, and institutional political arena and, using a series of case studies, showed how science and politics are deeply intertwined. In Part III we will examine how the work environment is structured by social and political conflicts that determine the outcomes for occupational health and safety.

In particular, we believe that the protection of workers and the environment from hazards depends on the ways in which "science" and its practitioners interact within a web of social relations. The material presented here suggests that the role played by science is never neutral and that scientists and professionals are trained in and act on a mistaken but specific conception of what their roles and functions should be. In fact, scientists and professionals have a series of related and often conflicting interests, their actions being largely determined by their training, social class, employment, career ambitions, and beliefs about the world.

In Section A, "Worker and Community Struggles," we look critically at what happens when conflict erupts over work environment issues, hoping thereby to suggest what types of tensions emerge from such conflicts and what lessons may be learned from efforts to ameliorate health hazards resulting from occupational and environmental exposure to toxic substances.

The role of the local community in acting against environmental hazards is described by Lewis in Chapter 19, on the impact of citizen organizations on controlling the hazards posed by polluting companies. The primary lesson of these stories, Lewis suggests, is that the community can be most effectual if it gains relevant knowledge from professionals and regulatory agencies concerning what is going on inside locally based companies. Using specialists provided by the National Toxics Campaign and information available as a result of the federal Community Right-to-Know law, citizens have in many instances mobilized to force companies to agree to stop their polluting activities by—as a first, critical step—establishing "Good Neighbor Agreements."

Parker and Morse, in Chapter 20, take issue with Lewis over what they consider an incomplete citizen action strategy for protecting the environment, incomplete because it makes no effort to include the workers within an industrial plant's walls in the collective efforts. For Parker and Morse this shortcoming has two major consequences. First, workers are often subjected to various types of "job blackmail" (i.e., with the employer pitting the environment against their jobs), as a consequence of which workers often side with the company against the concerned community to protect their jobs. Second, keeping pollutants from spreading *outside* a factory's walls does not consider what then happens to the workers exposed to these hazards *inside* the factory. The second half of the chapter takes up Lewis's reply to Parker and Morse.

The social conflicts surrounding the work environment include the tensions arising from the issues of race, class, and gender. In Chapter 21, Head and Guerrero assert that race, or racism, is a major problem confronting those who seek to improve environmental conditions. Minorities suffer *disproportionately* from hazardous waste in their communities, from hazardous working conditions, and from exposure to pollutants. Head and Guerrero argue that this environmental racism stems in part from the history of racism in the United States and elsewhere and exacerbates the problems faced by poor communities everywhere.

One solution to the problem, as Head and Leon-Guerrero note, is the activities of organizations such as the SouthWest Organizing Project (SWOP). In a similar vein, Buchanan and Scoppettuolo, in Chapter 22, describe the story of the Champion International paper mill in Canton, North Carolina, and efforts to combat the pollution and degradation of the work environment caused by this company. Buchanan and Scoppettuolo describe the possibilities for success and failure in building worker–community coalitions to combat pollution from a major industrial plant. The lessons from this case are very instructive. Building coalitions, fighting job blackmail, and ending occupational and environmental exposure are not easy tasks. To be successful, workers and communities must recognize that they have similar long-term goals. A successful coalition will depend on the willingness of labor and the environmental movement to work together democratically—to press for effective governmental regulation through OSHA and EPA, to strengthen worker and community rights, and to gain control over company investment strategies.

These themes are repeated in Baldauf's analysis (in Chapter 23) of the struggle against an attempt to site a dangerous ammonia plant in Texas. Baldauf documents how citizens organized successfully to defeat the attempt to locate this facility in their community.

Thus, Chapters 19–23 illustrate how political, social, and economic tensions come to the fore in struggles to protect both worker health and safety and the general environment. These case studies provide some pertinent examples of what can go right and what can go wrong in efforts to fight against corporate polluters. They especially illustrate how social forces interact around the bedrock necessity for production.

In Section B, "The Struggle for Occupational Health Clinics," these issues are illustrated by three chapters about the creation of occupational health clinics. Tuminaro's and Lax's articles examine the experience of the occupational health clinics in New York State, revealing initial hopes and a later, sober evaluation of actual progress. Dillard's piece is the reflection of an occupational physician on the trials and tribulations of working a market system.

Section C of Part III, "Programs for Change," offers comments on how these social and political conflicts may be resolved. It recommends the types of actions and programs that might work to improve working conditions and protect the environment.

Weil contends, in Chapter 27, that we must seek to reform the central federal regulatory agency, namely, OSHA. Weil argues that OSHA's enforcement strategy has been unsuccessful and that it desperately needs a new strategy. He views the absence of widespread unionization in American workplaces as central to the problem. Weil argues that OSHA should target the nonorganized large and small workplaces in American industry, those that are typically the most dangerous and where workers are not protected by unions.

Weil goes further and raises some controversial questions about the way in which OSHA sets standards, the role played by cooperative labor-management programs to ensure safe workplaces, and the policies OSHA uses to target certain industries. He concludes that OSHA needs more resources, a beefed-up inspection system, and that it must pay more attention to the large number of unorganized workplaces in the country.

Klitzman, Silverstein, Punnett, and Mock provide a detailed discussion of an extremely important issue for the work environment: the special problems facing women at work. Women make up an increasingly large proportion of the workforce. In Chapter 28 Klitzman and associates argue that women's special occupational health exposures and needs have not been addressed fully and that this is a particularly critical issue since women suffer from occupational segregation, lack of unionization, and the expectation of multiple roles in the home and the workplace. Key issues include reproductive hazards, employment discrimination, ergonomic problems, and stress resulting from the psychosocial conditions of work. All these problems have been inadequately studied, and little improvement has been made. Klitzman et al. conclude that meaningful improvements in working conditions must include much greater attention to a women's agenda in matters of occupational health.

As we have noted throughout this volume, efforts to improve working conditions must begin with the economic, political, and social structures that determine production. Nowhere is this more evident than in the often bitter tension between workers and environmentalists. The continual dichotomization of "jobs versus the environment" has prompted major political struggles. Chary, in Chapter 29, takes on this issue squarely, arguing that pollution prevention strategies and legislation designed to clean up the environment inevitably result in some loss of jobs and income for workers. The problem, according to Chary, is that workers and unions have never fully articulated how they expect environmentalists to push for meaningful environmental change while also considering all the demands by labor for job security and income protection. Chary challenges labor to integrate its political and economic struggles with those of the environmental movement, and she takes up in some detail the recent historical example of the Great Lakes United Task Force on Labor and the Environment.

Merrill, in Chapter 30, further takes up this challenge. Asking what role there will be for workers in the environmentalist's "brave, new green world of tomorrow," he finds the answers less than reassuring. Part of the answer, Merrill suggests, is a Superfund for Workers that will provide income protection and job retraining for workers displaced as a result of stringent environmental regulation. Environmentalists, he asserts, must find ways to incorporate protections for workers in their demands for environmental curbs or controls.

In Chapter 31,Geiser briefly traces the history of how we have sought to control environmental pollution, arguing that most of these strategies have dealt only with the outcomes of production—that is, with cleaning up the damage after it is done. A better strategy, Geiser asserts, is to not produce the polluting agents in the first place. Toxics use reduction measures seek to minimize production and use of hazardous chemicals before they can pose a threat to workers and the environment. Geiser describes the passage in 1989 and subsequent implementation of the Massachusetts Toxics Use Reduction Law (one of the earliest of such measures), which provides funding for relevant research, training, and promotion of a toxics use reduction strategy within that cutting-edge state.

SECTION A

WORKER AND COMMUNITY STRUGGLES

19

Citizens as Regulators of Local Polluters and Toxics Users

SANFORD J. LEWIS

Through data submitted by industry under the federal Community Right-to-Know law enacted in 1986, the public is now aware that as much as 22 billion pounds of toxic chemicals are emitted to air, water, and land each year. Citizens are learning, through this federal law, what portion of these emissions are caused by polluting companies in their own communities. They are also learning the amount of toxic materials stored at such local companies that could cause a chemical disaster like the one in Bhopal, India, in which more than 3,000 people were killed.

As a result of these disclosures, the public increasingly understands that our environmental laws and agencies do not guarantee the public a right against being poisoned through air or water by local factories. Some citizens are using the information gained to demand direct participation in decision making regarding local companies' use and management of toxic materials.

For example, residents of Berlin, New Jersey, won significant improvements in plant safety at the local Dynasill Company, which produces glass for high-tech applications, including laser and aerospace uses. An organization of residents known as the Coalition Against Toxics had become concerned about the facility after a fish kill in local lakes. Members wondered if chemicals discharged by Dynasill, which is upstream from the lakes, might be responsible for the water's acidity.

A DETAILED INSPECTION

In May 1988, with the company's permission, the Coalition Against Toxics conducted a detailed inspection of the Dynasill facilities with an industrial hygienist whose services were provided by the National Toxics Campaign. A report prepared by the hygienist after the tour made a number of recommendations for improving the facility's safety, such as:

- Completing the diking around storage tanks containing silicon tetrachloride, which, when exposed to water, can create heat and hydrochloric acid;
- Installing shower/eye wash stations;
- Training employees to be a company fire brigade.

The local Coalition Against Toxics had expected to negotiate with Dynasill about the proposals. But within one month of receiving the inspection report, the company implemented all of the recommendations.

Direct action by citizen regulators also was successful in Quinsigamond Village, a densely populated area in Worcester, Massachusetts. Residents were angered by years of odorous pollution by Lewcott, a local company engaged in resin coating of fabric. The residents also believed that the plant was responsible for sickness in the neighborhood. Through a series of meetings in the residents' living rooms, they launched an all-out campaign to get the company to clean up its act, or move out. After declaring to the company and the community that Lewcott would have to negotiate directly with its neighbors, the new Quinsigamond Village Health Awareness Group began to organize to win its demands.

The neighbors picketed the company. They inspected the company together with the National Toxics Campaign and published an inspection report. They held press conferences. They challenged the company's flammables storage permit, and won severe restrictions on the permit's duration. They pressured the state environmental agency to assess penalties for the firm's pollution. They lined a main street of Worcester with signs expressing their anger with the company's odorous behavior such as: *Don't ruin our summer again* and even *Lewcott get out*! They investigated the company's directors, and the environmental records of companies owned by the major shareholder in Lewcott.

As a result of these efforts, Lewcott executives finally agreed to sit down with the residents in a series of formal negotiating sessions. Through this dialogue, important information was disclosed and changes in the plant managers' thinking emerged. The company head began to realize that his plant was not properly located to begin with; it was simply too close to people's homes to be safe and a good neighbor. The owner concluded that the most economical and effective means of reducing the hazards to an acceptable level was to consolidate the Worcester plant and another Lewcott facility under one roof, with a new, more effective set of pollution controls.

A PRECEDENT-SETTING AGREEMENT

A surprising and precedent-setting agreement resulted. The company actually committed to relocate its plant, over a period of a year and a half, away from the neighborhood. For the interim period in which the plant would remain in the neighborhood, the agreement not only included a right of citizens to inspect the facility, but also provided for enforcement through a binding arbitration clause and designated penalties for violations.

There have been dozens of other examples of local campaigns such as these.

The genius of the "citizen-regulator" approach is that it places those with the biggest stake in safety at the front lines of the regulatory process. What the affected citizens can see for themselves, through direct inspections with their own expert consultants, can be far more reliable and extensive than what a government inspector may report to them. We do not have to be experts ourselves in order to make this strategy work.

Where citizens can get a firm to agree to immediate action or a written, legally enforceable "Good Neighbor Agreement," they may secure more protection than a standard government order to comply with existing (often inadequate) regulations.

Experience has shown that organized local citizens have as much, if not more, ability than government agencies to move local corporations to act. If a local company does not willingly negotiate, citizens must be prepared to wage a campaign to bring the company to the bargaining table, by organizing, picketing, or generating unfavorable media publicity. Community groups still may want to turn to government for assistance, but instead of filing complaints only with a single environmental agency, they may deploy dozens of agencies in fields as remote as zoning, banking, and insurance toward the goal of pressuring the company to negotiate. In some cases, citizens groups also file their own legal actions directly against the companies.

CITIZENS DEMAND PROTECTIONS

Using this strategic arsenal, determined local citizens often demand protections that the government environmental agency may be unwilling or unable to require of a corporation. For example, citizens in some communities are demanding that a company:

- Study and reduce its toxic chemical usage and waste generation;
- Provide technical assistance to the residents to review the firm's activities;

- Allow residents the right to inspect the facility periodically;
- Establish a comprehensive chemical accident prevention program at the facility utilizing the advanced measures the citizens' own experts identify; or
- Grant the citizens a right, along with workers in the plant, to ongoing participation in a company health and safety committee handling management decisions about toxics.

The direct democracy model is being deployed in pollution-prevention campaigns at about 100 factories across the nation today. The strategy is proving most effective in dealing with small- and medium-sized corporations. At big companies like Chevron in Richmond, California, and Exxon in Baytown, Texas, citizens are demanding detailed lists of safety reforms. But, so far, the concessions from these mammoth polluters have been few and far between.

The citizen-regulator strategy also can be applied in a number of related but different contexts. For example, at least 36 companies across the United States were asked to sign "Global Good Neighbor Agreements" committing them to reduce the use of chemicals that destroy the ozone layer. More than 1,200 local supermarkets signed agreements with the National Toxics Campaign to end, by 1995, their sales of fruits and vegetables with pesticides linked to cancer and other toxic effects.

In 1990, the National Toxics Campaign (NTC) mobilized as many as 1,000 local campaigns to get industries to sign "Good Neighbor Agreements" to curtail their pollution and reduce their usage of toxic and ozone-destructive chemicals.

The citizen pioneers in communities like Worcester, MA, and Berlin, NJ, have shown that the strategy of direct citizen regulation effectively supplements inadequate government regulatory systems that are already in place. There is literally no limit on the safety improvements and concessions that organized, creative, and determined citizens can win from local companies.

20

Involving the Community Inside the Factory Walls

ELISE PECHTER MORSE
JOAN N. PARKER

In "Citizens as Regulators of Local Polluters and Toxic Users," printed in the Spring 1990 issue of *New Solutions* [Chapter 19], Sanford Lewis described efforts by community groups to force factories in their neighborhoods to reduce toxic chemical emissions and to minimize the health risk to the community. As industrial hygienists employed by the Massachusetts Department of Labor and Industries, we often are in a position to see another perspective.

One of the strategies popularized in Mr. Lewis's chapter was the carrying of signs saying "Lewcott, Get Out," and the ultimate effect of this successful campaign was the closing of the Lewcott plant in Worcester, Massachusetts. Whatever effects these actions may have had on Lewcott employees, we can be certain that the whole story has not been told.

When the target for environmental change is an industrial workplace, the results always have significant impact on the employees of that workplace. Yet the article did not describe any attempts by the citizens group to contact workers to organize a more comprehensive community action, nor did it report any short- or long-term effects of their actions on the smaller "community" within the factory walls.

When a citizens group is successful in convincing a company to reduce toxic emissions to the outside, how can the group be sure that the production of toxic materials is actually reduced or eliminated instead of simply being kept inside where workers will suffer the ill effects? When a company responds to the pressures of a citizens group to close down or consolidate with another plant, how can the group know how the lives of workers will be affected by that change? In short, how can a citizens group pursue its own very legitimate concerns without adversely affecting the lives of some members of the very community it is trying to protect?

The only way to approach these questions is to recognize that the workers within are members of the larger community struggling with the actions of the company. When citizen activists act in isolation and fail to involve those employed by the offending factories, both groups suffer. The workers may be exposed to increased levels of toxic chemicals as company officials try to reduce pollution outside of the workplace. Jobs may be lost if the company feels pushed to eliminate certain operations instead of making them safer, and the activists suffer because they lose the most important allies they can have.

ECONOMIC BLACKMAIL

Community actions that target corporate activities can portray the workers employed there as part of the problem or as part of the solution. When these actions fail to differentiate beween workers and company officials, workers may be driven to defend the same company activities they might resist under other circumstances. When workers are pushed to side with and even take the first blows for the company for fear of losing their jobs, they are being subjected to the same type of economic blackmail that industry has traditionally used to resist implementing health and safety protections within the workplace.

Workers are given difficult choices. They are told by industry, "Force health and safety improvements in the workplace and drive us out of business or keep quiet and keep your job." They are told by community groups, "Keep working for the company and be part of the problem or stop working and support us." This pressure by community groups forces a strange and unnatural alliance between workers and employers—one in which economic security is the only stated goal. The distorted view of such an alliance depicts community activists as the enemy and their concerns as subjects of mockery. No one wins in the end; the community activists remain isolated and the workers continue in a hazardous environment.

PRODUCTIVE ALLIANCE

A far more productive alliance is one between worker and community groups where the strengths of the two groups are enhanced by each other.

Chemical pollution in a community is often symptomatic of more severe exposures and health risks inside the plant that uses or produces the chemicals. This fact points to a natural partnership among those who suffer similar consequences from the untoward actions of an irresponsible employer and business owner. Ultimately both labor and the community would be best served by the same solution—reduction in the use of toxic chemicals. If hazardous materials and processes are replaced with safer alternatives, the risk of occupational illness will be lessened and emissions to the air and groundwater will be reduced.

The common goals shared by labor and community activists can be most effectively achieved by joint action that employs the resources and strategies unique to each group. Where workers are organized into strong unions, alliances are easier to form because a structure is already in place for negotiating with plant management. It is important to note that uncontrolled use of chemicals is less likely to occur in strongly organized union workplaces.

WORKERS' RIGHTS

Probably the most insidious aspect of the failure to include workers in the struggle for community health is the implicit acceptance of the corporate view of workers. As soon as they enter the factory gates, workers give up many of their democratic rights. Freedom of speech is limited by threats of layoff or outright firing. Workers' rights are further limited by laws written to protect the rights of business owners. Democratic representation is only available to those with effective unions and stewards. The right to assemble cannot be exercised on company time. In fact, many believe that a paycheck buys workers' health as well as their labor. Hazards, occupational illness, and safety risks are considered part of the job. The term "occupational hazard" has come to imply an accepted danger, one for which no protection is expected. Workers' lives lose value. We have witnessed news reports that claim, "Two workers were killed but no member of the public was harmed . . ." If the environmental movement does not recognize workers as people, as members of the community with the right to live and work safely, the corporate view is reinforced and our collective strength is reduced.

GOAL IS HEALTH

The goal shared by environmental activists and workers is that of community health; this includes workers' health. If an independent action by a citizens group causes a factory to shut down, chemical use, worker exposure, and the public threat are simply shifted from one location to another, perhaps to a community that is not as well organized. As a result of this action, jobs are lost and a potentially powerful movement to force a reduction in chemical use and manufacture is undercut.

Sanford Lewis proposes several strategies for reducing community pollution, but mentions workers only in passing, instead of addressing them as the powerful allies they can be in this effort. He characterizes citizen-regulators as "those with the biggest stake in safety." Occupational illness claims the lives of more than 50,000 workers a year in this country. Who has a greater stake in safety? Some of Mr. Lewis's strategies are good strategies, but they are not complete strategies. They do not include labor as regulators who have everything to gain from the control of hazardous chemicals.

If the goal is to find the most effective methods of fighting for community health and upholding the democratic rights of all members of the community, we must stop fighting each other, sabotaging each other's positions, and offering up our weakness to the powers that pollute. Instead, it is time to build alliances between those who really share very common goals.

AUTHOR SANFORD J. LEWIS REPLIES

I agree that much more needs to be said about relations between community activists and workers in efforts to force local industries to become good neighbors. Despite the long-standing rhetoric of environmental and labor activists in support of worker and community alliances, the hurdles that stand between us are substantial and must not be ignored.

When pollution problems confronted by a community group are relatively small, it may be fairly easy to forge a worker–community alliance. There is plenty of common ground to be explored. Solutions such as toxics use reduction can meet pollution prevention demands without increasing worker exposure to pollutants. Workers may have information to share regarding company practices such as chemical accident prevention upon which the citizens might seek improvement.

Often, however, larger, seemingly intractable, pollution hazards are of concern. Local residents may favor cleanup, and yet be unsure of whether the only real solution is to close the plant. In such a context, neighborhood activists are generally unprepared to rule out shutting down the plant, in case the less severe approaches fail.

Residents, Workers Must Seek Alliances

As long as a potential plant shutdown lurks in the future, the unity of interests between workers and neighbors and their ability to join forces may remain tentative. Yet, residents and workers need to open lines of communication and seek alliances, however tentative they may be. There are strong and practical reasons for talking. Some of the issues on which community activists and workers may want to talk include:

- Identifying whether an acceptable end to pollution is possible without closing the plant, and enlisting workers in the cleanup effort;
- Demanding company financial disclosures to allow a realistic appraisal of the economic feasibility of environmental improvements;
- Preventing less experienced, nonunion workers from undertaking certain jobs where they may endanger the environment;
- Developing a complete community plan that will end polluting production practices while preserving the number and quality of jobs;
- Demanding periodic joint citizen/worker inspections for occupational and environmental hazards.

Out of talking points such as these, meetings of workers and neighbors may devise a list of common demands to corporate management.

Strength in Unity

To the extent that they are able to coalesce around such demands, the groups will find strength in unity. Together, they will be in a stronger bargaining position than they would be without the other group's support, and the unified neighbor–labor demands can be far more winnable than where neighbors act without labor support.

One recent example of the strengths of this approach occurred at the Sheldahl factory in Northfield, Minnesota. After disclosures required by federal law indicated that the company was the nation's 45th largest emitter of cancer-causing air pollution, local citizens organized two separate groups to end the pollution.

In that case, the citizens met with both corporate management and the union. The Amalgamated Clothing and Textile Worker Union, fearing a plant shutdown if citizen demands were not met, raised the community's concerns in its bargaining with management. As a result, pollution prevention was required in the union's collective bargaining agreement. The agreement calls for a 90 percent reduction in methylene chloride emissions by 1993, and a 64 percent reduction in toxic chemical usage by 1992. The development of a nontoxic alternative manufacturing process was made the top priority for the company's capital improvements budget for the subsequent two years.

To allow such a powerful alliance to emerge, it is probably advisable that some initial discussions between workers and community residents take place outside of any general company-community meetings. When management is present, they will often seek to drive a wedge between neighbors and laborers.

Aside from developing common demands to present to corporate management, other gestures by workers and community residents may help to build goodwill and prevent a rift from developing. For example, workers might consider joining the neighborhood group and actively supporting its environmental agenda. On the other hand, local citizens can support national legislation to create a Superfund for Workers, to ensure the economic security of workers when jobs are terminated on account of environmental concerns. Such legislation would pay displaced workers' income during a transitional period, and for their further schooling so that they can remain fully employable in our changing economy.

There is much more that could have been said about the Lewcott case itself, but clearly that case exemplifies the challenges noted above; it also raises other issues. (Readers may recall that citizens in that case in Worcester, Massachusetts, had faced 20 years of pollution from the Lewcott Company, a small resin-coating operation. They launched an all-out campaign, initially an effort either to clean up the company or to shut it down.)

Residents Wanted to Talk

These residents wanted to talk with the Lewcott workers, but since the plant was nonunion, they were unable to identify appropriate workers' representatives. Talking with a few isolated workers was not helpful in establishing an organized relationship for the citizens' bargaining efforts. Since less than 13 percent of America's private sector workforce is unionized, this poses a serious and commonplace obstacle to worker-community relations.

(A partial solution to this problem is suggested in proposed New Jersey legislation, the Hazard Elimination through Local Participation Act, which is supported by a coalition of labor and community activists. The bill would grant citizens a right to inspect, and also establish a semiautonomous workplace "hazard prevention committee." In a nonunion workplace, half of the members of this hazard prevention committee would be elected by the workers, the other half appointed by management. A question for workplace activists to consider is whether legislation is necessary to establish such committees, or whether they can also be organized on an ad hoc basis, perhaps in response to a neighborhood initiative.)

After months of intensive and bitter fighting with Lewcott's management, the citizens eventually gravitated toward a plant closure. They won a binding agreement from the company to shift all operations to the company's other plant about 10 miles away. Fortunately, the jobs impact was nowhere as severe as if the plant had closed and left the area entirely.

The neighbors of the second Lewcott plant, to which the operations were moved, also began to organize. As a result, advanced pollution controls were installed on the plant in their neighborhood.

I cited the Lewcott case in the preceding chapter because it demonstrates so graphically that organized local citizens can muster the power, even without a change in the law, to bargain as equals with corporate management. Although the point was mentioned, one can never emphasize enough that tripartite bargaining, between neighborhood groups, labor, and corporate managers, is the preferred framework for negotiating Good Neighbor Agreements. The challenge and opportunity is for workers and corporate managers to cooperate with local citizens to define the conditions under which safe neighborhoods and quality jobs can coexist.

21

Fighting Environmental Racism

LOUIS HEAD
MICHAEL LEON-GUERRERO

In September of 1989, the Albuquerque-based SouthWest Organizing Project (SWOP) and the Eco-Justice Working Group of the National Council of Churches conducted the Inter-Denominational Hearings on Toxics in Minority Communities. At the hearings, which were held in Albuquerque, an ecumenical panel of regional, national, and international church leaders heard several hours of testimony by people of color from Arizona, New Mexico, Texas, and northern Mexico who had been victimized by toxic poisoning.

Their stories exemplified the impact of environmental racism and toxics hazards on Third World people in the Southwest. To cite some examples:

Former and present Los Alamos National Laboratories employees testified that they, like other Chicano workers, have faced discrimination for years, have been forced to work in "hot" areas and jobs, and have been contaminated by weapons-grade plutonium on a regular basis. Said former worker Jerry Fuentes, "One supervisor contaminated the whole wing where I worked—he contaminated 12 people inside the wing, including myself. He was (subsequently) promoted from Assistant Section Leader to Section Leader."

Betty Griego of the Mountainview community south of Albuquerque spoke about how her child nearly died in her arms from drinking excessive amounts of nitrates in the water used to prepare his formula. She and others spoke about their current struggles (described later in this chapter), involving the SouthWest Organizing Project, to hold accountable potential groundwater polluters, including the highly suspect Kirtland Air Force Base.

THE HAZARDS OF MAQUILAS

Representatives from El Centro del Obrero and La Mujer Obrera in El Paso, Texas, testified about chemical spills and hazardous working conditions in maquila industry operations on both sides of the U.S.-Mexico border. (1) Often "runaway shops," these factories frequently employ teenaged women workers. The development of the maquila industries has been accompanied by marked increases in workplace hazards and the presence of toxics in local communities. Witness Guillermo Glenn said, "We feel that the location of this toxic zone is not a coincidence. . . . In fact, we believe that it is because this area is a low-income Hispanic community that city and state officials have allowed it to become a haven for unregulated hazardous material companies, and this also affects what is transported and brought back across the border. It is kind of a no-man's zone, a low regulation zone, where the authorities really do not understand what is being taken across and what is being brought back."

Virginia Candelaria testified that she and hundreds of other women workers are suffering from several forms of cancer and are dying after years of exposure to solvents in an Albuquerque General Telephone and Electronics (GTE) plant. During the resolution of a lawsuit brought by 465 plaintiffs against GTE, the poisonous section of the plant where Mrs. Candelaria had worked was moved to Juarez, Mexico. Employees there are paid extremely low wages for the same work in similar conditions.

Workers from the Fenix Junkyard in Juarez, Chihuahua, Mexico, told how they had been contaminated by a radioactive cobalt spill that also contaminated at least 500 people in the neighborhood surrounding the junkyard. The cobalt had been sent across the border from the United States. The lack of personnel at the Mexican National Nuclear Commission necessitated the cleanup of the junkyard and surrounding communities by the workers themselves, without adequate protection. The workers and their families were sterilized as a result of the contamination. In the words of one worker, "It affected our blood, it affected our bones and we are in a propensity in the future of having cancer or leukemia and other harm which can come from it."

Residents of the Rio Puerco area of the Navajo Nation described how they have been exposed to high levels of radiation from water contaminated by mining and milling of uranium used for nuclear weapons and fuel production. In 1979 a dam at a United Nuclear Corporation tailings pond broke, resulting in a 94-million-gallon spill that contaminated the water and bed of the Puerco River. The contamination has affected community residents, livestock, and the water supply of the local school.

Residents of Espanola, New Mexico, said Chicano public school employees in their district were forced to clean up asbestos in local school buildings without adequate protection. The district showed no concern about the impact of the dangerous material on either the schoolchildren or the workers who were removing it. An elderly maintenance man who was exposed to the asbestos was told by his supervisor that he need not worry about it because he was "not going to live 20 more years anyway." (2)

INSIDIOUS FORM OF RACISM

The systematic exposure of people of color to hazardous chemicals and wastes is one of the most insidious forms of racism today. It manifests itself in the disproportionate siting of polluting industry and hazardous waste dumps in communities of color, and the routine exposure of people of color to extreme health and safety hazards in the workplace. The underlying assumption in these practices is that people of color are expendable.

There is a growing body of documentation detailing the disproportionate impact of toxic contamination on communities of people of color. In 1983, a U.S. General Accounting Office report showed that three of four landfills studied in the Southeast were located in poor, African-American communities; included was Chem Waste Management's facility in Emelle, Alabama, the nation's largest hazardous waste dump. In 1987, the Commission for Racial Justice of the United Church of Christ released its landmark study *Toxic Wastes and Race in the United States*, which found race to be the variable most strongly associated with the location of commercial hazardous waste facilities. *Toxic Wastes and Race* also found that three out of five African-Americans and Latinos live in communities with uncontrolled toxic waste sites. Research by the Center for Third World Organizing in Oakland, California, shows that most of the 200 million metric tons of uranium tailings in the United States are located on Native American land. (3)

Two threads of history explain the disproportionate impacts of toxics on Third World communities. One is the long history of oppression of people of color in the United States, from indentured servitude, slavery, and colonization to newer forms of institutionalized racism. Another is the incorporation of people of color into heavy industry and agribusiness during this century. Historically, people of color have found employment in the most hazardous industrial and agricultural jobs, and have, for a variety of social, economic, and cultural reasons, lived in areas where polluting industry is most likely to be located.

In the pursuit of profit maximization, most industries seek to locate in communities that are economically depressed, politically disenfranchised, and therefore subject to economic blackmail. This is particularly true of waste disposal and other obviously hazardous industries. Poor communities are sold the notion that employment and economic development will offset any negative impacts caused by such activities. Instead, long-term devastation of lives and resources is the result, while few jobs are gained. When industry is challenged about its polluting practices, poor communities are threatened with job cutbacks and plant shutdowns, ostensibly the price to be paid for pollution control and/or cleanup. (4)

RESISTING ECONOMIC BLACKMAIL

In many parts of the United States, people are organizing to resist corporate efforts to environmentally exploit poor and Third World communities through such economic blackmail. One organization that illustrates this work is SWOP, a multi-racial, multi-issue community membership group active in New Mexico and throughout the Southwest.

SWOP's mission is "to empower the disenfranchised in the Southwest to realize racial and gender equality, and social and economic justice." The organization was founded in 1981 by a group of longtime community and student activists who envisioned the development of a locally based and directed organization that eventually would work at the state and regional levels, serving to develop strength through common experiences and conditions, as well as the diversities that characterize the Southwest.

SWOP's constituency and membership are composed primarily of working-class Chicanos. The organization also has worked in African-American communities and in coalition with Native American organizations and nations, and has a significant number of Anglo members. During the last 10 years, SWOP efforts have included massive nonpartisan voter registration, education and get-out-the-vote drives, organizing campaigns to obtain affordable housing and meaningful employment, and efforts to promote greater access to institutions of higher education by people of color. SWOP also maintains an international perspective on its work and engages in educational exchanges where possible with community and other progressive organizations outside of the United States. Much of SWOP's past funding has come from religious sources and some from foundations. Now, with some success, SWOP is emphasizing income-generating activities and the development of a contributing donor base.

In the '60s and '70s, some of SWOP's founders were active in addressing issues of race, poverty, and the environment, such as the disproportionate placement of waste treatment facilities in Chicano neighborhoods in Albuquerque; Earth Day 1970 actually focused on such issues in Albuquerque. In the early '80s, SWOP surveyed several low-income communities in Albuquerque. Residents were not very responsive to questions concerning "environment" or "environmental issues." However, when asked about "community" issues of concern, people ranked groundwater and air contamination high on the list, often in terms of the failure of state and local governments to do anything about these and similar problems.

Such experiences led SWOP to recognize environmental issues as issues of social, racial, and economic justice. Since 1985 when the organization established its Community Environmental Program, a major part of SWOP's resources has been directed toward this area of work. Although SWOP is not an environmental organization, its work on toxics issues has developed models for organizing to alleviate toxics and environmental problems that inordinately impact low-income communities and people of color in particular.

URBAN CLEANUP EFFORT

SWOP initiated its direct organizing efforts on toxic contamination issues in the Albuquerque area in 1986 by successfully taking on a local particle board–producing industry that had contaminated groundwater beneath the largely Chicano Sawmill neighborhood. Grassroots organizing, leadership development work, and effective efforts to promote accountability on the part of elected officials toward local residents resulted in the first groundwater reclamation to be forced on an industry by a community group in New Mexico. It also was the first time that any New Mexico industry had ever undertaken such a cleanup in an urban setting.

SWOP then took on a much bigger target: Albuquerque's Kirtland Air Force Base, which is strongly suspected of contaminating the groundwater supply of a local community with extremely high levels of nitrates and nitroglycerin. After several years of direct organizing, accountability work with the New Mexico 1st Congressional District representative, and endless hassles with the Department of Defense, United States Environmental Protection Agency, and New Mexico Environmental Improvement Division, an agreement has been reached that will guarantee community residents full participation in the investigation of the contamination sources and in the development of appropriate technology for cleanup. This marks the first time in U.S. history that a branch of the military has ever made such a commitment, one that should provide a meaningful precedent for others around the country who are confronted with military toxics hazards.

SWOP's organizing efforts have shown that environmental racism is very real and that environmental issues cannot be separated from the broader context of social, racial, political, and economic injustice. Ultimately, the challenge is to address a broad set of economic and environmental issues together proactively; SWOP is headed in this direction in the coming years.

Unfortunately, the environmental movement has been very slow to recognize such links. Indeed, some of the mainstream environmental organizations are being compromised by the very polluters who have been major contributors to the environmental crises that now threaten us.

The environmental movement enjoys the highest visibility within the spectrum of political activism today in the United States. Millions of dollars now fund the major environmental organizations such as those that affiliate themselves as the "Group of Ten" (for example, the Environmental Policy Institute/Friends of the Earth, National Wildlife Federation (NWF), Natural Resources Defense Council, and Sierra Club). Much of this money comes from membership in groups such as the NWF. Membership has skyrocketed over the last decade, and the organizations now claim millions.

FUNDING FOR "GROUP OF TEN"

However, a significant amount of the funding for several "Group of Ten" organizations comes from major corporate powers, including GTE, Exxon, Dow Chemical, DuPont, and others. The CEO of Waste Management, Inc., sits on the board of the NWF. Indeed, many corporations are now labeling themselves environmentalists. They provided much of the money and hype that made it possible for some Earth Day activities to become massive media events. However, this newfound and loudly expressed corporate environmentalism masks persistent industrial practices that subject workers and communities to unnecessary exposure to toxic chemicals. It coincides with deindustrialization at home and runaway shops abroad. This raises the question about whether the poor are going to continue to pay for past and present corporate environmental negligence in the United States and around the world.

As environmental groups enter the mainstream of U.S. society, they are pointing to what they claim to be a new and increasing interest of people of color in the environmental movement. As they see it, Third World people are finally "coming around" to recognize the importance of issues on which the environmental movement has been advocating for a long time.

However, people of color historically have been involved in struggles for environmental justice. The false idea that such involvement has not taken place is rooted in the narrow definition of environmental issues by environmentalists. People of color, on the other hand, have defined environmental issues as an integral part of community, labor, or civil rights causes. The pesticide struggle has for many years been spearheaded by Latino farmworkers. Occupational safety and health issues have been addressed by labor unions, or in the case of the border region, often by immigrant rights groups. Lead-based paint poisoning has been addressed by community-based housing organizations.

The reality is that the environmental movement, including but not limited to the "Group of Ten" organizations, has shown little willingness to recognize the legitimacy of struggles to alleviate the poisoning of communities of color. Environmental organizations across the board have only token involvement of people of color in their operations and policymaking bodies. When they have supported Third World and poor peoples' struggles, they have often brokered resources and attempted to politically control the activities of local groups. Finally, like the EPA, some mainstream environmental groups have taken steps in the field that have actually been detrimental to the interests of people of color.

For example, the Sierra Club, National Wildlife Federation, National Audubon Society, and others have sided openly with U.S. government agencies to further limit sovereignty of indigenous people (both Native Americans and Chicanos) over traditional lands in New Mexico. At the international level, conservation organizations advocate for, and participate in, "debt-for-nature swaps" in which minuscule portions of a country's debt are forgiven to allow for a conservation group to administer tracts of tropical rain forest lands. Such "swaps" hand over sovereignty to nonindigenous people, serve to legitimize the so-called Third World debt and furthermore involve, at best, token representation of indigenous people living in affected rain forest areas. Urban "debt-for-nature" conflicts also exist within the United States between affordable housing organizations on one hand and environmentalists on the other over the proper use of open space lands (housing or parks) that can be purchased at discount rates through the savings and loan bailout.

A CALL FOR DIALOGUE

In early 1990 these and related issues were raised in letters to the "Group of Ten" signed by SWOP and hundreds of other social and racial justice activists and organizations. The letters called for dialogue to take place between people of color and members of the Group. More recently, a formal call went out from the United Church of Christ Commission for Racial Justice for a national Environmental Leadership Summit to take place in the fall of 1991 that will bring together Group-of-Ten representatives and community-based and civil rights organizations. SWOP and others are working to ensure that the summit will be fully inclusive. In the meantime, model working partnerships with organizations such as the Natural

Resources Defense Council have been initiated by SWOP, which it hopes will set a trend for more equitable working relationships in the future.

The potential exists for a new understanding to emerge between the environmental movement and those organizations working for social and racial justice. The process will require better defining of how issues of environment fit within a social, racial, and economic justice framework, and what the relationship will be between environmental organizations and the growing social and racial justice movement in the United States.

The new grassroots social and racial justice movement which SWOP is helping to build is exemplified in the Southwest by the Southwest Network for Environmental and Economic Justice. Formed in April of 1990 out of a SWOP initiative, the network includes more than 75 activists of color and representatives of community and labor organizations from Arizona, California, Colorado, Nevada, New Mexico, Oklahoma, Texas, and Indian Nations in the region. The network reflects the fact that there is as much or more local organizing taking place today than there was in the '60s and early '70s. After surviving a decade of Reaganism, people are looking for mechanisms through which they can mutually strengthen their efforts.

These days, the work is quieter (read unreported) and is much less likely to be powered (read extinguished) by government funding. Issues being addressed include employment, workers' rights, affordable housing, education, municipal services delivery, immigrant rights, and a host of others. As a part of this broad agenda, almost unanimously, organizers from around the country point to the environment as being "the issue" of the '90s. Through broad coalition efforts, and a lot of honest struggle, it can be a unifying issue.

NOTES

1. "Maquila" industry refers to a variety of manufacturers located in northern Mexico that take advantage of extremely low wages in Mexico and favorable U.S. and Mexican trade and tax incentives. The firms involved are generally U.S., Japanese, and Western European multinationals that locate in Mexico in order to avoid having to pay decent wages and transportation costs, and to avoid environmental and other regulations that limit profit maximization.

2. See "Interdenominational Hearings on Toxics in Minority Communities Testimonies." Unpublished transcripts. SouthWest Organizing Project, 1990.

3. See "Toxic Wastes and Race in the United States: A National Report of the Racial and Socio-Economic Characteristics of Communities Surrounding Hazardous Waste Sites." Commission for Racial Justice of the United Church of Christ, 1987; "Siting of Hazardous Waste Landfills and Their Correlation with the

Racial and Socio-Economic Status of Surrounding Communities." U.S. General Accounting Office, 1983; and "Toxics and Minority Communities." Issue Pac #2, Alternative Policy Institute of the Center for Third World Organizing, 1986.

4. See Kazis, R., and Grossman, R. *Fear at Work: Job Blackmail, Labor, and the Environment*. The Pilgrim Press, 1983; and "Labor/Community Watchdog—A Three-Year Organizing Strategy to Impact the Los Angeles Clean Air Plan." Labor/Community Strategy Center, Van Nuys, California, 1989.

22

Environment for Cooperation

BUILDING WORKER–COMMUNITY COALITIONS

MILLIE BUCHANAN
GERRY SCOPPETTUOLO

Job blackmail is not a new tactic for industries facing pressure for change. Communities and workers, organized and unorganized, have for decades heard companies threaten to close their doors and move away rather than meet demands for changes in conditions within or outside the workplace.

The decade of the 1980s, however, brought both a resurgence of company demands for takebacks from workers, and increased business opposition to laws and regulations designed to protect the community and the environment.

Scientific research increasingly documents health and environmental risks from industrial processes and wastes. At the same time, new laws and regulations giving communities and workers the right to know about those risks are raising public awareness of pollution and its dangers.

Recently released reports of the risks from dioxin in bleached paper products, and toxic emission data showing that paper mills are among the most prolific polluters, place the pulp and paper industry in the forefront of environmental regulatory issues. The story of the Champion International Corporation paper mill in Canton, North Carolina, one of the first of the battles in this new wave of permitting, has ramifications throughout that industry. It may be instructive as well for other communities dealing with industry decisions.

WORKERS AND COMMUNITIES DIVIDED: THE CHAMPION STORY

In 1906, Champion Paper Company (now Champion International Corporation) built a large mill on a small river in western North Carolina. The mill, on the Pigeon River in Canton, has become both the biggest employer in the western part of the state and the center of a storm of controversy.

The Pigeon River begins as a native mountain trout stream near the Blue Ridge Parkway in North Carolina. It winds down through rural Haywood County into the town of Canton, passes through the Champion mill, emerges carrying the company's effluent, and heads toward the state border with Tennessee.

To discharge into the Pigeon River, Champion must have a federal National Pollution Discharge Elimination System (NPDES) permit. NPDES permits, under the Clean Water Act, limit industry discharges into surface waters, and require the companies to monitor certain pollutants. Like many states, North Carolina is authorized by the federal government to enforce the Clean Water Act and issue NPDES permits within its borders.

Champion's current permit expired in 1981, and has been extended while the new permit is negotiated. Between 1981 and 1985, the State of Tennessee attempted to have input into the conditions of the new permit. When North Carolina's Division of Environmental Management issued a draft permit in 1985 without consulting Tennessee's government or its citizens, Tennessee officially petitioned the federal Environmental Protection Agency (EPA) to veto the permit and take away North Carolina's authority for regulating Champion.

EPA complied, issued its own draft permit in 1986 calling for stricter color limits to comply with federal laws, and scheduled permit hearings for May 1987. (Hearings were later postponed and eventually held in January 1988.) That move signaled the start of a new, bitter battle in the long war over the Pigeon River.

CHAMPION'S OPPOSITION CAMPAIGN

Champion officials said the permit as drafted would force the mill to close, and launched an extensive and expensive opposition campaign. Company officials won't say how much they spent, but the effort included hiring a nationally recognized public relations firm to rally public support, busing supporters to hearings in North Carolina and in Tennessee, and purchasing dozens of newspaper and electronic media ads. "Don't Let Champion Fall" posters, sporting falling dominoes signaling the cost to the area should the mill close, sprang up all over the town of Canton and surrounding areas, and dire predictions filled newspaper pages and airwaves.

Tennessee community and environmental groups, with minor participation from some North Carolina and national environmental groups, began their own campaigns, less well-heeled but just as intense, in support of the draft permit. Tennessee supporters challenged Champion's contention that the mill would have to close if it had to clean up, citing the corporation's healthy bottom line. They charged that the job blackmail threats were designed to manipulate public opinion and gain a weaker permit.

The debate was consistently framed, both in the media and in the political arena, as "jobs versus the environment." The *Asheville Citizen*, the leading western North Carolina daily, ran a three-part, front-page "Champion Under Siege" series the week before the scheduled May hearings, examining the economic consequences should the mill close. The series included no analysis of the assumption that a stricter permit would indeed force such a closing. National media attention echoed the "jobs-versus-fish" debate. So did the comments of the hours-long parade of North Carolina federal, state, and local elected officials testifying on Champion's behalf during the hearings.

Following the hearings, but before EPA announced a decision, Tennessee, EPA, and North Carolina officials tried to hammer out a compromise. North Carolina officials say Tennessee agreed to back off on its proposed color limit; Tennessee officials say agreement was never certain. Whatever the truth of that debate, the compromise fell through after Tennessee's governor canoed the river from its clear beginnings to its dark and foamy entry into his state, and announced to cheering constituents his decision to stand firm.

THREAT TO 1,000 JOBS

After hundreds of speeches, thousands of letters, and months of contemplation, EPA officials came back in spring 1989 with a redrafted permit, with which neither side was totally happy. Champion officials said they could comply by cutting 1,000 jobs—about half the workforce. August 1989 hearings on the new draft were somber compared to the circuses of 1988. Tennessee supporters called for changes; Champion supporters blasted the permit, but wearily declared "enough is enough" and asked EPA to issue the permit as drafted.

In September 1989, EPA issued the permit with minor changes, together with a summary of comments and responses. Within that summary was an EPA hearing officer's response to comments attributing the 1,000-

job cutbacks to the stricter permit. "Regarding the loss of jobs resulting from this permitting action, EPA has no firm information that the announced job losses are related to the permit terms," the hearing officer wrote. "The company, however, some five or so years ago, indicated to EPA that regardless of the outcome of the permit, there would be fewer workers at the Canton mill. The stated reason was that the company desired to improve the mill's profitability and the best way to do that was through a major modernization process resulting in the need for fewer employees."

The variety of newspaper coverage of that story was instructive. The *Knoxville (TN) News Sentinel's* front page headlined "EPA: Champion lied about mill layoffs." The *Mountaineer*, Canton's hometown paper, also topped the front page with the story, headlined "EPA: Champion already planned cutbacks." The *Asheville Citizen* story, placed on an inside page, led and was headlined with Champion's denial of the charges. Champion officials said that they had no recollection of the EPA discussion and that cutbacks due to the modernization plans would have been much smaller.

It would be difficult to find anyone in either state who thinks the permitting process has been a good one. Following are the perspectives of some participants in the Champion permitting story:

STEVE AND JILL HODGES

Jill and Steve Hodges live downstream from the Canton mill, near Hartford, Tennessee. The Pigeon River runs through Hartford, its dark waters carrying a rank smell. Residents there believe the river also carries poisons. The small town is nicknamed "Widowville" because of what area residents believe is an unnaturally high rate of cancer in the community.

In contrast to prosperous Canton, Hartford is economically depressed. When the Hodges hear Canton residents talk about lost jobs and fractured families if the mill closes, they understand. Steve Hodges lost his job as a river rafter because no one wanted to raft the Pigeon. Hartford families are used to seeing younger generations move away for better opportunities.

Jill Hodges and Carroll Israel, president of the Canton paperworkers union local, met as panelists discussing the Champion battle during a spring 1989 conference on plant closings sponsored by the Highlander Center of Tennessee. Jill talked both with the union leader and with his wife, who fears her son (also a Champion employee) will have to move and she'll only see her grandchild on holidays. Jill sympathizes, she says, but can't translate that sympathy into agreement that the Tennessee residents should back away from their demands.

"Cocke County has been going through that for years," Jill says. "A lot of people have left home to make a living for their families, sometimes men leaving their families behind to go up north to get the better jobs in the auto industry. They come back here when they retire. Makes it real difficult. I can certainly sympathize with the workers and the community because we're already going through what they may be entering.

"We can see what not having a good tax base and not having a good employer can do to our schools. Our schools aren't accredited, our children can't get into college. Jobs—we haven't had jobs in a long time."

COLOR ISSUE VERSUS HEALTH ISSUE

The Hodges believe misperceptions have helped fuel the controversy. One is the contention in North Carolina that the river problem is only a color issue, not a health issue. They don't believe that industry didn't know about the dioxin in paper company discharge until recently.

"Being in the paper industry as long as they have, they certainly should have known," Jill says. "The people who live along the river have felt there was something wrong with it more than color for a long time, but felt who were they to say? In our area, so many people have lost so much they really weren't willing to stand up. Lots of people lost land to form the national parks and the national forests. This has always been a poor area. The fear was if they said anything, their food stamps would be cut off, then government checks would be cut off. Without jobs, that's about all they have."

The second perception that bothers the Hodges is the belief that those pushing for cleanup are elite nature-lovers, opposed to plain working folk.

"I don't think that most of the people fighting this issue in Cocke County consider themselves environmentalists, just people with a problem that's running through their backyard," says Jill. "But because Greenpeace came in and posted the river (with signs warning of dioxin in the fish), the perception of Greenpeace has changed here. People in Hartford are proud of those signs. They have told us that Greenpeace is the only one that came in for their protection, and that meant everything to them, because no one has ever cared enough to do that."

Both Hodges come from strong union backgrounds, and they blame the company, not the people or workers in Canton, for the trouble and for the growing bitterness between the areas. They believe the company is exploiting both communities, but are unsure whether things would be different had people from the two areas talked to each other long ago.

"The company has a chance to drive a real wedge, to fuel the hysteria—you side with the company or you lose your job," Jill says. "The bottom line is, that's worked well in Canton. The big comment in Canton is it (the Pigeon River) smells like money. In Hartford, it smells like cancer. It's kind of hard for us to find a common ground."

Cocke County's real fight should be with corporate Champion, says Steve Hodges, and Haywood County workers' more with "elitists who see dollars in front of their eyes, talking about the condos they'll build when it's cleaned up."

The Hodges and the Israels came to know each other tentatively during the Highlander Center conference. Jill isn't sure, however, how much difference more extended talks would make. "It's real hard for us to be objective," she says. "I don't think from that meeting we got any closer. There hasn't been any real contact since then. We have sent information to Carroll Israel, but we've never gotten any response back."

"NOT KNOWING, HARDEST THING"

"We do understand their stress—not knowing is probably the hardest thing," she says. "He [Israel] did say the accident rate at the plant is up. He attributed that to the stress on the workers, which is understandable. The company has all the control. I can't even imagine the stress the workers at Champion are under, to come out in support of the company one day [at permit hearings], and to fight for their jobs against the company the next [during contract negotiations, which coincided with the permitting process].

"If we could roll back time, I think if the people who live in Haywood County, North Carolina, and the people who live in Cocke County, Tennessee, had talked before we all tried to sit down and listen to Champion's corporate mumbo-jumbo, it could have been different. What happened was, we invited Champion officials here to talk to us. Basically, their message was, 'If you just leave us alone, we'll have that river partially cleaned up, better than you've ever seen it.' It was a waste of time. We've heard that before, and no one trusts what they say."

The Hodges also were disappointed that a proposal put forward by their group and others for a toxics reduction solution to the problem didn't generate any union support. Switching to unbleached paper products would reduce both the color and the toxic materials going into the river, but the Hodges don't believe the corporation is interested in that kind of answer.

"I've been seeing job blackmail for a long time where I've worked," Steve adds. "It used to be safe, then they gradually made you do more and more. It's just a way to break unions. My dad got asbestos in his lungs, and they lost all his medical records. He couldn't get any disability. He'd been big in the union. I told Mr. Israel about my and my dad's experience, but he couldn't associate it. He didn't see how all these big companies were doing the same thing."

"The sad thing that's happening," Jill says, "is that we're going to have to live here, and the people in Haywood will have to live there, and Champion will go its merry way. There will be bad feelings between Haywood and Cocke. It's unfortunate that companies can come in and pollute and get the hell out and leave such bad feelings behind."

CARROLL ISRAEL

Carroll Israel is the executive vice president of the Paperworker Local in Canton. He has been involved more than anyone in his union in dealing with the crisis at Champion Paper brought on by chlorine leaks and threats to close the plant. Carroll has strong feelings about the role of environmentalists in seeking the new discharge permit:

"We still have the problem. The people we have to deal with [the environmentalists] still want to shut us down. They don't want to compromise. I went to Chattanooga eight months ago to talk with them, but it didn't help. Greenpeace came up here putting up signs warning people about the water, but they're not from here. Only one or two of the environmentalists involved are from this area. The president of the Dead Pigeon River group is a real estate salesman and the vice president is a big landowner."

Carroll sees a double standard in the lack of reaction by environmentalists to the parallel but essentially invisible issue of health and safety problems in the plant. "There was a chlorine leak in here eight months ago that sent 13 people to the hospital and the environmentalists didn't respond to it at all," he says.

A joint labor-management health and safety committee in the plant has dealt with chlorine problems there, and Carroll sees modernization as the solution to the chlorine problems. "If the company modernized, it would help," he says. "We need to go from chlorine to chlorine dioxide to make a difference. The new permit would require us to cut back from 45 million to 29 million gallons of waste water a day. We will have three years to fix the problem. This cutback in output will mean layoffs."

Regarding the union's relations with Champion Paper throughout the ordeal, Israel admits that choosing sides on issues has been difficult. "The company has been helpful in allowing the voluntary transfer of 100 workers to other mills—to use attrition to avoid layoffs. As far as contract negotiations are concerned, it did weaken our position. It was hard to fight for a good contract when we had to be on the company's side part of the day to keep the mill open and then had to bargain with them that night."

CLARE SULLIVAN

Clare Sullivan worked for the United Paperworkers International Union during the Champion paper controversy, but it was not until the battle was raging that the International Health and Safety Department was asked to work with the Canton local. By that time, the lines were drawn, and the situation already serious.

The United Paperworkers Union at the international level has begun addressing the dioxin issue during the last 18 months, with two extensive articles in the union paper, *The Paperworker*. The International also participated in the Office of Technology Assessment's analysis of the dioxin problem for Congress, and worked with the National Institute for Occupational Safety and Health (NIOSH) on the health effects of dioxin on workers in the paper industry. When Clare talks about the issues at Champion, she focuses on dioxin, not color. When she talks about health issues, she thinks of both worker and community health.

"The union's position is that we know dioxin is a serious contaminant," she says. "We know the paper industry has been polluting for a long time and that it has to stop. Many management people in the industry itself know the time has come, maybe gone, and we're way past due for stricter regulations."

There are cases, Clare agrees, where the company is the obvious bad guy from both a worker and a community standpoint. She doesn't see the Champion fight as one of those cases. "Everything they [the company] are doing is really state of the art," she says. "The delay [caused by the long permit battle] if anything caused retraction in the commitment Champion has made to western North Carolina. That's understandable from a business perspective. It doesn't make sense to hold up corporate funds for investment, particularly in an old mill, if they don't know if they have a permit.

"GIVE INDUSTRY SOME SPACE"

"There's a tendency to make companies the bad guys, and they often are. But the paper industry didn't know about dioxin, and it's trying to be responsible, investing billions of dollars. The industry needs credit for that. At least give the industry some option, some space."

The health impacts of pollution are real, Clare agrees, but so are the health impacts of job loss.

"When we look at the impacts of massive job loss on people and their families, we know 100 percent of families will be affected healthwise, physically or mentally," she says. "What are the health effects of layoffs? We have statistical associations and anecdotal information, but it has not been looked at very carefully. Then, too, we really don't know yet what the health effects are of the pollution that's going on. There are many sources of toxic exposure, and it's hard to separate out the effects, particularly in a highly industrialized community.

"One thing I wish the Tennessee community could recognize is that the loss of a job is a health issue as well as pollution. And in many cases, you might be talking about just a little more time to let the company make changes. It's hard for people to deal with; they're angry over the past. But if it takes two more years, that's not a lot of extra pollution and it might mean a better solution and maintenance of jobs. I'd like to see a willingness to compromise, not on effectiveness, but on time. That's a position the community might start to think about."

She sees the Hodges' unbleached paper proposal as simplistic and unrealistic at that site.

"Champion Paper does not make unbleached paper," she says. "That (the unbleached paper campaign) is tantamount to saying Champion will close or sell out. If that would happen, the union could facilitate a buyout, and that may ultimately be what happens. I don't think that is what the local needed to hear. To them, it's a very simplistic solution to a really complicated problem."

Clare believes the environmental groups pushing for cleanup often represent neither workers nor the community, and the divisions often become divisions between classes and between outsiders and local people. Environmental groups can help the situation by community education and canvassing, by finding out what people's priorities really are. "Most people want to lead healthy lives and have jobs," she says.

TWO STATES COMPLICATE SITUATION

"The Canton situation was complicated by the fact that there are two states involved. The local really feels that if Canton had been in Tennessee, this never would have happened. Tennessee has many environmental problems as serious as Champion that need to be addressed. This would not have been done in this way if the mill had been in Tennessee."

Clare thinks the Canton fight is impossible to rectify at this late date. "It's just too late at this point to get people to understand the opposite perspective," she says. But she does see some hope for change in future battles.

At the international level, the union tries to make connections between worker health and safety and environmental issues, Clare says. "But it takes a lot of foresight on the union's part, and resources. Resources are the limiting factor. There's never anyone on staff with enough time."

She sees resources needed in many areas to help get beyond this point, resources for industry, workers, and communities.

"For industry, perhaps if there were money to help clean up, to assist a transition in the industry from bleached to unbleached products, incentives to make that happen.

"For workers, the new plant-closing legislation is a start, but it is limited. The economic situation in western North Carolina makes the possibility of a shutdown seem like the end of the world, because there are no equivalent jobs.

"For communities, economic development issues are overriding. As long as poor communities depend on one or two industries, the companies will be able to hold that threat over communities. We need to start broadening the issue somehow: not jobs versus environment, but community development AND clean communities."

NO COMPENSATION MECHANISM

"There also needs to be better legislation or funds available," she continues, "to help out people who are sick because of pollution. We don't have a compensation mechanism on either side, jobs loss or health. We need better support mechanisms and a better handle on what the effects really are."

Clare believes that, just as workers need to learn more about health and environmental issues, communities need more information about unions.

"Communities don't have a clear idea of what unions are supposed to do," she says. "The first responsibility of the union is to protect the jobs, health, and welfare of its members, and many have a broad perspective on what that means."

Unions have worked on strengthening regulations to protect workers inside their plants for years, she points out. Expansion to environmental issues must be based on workers' hierarchy of needs: "If we felt like we'd reached some level of success on the occupational level, maybe we could move on. But where no one is fighting that battle for them, workers have to concentrate on that. Maybe environmentalists also need to be looking more at occupational health to help win mutual support."

Some lack of cooperation "just might be lack of resources on both sides," she adds. Federal right-to-know legislation (Superfund Amendments and Reauthorization Act, or SARA Title III) giving communities more information about company discharges can be a tool, but the law fell short in not requiring labor representation on local planning boards.

SUSAN WILLIAMS

Susan Williams works with the Highlander Center's Economic and Environmental Interests Program. She has worked on similar issues in the Appalachian region for nearly a decade, and job blackmail is a familiar issue for her, one to which she knows there is no easy answer.

The few efforts that were made by the two sides to come together were too little, too late, Susan says. Early on, the Champion union invited members of the Dead Pigeon River Council to Haywood County for a meeting, but the council members declined.

"Perhaps there should have been a meeting in a neutral place," she says. "They [the union members] wouldn't have come to Newport, and the Dead Pigeon River Council wouldn't have come to Canton." Another late move by the union, the calling in of an independent expert to evaluate company figures and claims, would also have been valuable earlier, she says. "It's obvious the company was telling different people different things. Perhaps it could have been different if some neutral body—separate from the company, an agency of government or the politicians—had been able to intervene."

An independent group could help members of each faction clarify their needs and interests, what they are willing to give and what they might get from cooperation or negotiation.

Still, factors that weight the scales heavily against a coalition of workers and community—permanent long-term jobs with high pay in a company town—exist in Canton. Under those conditions, workers often feel they have to make an economic tradeoff, even in cases where it's clear their own health is at risk. It's possible no amount of early discussion could have changed that balance in Canton, but it would not have hurt to try since feelings could hardly have become any more bitter than they are.

ATTITUDES RELATED TO ECONOMICS

Attitudes can change quickly, Susan adds, when the economics change. "I remember one company in Marion County. The company supplied long-term high-paying jobs to keep out the union, and pretty good benefits. A couple of years ago, it shut down and reopened, contracting out for much lower wages and worse benefits. The strong base of the economy was suddenly ripped out from under people. When that kind of stuff goes on, it makes a difference in how people view the company. They're less naive, but more vulnerable."

Despite situations like the Champion one, Susan is optimistic that change is coming, partly because of economic problems. "Everybody's been touched by this economic downturn in some way," she says. "So many people have been laid off, are working for temporary agencies. Others are beginning to feel like it could happen to them."

Susan also believes there is growing awareness of the environment as an issue that affects everyone. "As many things as the press does wrong," she says, "it's gotten an awareness of the environment into the American conscience. Maybe it's some really basic fear that all this poison is touching off, that maybe we're all going to kill the world. People are acquiring a broader vision. They're more likely to see a bigger picture. It feels different than it did 10 years ago."

LESSONS FROM CHAMPION: BARRIERS TO COMMUNITY AND WORKER COALITIONS

What happened at the Canton paper mill illustrates the barriers that can come between groups like workers and community residents who might be natural allies. The economic and political power of corporations is increased when the natural opposition groups (workers, their unions, community residents, and environmental groups) are divided against each other.

Discussions with union leaders and environmentalists elsewhere in North Carolina show that what happened in Canton can happen anywhere when disunity replaces coalition. What are some of the barriers that confront communities faced with problems that lend themselves to job blackmail?

Corporate Control, Management Rights, and the Drive for Profits

In our economy, all corporate decisions regarding production processes and output level rest with the owners of the firm, the stockholders and their management team. The irreversible rules of competitive market economics drive these decisions.

In the paper industry, for example, any corporate outlay to reduce toxic emissions will be weighed against the firm's quarterly dividend statement and the need to be competitive with other paper firms. In economic terms, paper mills like Champion, Weyerhauser, and Federal Paper (all operating in North Carolina) all decide to reduce marginal costs and maximize marginal revenue.

Paper mills have been operating at high capacity in the last few years as demand for products has increased. Increased production levels alone can account for increased toxic emissions in the absence of other factors. For instance, the paper mill in North Carolina with the biggest increase in toxic emissions between 1987 and 1988 was the Weyerhauser plant in Plymouth. According to a Weyerhauser representative, the increase in chlorine emissions resulted from an increase in production. (1)

This is not to suggest that production levels alone determine the level of pollution. In that same Weyerhauser plant, new production processes will be implemented in coming years to reduce chlorine usage. (The mill produces bleached paper.) This cost, and $8 million spent recently on a pipeline to divert wastewater from a creek to the Roanoke River, constitute investments for Weyerhauser that put the company at some competitive risk. Under market conditions, some firms cannot remain competitive and make these outlays. The Canton Champion paper mill may be an example.

Even if workers and their unions want to pressure a company to change production processes, or to move to a less polluting product, such a demand would fall under the heading of "Management Rights" and is forbidden under the National Labor Relations Act (NLRA). This is not to say that unions cannot act outside the realm of the collective bargaining session and apply pressure elsewhere. But management has no obligation to negotiate any changes at contract time. In other words, workers and their unions have fewer rights than might be generally known when it comes to some aspects of environmental change.

The toxics use reduction movement now underway among environmental and community organizations is meeting strong opposition from corporations even when it makes economic sense to reduce, partly because the debate of this issue in the public domain threatens traditional management prerogatives and powers.

Lack of Communication across Community, Worker, and Environmental Lines

In the Champion Canton mill battle, community and environmental representatives lost a golden opportunity to show their concern for worker health and environment when the company's chlorine leak hospitalized workers. It is possible that some of them weren't even aware of the problem. Clearly, too, the Hodges have failed to make clear to Carroll Israel and the other union members the distinction between their concerns and those of other players on the Tennessee side. No community, environmental, or worker group that we could identify has set up regular communication with groups having different primary interests. Unless and until the factions communicate regularly and clearly, barriers will remain in place.

Lack of Ability in Communities to Obtain Information about Corporate Options

While workers have no legal right to a voice in "Management Rights" decisions, they may at least have considerable information about processes, toxics use, and company economics. Communities, on the other hand, have little such information. Community Right-to-Know provisions of SARA Title III provide some information about toxic emissions, but much more is needed to analyze thoroughly corporate records on pollution and suggest informed options. Most company economic and production information is closely guarded. When a company makes the claim that a new permit limit will mean 1,000 layoffs, it is difficult for communities or workers to analyze the reality of that threat.

Lack of Ability at the Community or Worker Level to Set the Agenda for Public Debate

The massive public relations resources of corporations give them the power to frame the debate. Our society, including mass media and governmental agencies at all levels, is enamored of expertise. The corporation has or can hire the experts. Community, environmental, and worker spokespersons relying on common sense, fairness, and logic have trouble competing, especially when they have no access to hard figures. Even though their opinions may be reported in the media, the questions they must address are framed by the corporate and regulatory agenda. In the Champion permit battle, for example, the issue most discussed was units of color in the water; the issue that most concerned community residents was dioxin. Aesthetic concerns like color are much easier to trivialize than health concerns.

Once the media have accepted the parameters or conditions of the debate, changing those parameters is extremely difficult. EPA's claim that Champion planned to cut jobs regardless of the permit was a one-day wonder in most media; within days, the phrase "the permit, which will cost about 1,000 jobs" was routinely appearing in stories about Champion once more without caveat or examination.

The Assault on Labor in the 1980s

North Carolina, like all states in the South, is a "right-to-work" state. Workers cannot be required to join a union. This fact, along with the extremely low unionization rate in the state as a whole (about 4–6 percent), accounts for a relatively weak labor movement.

However, one should be careful not to make too much of this. Southern manufacturing plants—especially in heavy industry—often have high unionization rates if they are unionized at all. The paper industry is largely a unionized industry. So where does economic pressure on unions come from? Is it plant flight to nonunion settings? Is it basic economic competition between unionized firms? An industry analysis is beyond the scope of this paper. However, it should be taken as a given that at the very least the threat of layoffs is real, whether it comes from competition from other paper firms, from forced environmental cleanup costs, or from other factors affecting corporate decisions.

If we want to assume that there is not a union/nonunion wage differential applying pressure on competing paper mills (because wages have been "taken out of competition"), then it is logical for unions to assume that environmental costs may pose a threat to members' jobs. In the Canton example, this threat was enough to create a situation where the union local found itself confronting management at the bargaining table during the day and siding with that same management against the environmental/community groups at night.

Does this weakness make the unions "company unions," in a situation where the problem is defined as "jobs versus environment"? Are we expecting too much of unions? Clearly unions in the South must have some strength or they would not exist. That they continue at all indicates a certain resilience and grit that may go unappreciated. That they survived in the general assault on labor during the Reagan years indicates that this strength has been maintained at great costs. What are these costs and do they have anything to do with the tensions that we've seen between workers and environmental groups?

The reactionary labor policies of recent years (coupled with the reduction of living standards measured in real terms for the working class) have moved the compass of the debate to the right. Where once labor fought for the closed shop against the union shop, it must now be considered a victory to have any union at all in right-to-work states. In the era of "givebacks," unions now count themselves lucky to maintain wages at current levels, when once they fought for cost-of-living increases. In this new environment, is it any wonder that workers and their unions would choose not to fight for environmental concerns? Many paper mills with unions do not even have a health and safety committee to protect workers in the plant, let alone energy to worry about residents outside the plant. Presidents of North Carolina union locals have varying views on these circumstances.

Eugene Brown is president of United Paperworkers International Union (UPIU) Local 738 in Riegelwood, NC (Federal Paper). His members make bleached paper; in 1988 that plant was the second highest emitter of toxics in North Carolina with emissions totaling 3,177,220 pounds.

"The company has a health and safety committee with members of the local on it; it is not a union-organized committee," Brown says. "The company is doing a better job with environmental problems and is spending large sums of money dealing with these problems. There have been environmental groups putting pressure on the company. We should have both jobs and a suitable environment."

Another local union president, Emmett Parnell of UPIU Local 1167 at the Weyerhauser Mill in New Bern, NC, declares that he would be willing to talk to environmental groups. Like Brown at Riegelwood, he believes, "Dioxin is not a problem, because it exists in such small levels." The Weyerhauser plant was the 13th highest emitter of toxics in 1988 in North Carolina, and the union has a health and safety committee in the plant. Parnell believes that the company is "moving to chlorine-free production because the European market dictates it." Unbleached paper is more established with European consumers.

Discussions with the president of UPIU Local 1423 at the Plymouth Weyerhauser paper mill suggest the same kinds of tensions that confronted the workers at Champion Paper in Canton. The union president at Plymouth is C. B. Robinson.

"We talk to management about chlorine but haven't had any communication with environmental groups," Robinson says. "A month ago Greenpeace came around putting up flyers so we went out with the police and tore them down. We don't know anything about Weyerhauser switching to unbleached paper and we don't want them to switch to unbleached paper anyway because there is hardly any chlorine in the plant anyway. I just went fishing in the Roanoke River yesterday (where the company discharges toxic wastes in the effluent) and I don't see that there is any problem."

Clearly there is a range of sentiments among paper union locals as well as an openness to communicating with environmental groups. Defined union interests are in "bread-and-butter" demands. The wage bargain—and to some small extent, health and safety issues—are the parameters of concern. Wider interests are sacrificed to this standard. This is not to say that labor unions are always so self-contained. In truth, they are often involved in larger social and class issues—when they act as institutions. But when they bargain, at the plant level, there is often no room for these considerations.

Workers, Environmentalists, and Community: Confusion about Who Speaks for Whom

Community residents and workers in the mill may share the same background and interests, yet still clash over toxic pollution, no matter how much the fear of unemployment is commonly acknowledged. In the Canton example, it was the state of Tennessee that actually blocked the permit, causing the confrontation. It is Tennessee and national environmental organizations who continue to protest any compromises in the permit. As Carroll Israel points out, the group that is continuing the battle is made up largely of middle- and upper-middle-class individuals (real estate developers, Chamber of Commerce representatives). Factory workers and immediate neighbors of the polluted portion of the river are represented sparsely, if at all, in those groups. The Hodges make a clear distinction between their goals as community members and those of the developers, but Israel makes no such distinction.

How would environmental groups in North Carolina be expected to react when North Carolina plant permits come due? How should we characterize these environmental groups? Whom do they represent? What do they share with the workers in the mills?

The North Carolina Coastal Federation may attempt to block or modify permits in some of the paper mills on the lower coast (all unionized and represented by the UPIU); it is crucial to understand their role in this struggle. They represent 55 citizen groups in 20 coastal counties with a total membership of 1,800. Three to five percent of their members are fisherpeople whose interests most clearly could be said to correspond to the mill workers. The NC Fisheries Association, an industry group, is the biggest member. Recreational fishers are represented also. The Coastal Federation is examining the pollution from Federal Paper at Riegelwood, Weyerhauser at Plymouth, and Weyerhauser at New Bern. The union presidents and representatives from the Coastal Federation have never talked to each other. The Coastal Federation is currently preparing for a possible challenge to the permits for these mills.

In July 1989, Todd Miller of the Federation explained his group's work: "We look at the water discharge sheets and assess the water quality standards—in this case for the two Weyerhauser plants and the Federal Paper plant. Permits are for five years. Realistically, it's hard to shut a plant down, but we can negotiate on the strength of the permits. There can be many people opposed to a strong permit, and the company can turn them out at a public hearing that would have to be held as part of the process. We would want to avoid these kinds of hearings. It would be better to have communication with the workers outside these kinds of high-visibility meetings."

Miller agrees with union officials that communication would be helpful, but goes on to say that the workers' fear can be exploited in this kind of situation: "The biggest roadblock is misinformation about losing jobs if the permit is threatened."

If it is clear that at least some of the members of the Coastal Federation live in the same community as the mill workers, there are also differences. Many Federation members are small business owners and recreational enthusiasts who may come from more privileged backgrounds than the mill workers. And the workers' fear may be based on more than "misinformation"; their bargaining position may be threatened and jobs may really be lost.

"Assessing the possible impact on workers is not a formal part of our decision-making process," Miller says. "But we do care about getting everybody behind us. We feel that the existing regulations can be made to work short of shutting the plant down. Communication with the affected workers would be fairly important."

Again we hear agreement on the need for more communication before a problem develops and awareness of the need for involvement from the affected workers. However, it would appear that thus far there is no effective organizational form or process to bring workers and environmentalists together. And although the Coastal Federation is more than just a lobbying organization and does indeed represent some workers whose jobs are threatened, its organizational forms, professional, public policy orientation, and at least partial control by business groups distances it somewhat politically and even culturally from the manufacturing workers who work in the mills. This distance could be shortened.

The Role of Federal Regulatory Agencies in Dividing Workers and Community: EPA, OSHA, and NLRB

The several federal regulatory agencies that are involved in community and labor health and safety pose their own barriers to effective coalition building. The Environmental Protection Agency (EPA) regulates industry pollution, and the Occupational Safety and Health Administration (OSHA) oversees health and safety in the workplace. There is no coordination between agencies, and the regulations of one might conflict unnecessarily with the interests of the other. EPA is supposed to regulate what goes on outside the plant, while OSHA regulates inside the plant.

If there is one thing that both environmentalists and workers would agree on, it would be the ineffectiveness of these agencies. Both are painfully slow in issuing minimum standards, and neither acts unless constant organized pressure is brought to bear by lobbying groups, unions, and other advocates. While both agencies resulted from the activism of the '60s and '70s, the movements that spawned them, especially the labor movement, are now under attack and fighting to hold on to the gains of another era. EPA and OSHA, by their ineffectiveness, understaffing, and underbudgeting, now can be said to protect corporate interests from workers and community environmentalists rather than the other way around. Just as environmentalists have a public record of struggle with the EPA, so do unions in their dealings with OSHA.

But there are also built-in differences in how advocacy groups may deal with their respective agencies. EPA is very slow in responding to citizen pressure, but at least in theory anyone can lobby it and bring pressure for change (including workers at the polluting plant). The same is not true for OSHA. If there is a health and safety problem, only the workers and/or their union can file a complaint to get an inspection. For example, dioxin in paper plant effluents is actionable by environmental groups and other lobbies, but if there is dioxin in the workplace, only the workers in the particular mill may initiate a complaint procedure. Thus, while anyone may join a Local Emergency Planning Committee (LEPC) to monitor and have effect on plant pollution, only a relative handful of people can oppose being placed in a dangerous health situation inside the plant. Help for workers breathing dioxin is hard to come by. This may lead to a feeling on the part of workers that a double standard exists. Through regulatory channels, environmentalists can't help workers who are being poisoned, even if they want to.

EPA and OSHA regulations also measure and evaluate the health effects of pollution differently. For example, currently there is no OSHA standard for inhalation of dioxin in the workplace. Until there is a standard, a paper manufacturer may expose mill workers to unhealthful dioxin levels without fear. On the other hand, the EPA soon will issue a standard defining permissible community exposure to dioxin. The EPA is posting warning signs on rivers and waterways downstream from paper plants, but there is no equivalent warning sign inside the plant where workers are being exposed.

Here are EPA's standards for safe intake of dioxins and furans (ppt = parts per trillion): nosebreathers, 158 ppt; nose/mouthbreathers, 143 ppt; and mouth breathers, 28 ppt. The EPA/American Paper Institute five-plant survey conducted in 1987 turned up dioxin levels in pulp of 3–51 parts per trillion. This would expose workers without protective masks who might be breathing through their mouth that day to harmful levels. Worse still, in the June 1989 EPA study of 104 paper mills, the Plymouth Weyerhauser plant had dioxin/furan measurements in pulp between 82 and 318 parts per trillion, well above the EPA standards. This is the plant where the union president joined the local police in tearing Greenpeace flyers off telephone poles.

There are other problems as well with OSHA, the presumed "EPA for workers." The general assault on the labor movement discussed above makes workers in nonunion plants afraid to ask for OSHA inspections for fear of discrimination and termination. The protective language in the act, Section 11c, is rarely enforced, and is so weak as to be considered not a deterrent at all. Therefore, workers in plants that pollute, when asked by environmentalists to support (or at least not oppose) strong discharge permits, are being asked to protect the community outside when there is no standard for the workers inside and when workers are not free to raise health and safety demands without fear of economic retribution. Workers pay a cost for merely opposing exposure to dioxins at the workplace, while environmentalists do not bear similar burdens when organizing against exposure in the community outside the workplace.

The Role of State and Local Governments in Job Blackmail Issues

Competition for industry among states and localities leads to a race to see who can provide the best "business climate," and that race has been prominent in the South in recent years. Too often, however, what's viewed as a good business climate translates into a chill one for workers and community: discouragement of unions; promise of low-wage labor pools; lax environmental, health, and safety regulations; and general government laissez-faire. Part of the effort for those concerned about labor and environmental issues is to maximize the advantages of the increased industrialization of the South (high-paying, skilled jobs and the commitment to training people for them) while minimizing the disadvantages (environmental degradation and the influx of jobs with low pay, benefits, and job security).

STRATEGIES FOR THE FUTURE

Some powerful basic forces in our society work against coalitions between workers and communities concerned about the environment outside the plant. Ironically, these seem most likely to be broken down only after the damage is done—when jobs are gone and the people and/or the environment are already suffering.

The financial and political strength of corporations gives them the ability to frame the debate and the scope of the discussion. The workers, unions, community groups, and environmental organizations are typically underfunded, understaffed, and their limited resources are stretched to the limit. Without basic societal change, the pressures that make job blackmail effective will remain. Still, some strategies for overcoming the historic barriers between these groups suggest themselves.

a. *Unions and environmental groups must set up formal lines of communication before crises develop.*

There was nearly universal agreement that communication was not going on but that it should—and before a problem gets started, not after. Groups must talk to each other before a crisis, not yell at each other in the middle of a disaster at public meetings. Once all parties can meet together, face to face, people will be in a position to hear and appreciate legitimate grievances. This can lead to a rise in consciousness about the broader aspects of economic/environmental problems.

b. *Groups must learn about each other's needs and goals and develop a set of shared interests.*

Environmental and community representatives need to understand and respect the role and limitations of unions and the needs and concerns of workers. Workers need to acknowledge the legitimate concerns of the community. Once communication and networking are established, perhaps each group will agree on a specific proposal and consensus can result. Thus, perhaps union and environmental lobbies can compromise on the level of discharges they will oppose in a permit procedure that will satisfy both parties. This consensus might be easier to achieve when unions and environmental groups are clearly rooted in the community.

c. *Questions of organizational forms, self-determination, and decision making must be settled before coalition building can occur.*

Matters of process must be settled before strategies and policies are chosen. Shared decision making, with community control and democratic process the guiding principles, can build trust and mutual respect. Together, the parties must try to develop an identification as a community with shared interests and concerns. This decision-making process will be complicated by the fact that both unions and environmental groups are often part of larger structures.

d. *The corporate profit motive must be acknowledged as central to any analysis of the problem*.

Workers may align themselves temporarily with the polluting company. Environmental groups may represent small businesses that operate according to their own profit motives (to protect small business profits being threatened by big business profits). These conflicts and contradictions must be met with an analysis that identifies market competition and the drive for profits-at-any-cost to worker or resident as central to the problem. This analysis should make reference to issues of race, gender, and class; capital mobility/plant flight; international market conditions; union/nonunion wage differentials; and interstate rivalries to create a "good business climate." Such an interpretation can help worker, farmer, community resident, and small business person alike share a sense of the manipulative forces to which they are all exposed.

e. *Everyone involved must acknowledge the conflicting pressures each faces, and provide support for difficult choices*.

Workers may live in the community being polluted and fear for their families' safety at the same time that they must worry about putting food on the table. The community as a whole needs to take charge of decisions and set its own agenda, not letting internal hierarchies of any outside group control the debate.

f. *The groups must work together for public policy reform within OSHA and EPA, and for more protective laws and regulations for worker and community.*

If groups can successfully address the organizational and political challenges that have divided them, they can then move on to struggle against some of the institutional barriers. If groups can unite in working for these changes, basic reforms that will reduce the chance for successful economic blackmail may occur. The needed changes are legion, but the following are examples of specific legislative and regulatory changes toward which all groups can work.

- Changes in SARA Title III to require a labor representative on all Local Emergency Planning Committees (LEPCs). These groups are charged under the 1986 federal law with developing plans to control chemical emergencies and with administering the local community right-to-know about hazardous chemicals. The LEPCs have broad representation throughout the community, except for labor. They can be natural meeting grounds where community and labor can address shared safety goals in a nonthreatening setting.
- Stronger right-to-know laws that expand into areas now considered corporate prerogatives—production information, toxics use and toxics reduction information, internal safety and health audits, and analyses of product and process environmental impact.
- Requirements that each EPA industry standard for a hazardous chemical be matched by a corresponding standard for the workplace in that industry, and that both agencies accelerate setting of minimum standards and reevaluation of standards as additional health data are received.
- Strict, across-the-board enforcement of environmental regulations. If environmental costs are indeed a factor in company decisions, their effect will be minimized if all competitors face similar costs. As long as some companies are allowed to avoid spending to decrease their pollution, others will be at a competitive disadvantage if they spend money to clean up.
- Strengthening of OSHA's section 11c to provide real protection for workers who bring complaints about unsafe conditions.
- Reform of the National Labor Relations Act to allow unions to bargain collectively over "management rights" (for example, whether or not to convert to unbleached paper or other nonpolluting technology).
- Strengthening of the plant-closing law to forbid a plant to relocate until it cleans up its pollution outside and inside the plant.
- Enactment of mandatory toxics use reduction laws tied to issuance of new or renewed environmental permits. Many of these chemicals can be eliminated, making the current toxics shell game unnecessary.

SUMMARY

Workers, communities, and environmentalists share long-term interests, goals, and needs, although short-term interests and needs among the groups may conflict. Strong institutional and societal barriers against cooperation among these groups exist and are unlikely to disappear without basic societal change. Industry, and too often government, emphasize and exaggerate those differences to the detriment of both worker and community.

It may be possible, however, to overcome those barriers in specific settings. Precrisis communication and sustained effort at developing shared goals are the keys to coalition building. Work together on issues of mutual concern could begin to build the coalitions among communities, workers, and environmentalists that are in turn the keys to more basic societal changes.

ACKNOWLEDGMENTS

The authors wish to thank the Bauman Family Foundation for its support of this project. The Clean Water Fund of North Carolina has granted permission to print this chapter.

NOTE

1. Toxic Air Emissions in North Carolina, NC. Environmental Defense Fund, p. 9.

23

Texans United Scores Historic Victory against Ammonia Facility in Houston

JIM BALDAUF

It could only happen in Texas, where everything—including industrial arrogance and governmental neglect—sometimes seems bigger than life.

It occurred in the fall, in October of 1990, after a series of catastrophic industrial accidents and oil spills struck the Texas Gulf Coast, where 30 percent of the nation's petrochemical output makes the Lone Star State the leading producer of toxic waste in the nation.

It occurred just weeks before an environmentally ravaged electorate voted to replace Republican Governor "Oil Spill Bill" Clements by electing progressive Ann Richards over the Clements heir apparent, Clayton "Lay Back & Enjoy It" Williams.

Governor Bill Clements, an oil man himself, earned his nickname by telling reporters that a massive Gulf Coast tanker spill was a "so-called oil spill that will probably evaporate in a couple of days." And Williams became a nationwide bad joke during his campaign against Richards when he told reporters that rape was like bad weather; "If you can't do anything about it, you might as well lay back and enjoy it."

Only in such an insensitive environment could a major industrial firm hope to influence a Clements-appointed regulatory agency by arguing that a proposed site for an extremely dangerous ammonia facility was "appropriate" because the neighborhood consisted mainly of "small, poorly maintained houses . . . small junky businesses . . . and very low quality housing." The quotes are taken directly from the company's formal permit application as submitted to the Texas Air Control Board.

Incredibly, LaRoche Industries wanted to site an extremely hazardous 105,000-gallon ammonia storage and distribution facility in a low-income residential neighborhood called Cloverleaf located just east of Houston. The community is near the infamous Houston ship channel (once declared a fire hazard) in a heavily industrialized (polluted) part of town. It's a blue-collar, multiethnic neighborhood containing small homes and businesses as well as schools, clinics, and a hospital. In spite of the dense population, the company's ammonia site application was moving right along the permit pipeline when local residents got wind of the scheme.

That's when neighborhood leaders called Texans United (TU). As special projects director at the Houston headquarters office, I took the call. In our three years of operation, we've seen many outrageous cases, but this turned out to be one of the worst.

Texans United is a 70,000-member citizens' environmental organization that works to help grassroots community groups organize to protect their rights against the powerful Texas industrial/political complex. In addition to organizing, we engage in legislation, litigation, and public information work. We have worked for years to enact environmental reform legislation such as Texas's recent moratorium on waste disposal facilities, a pending waste reduction law, criminal penalties for corporate polluters, an oil spill response program, and other overdue laws.

We are currently engaged in lawsuits against Formosa Plastics Corp., the Texas Chemical Council, Exxon, the Texas Water Commission, and the Environmental Protection Agency. We have sponsored a nationwide Environmental Health Conference and a statewide Grass-Roots Environmental Conference, and have traveled from the Texas Gulf Coast to the island of Taiwan investigating the international polluter, Formosa Plastics.

Whether you're in Texas or Taiwan, pollution solutions begin with people and with grassroots organizing. After we got the call on the Cloverleaf ammonia facility, neighborhood residents and TU organizers obtained a copy of the permit application and used it to very good advantage in a series of public meetings conducted to educate and mobilize community opposition.

"The arrogance of the language in the formal permit application reflects LaRoche's lack of concern for the community and its safety," said Rick Abraham, Texans United's executive director. Community leaders and TU staffers decorated the meeting halls with wall posters quoting the offensive language to help show the attitude of the company toward the community.

The first community meeting, held on October 3, turned into a heated exchange when a packed hall of irate residents greeted the unexpected (and uninvited) arrival of Texas Air Control Board officials from Austin with hostile questions and objections to the ammonia scheme. Residents cited an earlier ammonia tank truck disaster on a Houston freeway where a single 3,000-gallon tanker wrecked and ruptured, killing seven people, hospitalizing 50, and requiring the evacuation of more than a thousand others. The citizens also questioned the Air Board about LaRoche's insulting permit application, asserting that "small, poorly maintained houses" were not good reasons to locate the facility in the midst of their residential neighborhood.

Karla Land, the Cloverleaf resident who spearheaded the local oppostion, said, "I believe the Air Control Board officials flew into our first meeting expecting to smooth things over with a few fearful homeowners. But when they looked around the hall and saw those posters, they knew they'd walked into a hornet's nest of organized opposition. The Air Board seemed a little stunned but promised to consider our objections and then headed back to Austin."

The second public meeting proved to be even more interesting, she reported. "We invited LaRoche Industries to tell their side of the story—along with the Air Board and the community leaders—at a meeting of neighborhood residents, school board members, politicians, and state and county officials who we organized to support our opposition."

When the Air Board and LaRoche first tried to seize control of the planned meeting, and then to back out entirely, community leaders and organizers insisted that the meeting would be held as orginally planned.

Abraham said, "We told them that if they wanted to boycott the meeting and be represented on the stage with empty tables conspicuously labeled 'Air Board' and 'LaRoche Industries,' well, that was up to them, but it wouldn't look very good."

Tension and uncertainty mounted as the day of the meeting approached. "To our surprise and delight," said Mrs. Land, "after a flurry of last-minute letters, faxes, and phone calls, the Air Control Board arrived on time and asked permission to read a letter it had just sent to LaRoche, a letter saying it would not approve the permit because the idea was just too unsafe!" We had won the battle!

The crowd rose to its feet and cheered the victory. They even cheered the Air Control Board members whose sheepish grins said that public approval was a whole new experience for them. Then Karla Land read a letter that LaRoche had faxed to her, just an hour before the meeting, saying it was formally withdrawing its application and was giving up on the project. More cheers . . . with the Air Board members joining in! We had actually won the war!

The community meeting turned into a victory celebration! And what an historic victory it was! Texas Air Control Board officials admitted that never before in the agency's history had they turned down a permit application. The time was ripe for a victory . . . change was blowin' in the wind. The people of Texas were sick of the rape and plunder of their state's environment. The "good ol' boy" Republicans in Austin were about to be voted out of office. The state regulatory agencies knew that "bid'ness as usual" was a thing of the past. The handwriting on the wall bore the new signature of a progressive reform administration. And grassroots environmental organizers put together just the right combination of experience, strategy, and tactics to make all these factors work for the people of Cloverleaf and the Texas environment.

As the victory meeting broke up and folks headed home, I reflected that the early frontiersmen and cattlemen who founded the Lone Star State would have been good members of today's grassroots environmental movement. In their day, it was a "hangin' offense" to poison a neighbor's water well. As I began to wonder if we had let the Air Board and LaRoche Industries off a little too easy, I realized I wasn't thinking clearly and needed to get to bed. I went back to my own low-quality housing and had a high-quality night's sleep, reassured that the power of organized people could defeat the power of organized money.

SECTION B

THE STRUGGLE FOR OCCUPATIONAL HEALTH CLINICS

24

Organizing for a Statewide Network of Occupational Disease Diagnostic Centers

DOMINICK J. TUMINARO

Today there is a coordinated network of seven occupational health diagnostic clinics in New York State, strategically located to serve workers in every geographic area. Five are located in major cities: New York, Buffalo, Rochester, Syracuse, and Albany. The remaining two are at Stony Brook, Long Island, and Cooperstown. The latter is aimed at injuries and illnesses experienced by farmers and farmworkers.

Funding for the clinics is provided by the state of New York through an assessment on workers' compensation insurance premiums paid by the state's employers. Current state support is nearly $3 million annually. It is expected that the clinics will derive some portion of their income from workers' compensation reimbursement for medical expenses, union and employer insurance, and fees for service. But no worker without a means of paying will be refused access to clinical services.

Establishment of the clinics network culminated a two-year campaign by a coalition of labor and public health advocates spearheaded by the New York Committee for Occupational Safety and Health (NYCOSH). These activists recognized that, while many workers were suffering from occupational diseases and needlessly dying in large numbers, efforts to identify and protect worker populations at risk were virtually nil. Given the failure of the market to prevent occupational disease, it was critical that the state step in and fund occupational health services.

The safety and health activists who initiated the clinics campaign believed the establishment of a network of clinical services would further a number of important occupational health and political objectives.

AN INVISIBLE PROBLEM

First, the campaign itself would increase public awareness of what was then largely an invisible problem. Elected officials, the general public, and even trade union officials were only beginning to understand the magnitude of workplace illnesses and to recognize the possibility of preventing the exposures that produce disease. Organizing for clinical services might help give the issue of occupational disease greater priority on the agenda of organized labor in the legislative arena.

Second, by identifying industries and work processes that were generating harmful exposures and disease, the clinic network could contribute to early detection and prevention of occupational illness. At workplaces where workers had been diagnosed with illnesses related to occupational exposure to toxic substances, the clinics could contact both labor and management and offer to provide industrial hygiene and education services to prevent exposure. In addition, the establishment of a uniform statewide data collection network would make it possible to determine the extent and dimensions of occupational disease among different worker populations. This would lay the basis for targeting of educational and regulatory efforts by the clinics and the state.

Having identified a need for specialized clinical services to diagnose occupational illnesses and target preventive efforts, the New York City–based NYCOSH convened a breakfast in November 1985 and invited occupational health advocates from unions, government, and academia to begin to discuss what a network of clinics would look like and to devise a political-strategic approach to achieve its realization.

Approximately twenty-five professionals and activists, including industrial hygienists, attended the initial planning meeting in New York City. Representatives from other COSH groups participated. In New York State, active COSH groups exist in Buffalo, Rochester, Syracuse, and Albany, as well as New York City.

A subcommittee was established to meet regularly and continue planning and organizing. An early phase in the effort focused on obtaining $150,000 in state funds for a study of the magnitude of occupational disease in New York State. The study was carried out by the Division of Environmental and Occupational Medicine at Mount Sinai School of Medicine of the City University of New York under the direction of Philip J. Landrigan, M.D., and Steven B. Markowitz, M.D. The final study report, which included a proposal for a clinics network and estimated budget for a five-year demonstration project, played a critical role in legitimizing the clinic campaign by documenting the scope of the occupational health problem in New York State.

A central organizing feature of the campaign was the establishment of a large advisory board composed of nearly 100 labor union representatives, occupational health physicians, industrial hygienists, and governmental representatives. Many labor and occupational health professionals met and got to know one another within the framework of the advisory board which met monthly over a period of 14 months. They participated in discussions that elaborated the particular needs the clinics would serve and addressed issues of financing and governance.

INFRASTRUCTURE FOR A CAMPAIGN

After the Mount Sinai study was completed, the political organizing phase became a central focus. The advisory board members, particularly the union representatives, plainly constituted a political infrastructure for a campaign to persuade the legislature to approve funding for the proposed statewide clinic network. A second network of activists around the state who were not participants in the advisory board was coordinated by NYCOSH and other COSH groups. This coalition of labor and occupational health professionals carried out a systematic lobbying campaign that succeeded in making the clinics a priority on the agenda of the New York State AFL-CIO and ultimately on that of the legislature as well.

The establishment of the clinic network in two short years is a testament to the effectiveness of the coalition of trade unionists, occupational health activists, and professionals. The presence of the COSH network throughout the state provided an important grassroots foundation for the campaign and generated the political momentum that transformed the legislative study on occupational disease into the reality of the statewide clinics network.

The clinics campaign had additional benefits for the COSH movement. Grassroots activists had an opportunity to become more sophisticated about the legislative process and to learn to deal more effectively with government regulatory agencies, members of the state bureaucracy, and the labor hierarchy. In short, we learned a lot that will help us grow and develop politically.

It is critical that the coalition that established the clinic network remain active to ensure that its development continues in ways that serve the interests of the workers for whom the network was intended. We can anticipate that corporations will pressure the legislature to cut the clinics' funding as workers discover that their illnesses are occupational diseases. There is also a danger that demands for financial self-sufficiency could force the clinics to take on programs for employers, such as preemployment physical and drug testing, that would change the mission of the clinics. Only continued vigilance by the coalition of labor and public health advocates can ensure that diagnosis and prevention of occupational disease remain the principal objectives of the clinics.

25

The Occupational Health Clinical Center Network in New York State

ACHIEVEMENTS, PROSPECTS, AND CONSTRAINTS

MICHAEL LAX

In New York State, the demonstration of a significant burden of disease and cost associated with workplace hazards, coupled with a clear lack of occupational health resources, made a publicly funded Occupational Health Clinical Center (OHCC) network possible. (1) The impetus to study the problem and, subsequently, the successful push to obtain funding through the state legislature came from the occupational safety and health movement (with the COSH groups in the lead) and the trade union movement. The state AFL-CIO made funding a priority and continues to carry the clout necessary to make its weight felt in the halls and backrooms of the legislature. (2)

The OHCCs were to fulfill several functions including:

a. Accurate diagnosis of occupational illness by practitioners trained in occupational medicine;
b. Collection of data to identify high-risk industries and the development of intervention strategies to address the needs of workers in those industries;
c. Education of other health care providers, raising awareness and skills in occupational health.

Acute injury care, preemployment physicals, and routine compliance examinations would be left to the industrial clinics and other entrepreneurs. The establishment of an OHCC network was a major achievement, opening up the possibility that, for the first time, workers would have access to statewide occupational safety and health resources independent of management control. The dominant vision of the OHCCs saw independence from management dominance, universal accessibility, and technical expertise as the cornerstones upon which the OHCCs would be based. This vision sprang from a broader perspective on workplace relations between management and labor that acknowledged conflict and a need for workers to struggle on their own behalf to improve workplace conditions.

Historically, management has shown that, if left to its own volition, it will not reduce hazardous exposures at the workplace. From Gauley Bridge through the asbestos cover-up to the recent revelation of massive tampering with environmental monitoring results by coal operators, management has consistently sought to deny or minimize workplace hazards and to shift the costs of occupational disease to workers and their families. (3–5)

Government, under pressure from labor, has recognized this history and has acted to curb at least the most flagrant abuses. These efforts are undermined by inadequate laws, underfunding of government agencies and grant programs, and understaffing of regulatory agencies. As a consequence, government efforts are not protecting workers from occupational injury and illness.

Because of management and government's failings, the burden of pressing for safe and healthful working conditions has fallen, and continues to fall, on workers and their unions. Labor efforts have been crucial to every positive step made in occupational safety and health, including the OSH Act, the Mine Safety and Health Act, Right-to-Know laws, and the establishment and support of the Councils on Occupational Safety and Health (COSHes). Day to day on the shop floor, workers recognize hazardous conditions and symptoms of toxic exposures, and struggle to put controls in place.

Given this history, it is clear that safer and more healthful working conditions depend upon knowledgeable workers who are organized and possess the power and will to persist in pursuing their goal. Unions, COSHes, and OSH activists have recognized this reality and have organized themselves to provide safety and health resources to workers. Many of these groups have suffered, however, from a lack of access to technically trained, professionally recognized resources, including occupational medicine physicians and industrial hygienists. The OHCCs were meant to fill this void.

This history of occupational safety and health led the coalition to seek public funding for the OHCCs. The coalition reasoned that independence from management would allow OHCC staff to make diagnoses and recommendations free of the financial pressures inherent to industry-based practice. Public funding created the possibility of an independent setting. Practitioners would be freer to make work-related diagnoses, and to aggressively pursue preventive activities. Unions and COSHes saw themselves interacting closely with the centers. The OHCC was seen as an institutional resource of the occupational health and safety movement, acting within the public health tradition, and sharing the movement's belief that improving workplace health and safety depends upon worker knowledge and power.

It should be noted that the coalition backing an OHCC network was not unanimous in its vision of the future network. The dominant view of the goals and role of the network was outlined above, and generally continues to serve as the basis for network practice. However, dissenting opinions exist and can affect the practice in significant ways. Important dissenters include business interests who were not active in the coalition, but whose concerns have modified OHCC practices nonetheless, those (mainly professionals and politicians) who view occupational safety and health as a technical problem best left to the experts to solve, and trade unionists who see the network simply as a union benefit. For them, the OHCC network is a service for union members only, not a tool for the empowerment of all workers whether organized or not. The impact of these opinions will be described in the various sections that follow.

Are the OHCCs living up to their promise, or at least making strides in the right direction? This paper will explore this question with a broad overview, and from the particular perspective of one of the OHCCs. It will argue that OHCC development has been uneven throughout the state, and that four major constraints prevent any of the OHCCs from completely fulfilling their original vision. These constraints include:

a. Socialization and education of OHCC professional staff that often serves to conservatize and depoliticize individuals and their practice in ways detrimental to worker health.

b. The agenda of each OHCC's parent institution, which may conflict with the goals of the OHCC.

c. A political economic context characterized by economic deterioration and a management assault on labor that has greatly weakened the labor movement. The consequences of this context are negative for efforts to prevent occupational illness and injury.

d. The dependency on public funding creates a positive space for OHCCs to operate, but also puts OHCC funding at the mercy of state politics. Maintaining political support requires a rhetoric of harmonious existence in a context of labor–management conflict. The result is a decrease in OHCC aggressiveness in an effort to mute political opposition.

Recognition of these serious constraints should not obscure the significant achievements of the OHCC network. Confronting the constraints is necessary, however, if we hope to overcome them. To the extent the OHCC network is able to overcome them, its original mission will come closer to realization.

WHAT DO OHCCs DO?

As envisioned in the study by Mt. Sinai, OHCCs would provide both diagnostic and preventive occupational health services. A multidisciplinary team of physicians, nurses, industrial hygienists, educators, social workers, and administrators would work together to address these areas. These teams would operate under the guidance of a broad-based advisory/governing board, on which representatives from labor, management, and the community would sit. What is the reality?

(The OHCCs that were part of the original network are located in Buffalo, Rochester, Syracuse, Albany, Mt. Sinai in New York City, and Stony Brook on Long Island. Two other OHCCs currently part of the network are: an OHCC in Cooperstown that focuses on agricultural hazards and has a different history and orientation. It is not included in this discussion. An OHCC at Bellevue Hospital in New York City very recently joined the network. Because of its newness, it also will not be discussed.)

Uneven Development

For reasons to be discussed below, the OHCCs have developed quite unevenly. Certain OHCCs lack staff, even including occupational physicians and industrial hygienists, with a resulting inadequate level of services. The level of compensation claims varies quite widely from OHCC to OHCC, reflecting different clinical practices and possibly different approaches to occupational medicine. Some OHCCs have failed to develop multidisciplinary staffs or have not been well integrated into the work of their local COSHes and labor groups. While the focus here is on the Central New York Occupational Health Clinical Center (CNYOHCC), the challenges identified are relevant to other OHCCs in New York and nationally.

Compensation

The overwhelming majority of OHCC staff time is related to diagnosing, following up, and billing individuals with possible work-related disease. Identifying occupational illnesses and helping workers through the legal morass of Workers' Comp has been the OHCC's most obvious and immediate success. The several hundred patients who have been evaluated at the Syracuse OHCC and who have filed Workers' Comp claims will attest to the difference the OHCC has made in establishing their cases. To many of them, establishing their claims for compensation means being able to feed a family, stave off a mortgage foreclosure, or obtain needed medical care. Consequently, the concrete effects the OHCCs have had for individual victims of toxic exposures and hazardous conditions should not be minimized.

Prevention

Victims of occupational illness must be compensated and need medical support. But these victims should never have gotten sick in the first place. The prevention of occupational illness requires the improvement of workplace conditions. This is the task that requires massive infusions of time and resources. Ironically, at the OHCCs, it is the task given short shrift, one whose results are hardest to see, especially in the short term.

Industrial hygiene, education, and organizing are the activities central to prevention. All of these activities require access to workers, and many require access to the workplace. To be successful, an intervention with preventive goals cannot be carried out by an OHCC alone.

The CNYOHCC has been most successful in gaining access to workplaces as a result of individual patient evaluations. An employer is told that a work site evaluation is necessary to aid the diagnostic process, and also as part of making return-to-work recommendations. Understanding that one visit is all the hygienist usually will be allowed, the inspection is designed to focus on the specific patient issue, but with an eye out for broader hazards as well. Recommendations resulting from these inspections typically address concerns for workers beyond the original patient.

Access to the workplace also has been attained through unions coming to the CNYOHCC with specific issues that need investigation. In these situations, CNYOHCC serves as the union's expert put forward to evaluate, and sometimes counter, the work of experts employed or hired by management.

Educational activities do not necessarily require access to a workplace. CNYOHCC has participated in a wide variety of educational activities, the majority of which have been aimed at workers and organized by the local COSH group (CNYCOSH), labor unions, or other labor-connected groups. For the most part, these have been issue-oriented (for example, indoor air quality, cumulative trauma disorders) and designed to appeal to a cross-section of workers from a variety of work settings. The other broad-based group addressed by CNYOHCC's educational efforts has been health professionals in practice and in training.

Part of the reason for the relative short shrift given to preventive activities was suggested above: the burden of evaluating individual workers and the limitations of staffing. Another part of the problem is the need to justify the OHCC network to legislators who control the funding. Though the State Health Department has not noted it to be a major legislative concern, counting patients served is perceived by many OHCCs as the most tangible way to justify their existence. Even if these parts of the problem were overcome, a key constraint to prevention would remain: the OHCC's lack of power to gain access to workplaces and its lack of power to enforce the recommendations made.

If, as proposed earlier, workers themselves are central to any effort to improve workplace conditions, then OHCC preventive activities are dependent on workers for implementation. This requires workers ready to take collective action around health and safety issues, which in practice almost always means workers who are unionized.

But, workers require more than union membership to reduce hazards. They need a union willing to make health and safety a priority, and strong enough to pressure management into taking positive action. Given the current economic climate, union strength has ebbed and occupational health is a lower priority than survival. Consequently, CNYOHCC's preventive activities have had limited obvious short-term success. Unions and workers have been successful in implementing some of CNYOHCC's recommendations, but these usually have been on issues of narrow focus and modest economic impact on the employer. The CNYOHCC has provided technical resources to several union organizing drives. An organizing drive where health and safety was a central issue in one small workplace was victorious, but in another workplace the election was lost, and in a third the drive stalled in a very early stage.

Unions have typically not been able to persuade employers to hire CNYOHCC to provide services such as medical surveillance or industrial hygiene. Nonunion employers who have approached CNYOHCC have not elected to use CNYOHCC, except in a few instances to provide OSHA compliance examinations. Consequently, in a number of instances, the role of the CNYOHCC has become more that of a critic of existing or proposed safety and health programs, and less of an actual provider.

Given the choice, employers will choose occupational health "consultants" marketing their services to corporations, over an OHCC perceived as sympathetic to workers. In central New York, entrepreneurs filling this niche are readily available to business. Engineering companies provide air sampling; industrial clinics contract with employers to provide acute injury care and routine surveillance exams; physicians under contract to Workers' Compensation insurance carriers examine workers pursuing compensation claims; and rehabilitation specialists assess work capacity and recommend work placements for injured workers at the request of employers and insurance carriers. The CNYOHCC is often called to critique the analysis being put forth by these entrepreneurs. On other occasions, the CNYOHCC may be initially involved in a health and safety situation, usually at the request of a union or an individual worker, and management subsequently hires a "consultant" to counter CNYOHCC's perspective.

Clearly, prevention does not occur overnight, and the effects of prevention may not be evident for a long time. Evaluation of prevention should focus on process as well as outcomes. Seen in this light, CNYOHCC activities have certainly disseminated information to a wide range of workers, and have stimulated thought, discussion, and the beginnings of activity in many workplaces. Additionally, ongoing relationships with a number of local unions are being nurtured. These relationships hopefully will bear preventive fruit. None of the preventive activities aimed at workers could be successful if the CNYOHCC did not participate as part of a network including CNYCOSH and local and regional labor organizations.

Environmental Exposures

Illnesses resulting from nonoccupational toxic exposures clearly share many characteristics with work-related illnesses, and the expertise of the OHCCs could be utilized to address this issue. The OHCC network has no explicit policy regarding environmental disease, but there is a tacit understanding that individuals with an environmental illness are appropriately evaluated by the OHCCs. The OHCCs have varied in their response to this understanding, leading some in the environmental movement to be disappointed by the performance of the network. This problem needs a full and open discussion by the network with a commitment to serve the needs of individuals exposed to environmental toxins.

THE SOCIALIZATION AND EDUCATION OF OCCUPATIONAL HEALTH PROFESSIONALS

Most occupational health professionals would deny they are biased or that their work is anything except based on objectivity. The belief that health professionals become *scientists* separated from, and unaffected by, their economic, political, and social milieu is central to the way these professionals see themselves, and the way they portray themselves to others. *Objectivity* and specialized training are the critical props to these professionals' prestige and credibility, from which their power flows. In reality, however, professionals spring from the same context that produces managers and workers. They are socialized and educated in that context, and they occupy specific niches in the class, gender, and race hierarchy of that context. As a result, many of their activities have the effect of supporting and perpetuating the existing economic, political, and social structure, and as such are often inimical to worker health.

Class Identification

Occupational health professionals are privileged, many by their class, gender, and ethnic background, all of them by their training. After being processed by their education, they are all privileged in the sense of having relatively more choices than the vast majority of other people in this society. They are relatively better paid, have more job security and mobility, and more autonomy. As a consequence, their work experience is far removed from the worker on the assembly line or at the VDT keyboard, and their understanding of, and empathy for, those workers is reduced.

Secondly, for most professionals the goal, or at least the outcome, of their education is a corporate or corporate-connected job. Recognizing this reality, most training is aimed at building skills for the corporate setting.

An important underpinning to professional training is the ethos of scientific certainty. Scientific certainty requires replicated studies, statistical significance, elimination of all confounding variables and biases, and impeccable exposure data before it can be suggested that an exposure is causing a disease. The result is a reluctance to impute causation in individual cases of possible occupational illness, or to take preventive steps to control hazardous exposures. Uncertainty becomes a justification for inaction, and the consequence is that workers continue to get sick and die waiting for the definitive study to be done.

Another essential part of the training instills an elite sense of knowledge ownership among each professional group. Each profession expends major energies in defining its own unique role and protecting that role from incursions both by other professionals and by nonprofessionals.

As a consequence of their privilege, their training, and their dependency on corporations for their livelihood, most professionals tend to identify, either overtly or unconsciously, with managers and corporations rather than workers and unions. The result is an approach to occupational health that allows corporate costs and profitability to become inordinately important in health decisions. Workers are viewed mainly as objects that management controls and manipulates, and who often need reassurances about their misplaced and exaggerated occupational health concerns. Professionals see themselves as creating and controlling occupational safety and health knowledge. It is their job to train workers, and it is the worker's job to learn and put into practice what the professional tells them. This view of workers as a necessary nuisance leads professionals to ignore the collective knowledge and skills workers develop at their jobs about their work.

And it leads to a failure to appreciate the critical role workers play in identifying and controlling workplace hazards.

The training, corporate connections, and belief in scientific certainty lead to a clinical practice that tends to minimize workplace hazards, shift responsibility for ill health from the workplace on to individual workers, and a less-than-aggressive pursuit of preventive measures in the workplace. Sympathy for management often translates into a profound and automatic distrust of patients who are seen as malingerers or out for some secondary gain, especially when benefits such as Workers' Compensation or disability are an issue. Physicians' sympathy for management and their adherence to the credo of scientific certainty leads them to place an often insurmountable burden of proof on the worker to show s/he was exposed to toxic conditions at work, and that those exposures have been "scientifically proven," beyond a shadow of a doubt, to cause the effect the patient is claiming. Lifestyle is often invoked to explain a worker's illness, diverting attention from the workplace. Industrial hygiene data are used to show that no exposure occurred or that the exposure level was too low to possibly have any adverse effects. A lack of literature may be used to deny an association between the workplace and an exposure, when the lack of literature really means the question has not been studied. All of these arguments are used to deny that workplace hazards exist.

Academics and Government: A Little Space?

Professionals in academia and government may have more room to deviate from the practice described above. However, they share the same background and training as their corporate and consultant counterparts, which keep them from straying too far off the beaten path. Job security in academia comes from attaining tenure, and this requires achieving credibility among one's academic peers. Professionals who challenge the idea of a professional monopoly of expertise, or who apply their knowledge to activities that corporations find threatening, do not generally build peer credibility. They are ostracized as ideologues deviating from the mainstream. Additionally, they have difficulty finding support for research leading to the publications tenure requires. Consequently, aspiring academicians find it hard to resist the conservatizing influence of academia, and their occupational health practice reflects that influence. This description of academia should not be construed as a claim that worker-oriented professionals cannot find a home in academia, because many have. But it should be recognized that, despite the rhetoric of academic freedom, there is an academic "party line" that is overtly challenged at the risk of one's academic career.

Government offers a niche for some worker-oriented professionals, but there are significant constraints in these settings as well. As the economy continues to stagnate, government increasingly sees its role as that of stimulator and protector of business interests. The efforts of the Office of Management and Budget (OMB) in obstructing stricter OSHA standard setting throughout the Reagan–Bush years, and more recently the antiregulatory activities of Dan Quayle's Council on Competitiveness provide clear examples of this role. The protection of business through deregulation, coupled with the fiscal crisis of the federal government and many states, combines to reduce the resources allocated to occupational health, and the aggressiveness with which new regulations are sought and existing ones are enforced.

Government occupational health services are at the mercy of their political masters. An agency or service can be built by one regime and then gutted by the next. Political priorities of the federal and state administrations determine more than staffing levels at the various agencies. Agendas, priorities, and aggressiveness are also strongly influenced by the administrations, often making it difficult for professionals in an agency to pursue what they would consider a public health agenda.

Bias, Objectivity, and Advocacy

The discussion above illustrates the milieu all OSH professionals operate in, and demonstrates the impossibility of fulfilling the lament often heard from the occupational health professional: "If only we could get rid of the politics and concentrate on science." The problems OSH professionals deal with are not just technical problems with technical solutions. In the United States, they originate in the social and economic structure of capitalism, and their solutions have effects on that structure. When a worker gets asbestosis or lead poisoning, the "cause" of those health problems is not mere exposure to asbestos or lead. An employer chose to use those materials, chose to control or not control the exposure in a particular way, and chose how fast the material was to be used. "Knowledge" about the level of hazard these exposures represent is heavily influenced by corporate producers and suppliers, and as a consequence OSHA regulates exposures with employer profitability concerns in mind. (11) Solutions in this context have implications for the foundation of the capitalist system, threatening profitability and management control over the workplace. (12)

Every professional approaches the problem and offers solutions based on his/her world view and the constraints of his/her position in the social structure. These are the "biases" with which every professional operates, and their diagnoses and solutions to preventive problems betray these biases, bringing them to light for inspection. The biases of the corporate-connected professional are usually transparent. The bias of the technocrat is more opaque but present nonetheless. The technocrat demands exhaustive study for every problem and is reluctant to take action until *scientific certainty* is achieved. S/he is further guided by a belief that "good science" sooner or later is translated into "good policy" by public health-minded government or by enlightened employers. By failing to recognize the underlying dynamics of capitalist society, this approach may produce results positive for worker health, but they are usually too little, too late. The union-linked professional is biased in the direction of increased sensitivity to union and worker concerns.

Instead of pretending that biases do not or should not exist among them, occupational health professionals should acknowledge their biases and recognize their function within their social context. (13, 14) This finally brings the discussion back to the OHCCs. An OHCC can and should diagnose occupational illness and contribute to the prevention of occupational illness. But it should do it with a recognition that workplace illness is an outgrowth of a specific social structure. That structure is dominated by management, and solutions to safety and health problems require knowledge, participation, power, and collective action on the part of workers. The role of the OHCC is to provide resources and to serve as an advocate of workers in this struggle.

On a clinical level, the resource/advocacy role requires a basic trust in worker/patients. The beginning assumption is that workers are telling the truth about their illness. The burden of proof should be on the employer to show that a worker was not exposed to a hazard or is not being truthful about some aspect of his/her illness. Clinicians at the OHCCs are charged with observing and suggesting new connections between workplace conditions and illness. The clinician should not confuse scientific certainty with legal certainty or with public health certainty. Protecting worker health requires action before illness occurs, and usually means acting in a context of significant uncertainty. In this context, the benefit of the doubt should go to the worker, not to the exposure, the workplace, or the employer. (15)

On the preventive side, worker participation and empowerment at the workplace require that OHCC education be made a priority and that education and research be carried out *with* workers, and not for or on behalf of workers. This model of participatory education and research is in distinct contrast to the usual mode where experts generate knowledge and transmit it to passive and unknowledgeable students. The participatory model recognizes workers as subjects instead of objects, capable of understanding their circumstances and of changing those circumstances. (16, 17)

Critics may charge that this is hardly an "objective" role for the OHCC to play. In light of the entire previous discussion, it should be clear that there is no contradiction between objectivity and advocating a particular position. Objectivity and lack of bias are often defined as synonyms when they really have different meanings. No one lacks a bias, but despite that reality, one can still be objective. An objective professional is one who approaches all worker/ patients the same way, and who consistently uses the same techniques in analyzing hazardous situations. When those chosen techniques do not give him/her the expected or desired answer s/he says so. It should also be clear that objectivity is not synonymous with scientific certainty. Occupational disease prevention demands action in the face of uncertainty. Objectivity implies a consistent approach to safety and health. It does not imply inaction while waiting for the perfect study to be published. Following these tenets, an OHCC can serve in the role of worker advocate while still maintaining objectivity.

INSTITUTIONAL INTERESTS

While the state legislature and the Health Department shape the contours of the OHCC network in important respects, real day-to-day control of individual OHCCs is located in their staffs, their parent institutions, and their advisory/governing boards. The institutional location of the OHCCs varies, with four residing in academic centers (medical schools), one in an HMO, and one in a free-standing union-based structure. All of these institutions possess their own agendas and interests that can impact on OHCC activities in major ways.

Academic OHCCs

Academic settings offer both pluses and minuses to an OHCC. One major advantage to an academic center is the resource base that can help support an OHCC. This resource base may include: clinic space, salaries, a network of diagnostic and treatment services, opportunities for research collaboration and funding, and the potential development of a group of colleagues to interact with. The academic setting allows access to professionals in training, who can be reached with OHCC educational efforts. The prestige of an academic affiliation is helpful to the OHCC in situations where it can add weight to OHCC statements or recommendations. Academic prestige is also obviously useful in obtaining grant funding. Occupational physicians are much easier to recruit to an academically affiliated OHCC. As noted earlier, academia, despite the limitations, offers more potential space for worker-oriented activities than most other settings.

The potential negatives of residence in academia should not be overlooked. Academic programs are not built or expanded by challenging prevailing orthodoxies or powerful interests. Controversies raising the ire of the business community make academic institutions nervous and can result in subtle or obvious pressure on OHCC practices. As discussed earlier, the pursuit of tenure adds pressure to dwell within the mainstream. Beyond these larger issues, academic institutions are notorious for their personal and departmental institutional battles. Becoming embroiled in this type of academic combat can be disastrous for an OHCC.

The balance sheet for the academic OHCCs is mixed. Some OHCCs have benefited from their academic connection, and any outside pressure has been kept low-level. Mt. Sinai under Dr. Selikoff built an institutional academic reputation that is long-standing, making its program much less vulnerable to attack. Other OHCCs have suffered from institutional resistance to accountability to a community-based advisory board. Intraacademic politics also have played a negative role with major consequences for one OHCC. The result of these difficulties has been one academic OHCC that has failed to build a functioning program and a second OHCC with an originally viable program now struggling to survive.

OHCC in an HMO

Residence in an HMO results in a somewhat different set of constraints than residence in academia. The advantages to the HMO in housing an OHCC include the ability to market itself as a more comprehensive health care program with a community orientation. Revenues can be generated through Workers' Compensation claims. What is most threatening to the worker-oriented OHCC in an HMO is that HMO services are marketed to employers offering the plan as a benefit to their employees. As a consequence, HMOs would not be overly eager to offend employers by filing too many Workers' Compensation claims or by pursuing an aggressive preventive program stressing worker participation. The HMO housing the OHCC in New York State has resolved some of this inherent tension by creating two occupational health programs with separate staffs. One program markets services to business offering acute injury care, compliance examinations, and preemployment physicals, while the other program is part of the state-funded network.

Union OHCC

A union-based free-standing OHCC theoretically would offer the most advantageous institutional setting for an OHCC to chart a worker-oriented course. The network's union-based OHCC started strongly, but interpersonal, interunion, and board/staff conflict has driven staff away, eliminated interaction with the local COSH, and severely limited the effectiveness of the OHCC. A major source of the problem was that governing power was concentrated in the hands of a few individuals representing a limited number of unions. Their vision limited both who would benefit from the OHCC and the nature of the services provided. Occupational health, in this view, became defined as low-cost medical care for union members. As a consequence, little attempt was made to direct services to high-risk groups or to unorganized workers, and preventive activities lost out to indiscriminate mass screenings. Little emphasis was put on empowering rank-and-file workers through training, work with the COSH, or with local health and safety committees.

Another consequence of the tensions at this OHCC has been difficulty attracting and keeping a professional staff trained in occupational health. Part of the problem has been the paucity of such individuals, a problem plaguing the rest of the network as well. Another major part of the problem has been staff relations with the OHCC's governing board, with tensions coming to a head over staff objecting to board micromanagement, and the board fearful of losing control of the OHCC. A third part of the problem that should not be underrated is the reluctance of many professionals to be identified with an explicitly worker-oriented entity. A union-based OHCC is dismissed by business, academics, and other professionals as "biased and unobjective," and therefore not a congenial place for the "unbiased and objective" to practice. Fear of credibility loss as a result of this ideology keeps even professionals sympathetic to labor away.

Advisory and Governing Boards

All of the OHCCs have community-based boards that are either advisory (all of the OHCCs except the union-based one) or governing. The original intent of these boards was to provide the constituencies served a voice in OHCC functioning. The composition of the boards, their specific role, and their power were, and remain, subjects of debate and struggle. A strong labor/ community-based advisory board is crucial in protecting the mission of the OHCCs and in providing a counterweight to any pressures emanating from the OHCC's parent institution. Such a role presupposes a board that not only is dominated by labor and worker-oriented community groups, but also a board that has organized itself into a cohesive body, ready and able to speak and act collectively. Such a role also requires that the board feel ownership of the project, that actions taken by the board matter, both in the narrow sense of influencing OHCC activities and in the broader sense of building a program that is important for workers. There is an ambiguity inherent in the role of the advisory board, particularly concerning its power. Though advisory board members may have had aspirations to govern, it became clear early on that the OHCCs' parent institution, not the advisory board, ultimately controlled OHCC operations. Whatever influence the advisory board was to have over budgets, personnel decisions, and policy would have to be carved out via negotiation and/or confrontation with the parent institution, mediated through OHCC staff.

Advisory boards do not spontaneously organize themselves. Individual members bring varying levels of occupational health knowledge and experience to the board, and it requires a conscious effort to bring the group to the same level. That force may come from OHCC staff, the local COSH group, experienced board members, or some combination of the three. Once organized, the board will be able to play an important and ongoing role in directing OHCC activities. It also will be in a better position to respond decisively in times of crisis. The exact limits to the board's authority will vary from OHCC to OHCC and depend upon the board's ability to organize itself.

In order to play its role, the board needs to remain independent of OHCC staff and the OHCC's parent institution. Obviously, the board should maintain close contact with the staff, and staff should have input into board deliberations. Ideally, the board should develop a relationship with the parent institution that is independent of the staff. Contacts will force the parent institution to recognize the board as a force with authority that must be consulted before unilateral decisions are precipitously made by the institution.

A labor-based board with power, even governing power, does not guarantee a healthy OHCC. The labor movement is not monolithic and does not share a unanimous conception of how the OHCCs should function. The consequences of this dissension can be disruptive for an OHCC. The lesson to be learned from New York's experience is to make sure that even a labor board is broad-based and power is not concentrated in the hands of one or a few individuals. Broad-based labor participation is essential to allow an OHCC to grow and to serve the needs of the entire working population, not just a portion of it.

Statewide Advisory Board

While the OHCCs each have their own advisory boards, the network as a whole is to come under the scrutiny and guidance of a statewide advisory board. Members of this board will be appointed by the governor, legislative leaders, labor, and business. Though the network has been in existence for more than four years, no statewide board has been assembled. Apparently, political wrangling is the cause for the delay. There is a real possibility that a politically appointed board will not strongly represent workers. Such a board may well function either as a figurehead without much activity or influence or as a place for business to make its concerns felt.

The OHCC network needs an overall watchdog with a composition and role similar to that of the individual advisory boards. The committee that directed the Mt. Sinai study giving shape to the OHCC network could form the nucleus of a watchdog group. The committee could be supplemented by individuals who have since become active members of the various OHCCs' boards. That body would more accurately represent the working community than a politically appointed advisory board could, and could build an unofficial authority that may prove crucial to the worker-oriented success of the OHCC network.

POLITICAL ECONOMY

"Rapidly changing" and "in crisis" are terms used by many commentators to describe the current state of U.S. economics. The state of the economy and the direction and content of economic trends have profound effects on occupational safety and health. A changing economy produces a changing spectrum of workplace hazards and workplace illnesses. An economy in crisis produces a context in which occupational safety and health issues become more difficult to confront, as worker concerns focus on economic survival and workers' power declines in the face of a management offensive. For an OHCC, this presents a challenge, both in terms of maintaining its own funding and maintaining a role with positive effects on disease prevention amidst a retrenched occupational safety and health movement.

A Changing Economy and Changing Hazards

The decline of manufacturing and the expansion of the service sector have been noted almost universally as a major aspect of the United States's economic transformation. As service workers in both the private and the public sectors continue to increase their numbers and their role in the union movement, newer occupational safety and health concerns are brought to the fore. Cumulative trauma disorders now make up more than 50 percent of all Workers' Compensation claims and are epidemic in a wide range of work settings. Indoor air quality has emerged as a major topic of concern, particularly in offices and schools. Occupational stress and its links to mental health as well as to physical ailments is receiving more attention as a workplace hazard. Communications workers, among others, are concerned about the potential health effects of electromagnetic radiation. Health care workers face the full gamut of hazards including AIDs and bloodborne diseases, chemicals such as ethylene oxide and formaldehyde, and musculoskeletal injuries. Solvents, present ubiquitously, command the attention of many working groups. In the construction trades, in addition to the traditional hazards, opportunities are opening up in the cleanup of industry's detritus (hazardous waste work) or in the abatement of widely distributed hazardous materials previously deemed safe (asbestos and lead) with potential health consequences. (19)

Exploitation of Nonunion Workplaces

A second major change in the U.S. economy is the decreasing proportion of unionized workers, which now stands at about 16 percent nationally. The growing nonunion sector is the traditional scene of increased exploitation and exposure to high-risk hazards, the fire killing 25 workers at the Imperial Foods chicken processing plant in September 1991 being a recent tragic example of these conditions. Workers in these unorganized, high-risk workplaces are disproportionately women and people of color, but the concerns of these groups have long been hidden and ignored. Often, this population is not only a victim of workplace toxic exposures, but of toxics in their communities as well. Toxic waste dumps and Superfund sites are found much more frequently in low-income communities of color than anywhere else, revealing a pattern that can only be called "toxic racism." (20)

Falling Profits, Job Loss, and Increased Hazards

For much of U.S. business, the era since the mid-1970s has been marked by declining profits. Business has aggressively sought to maintain profits through a variety of means, all of which are aimed at increasing management control over capital, labor, and the workplace. Business strategy has included the exporting of jobs, continuing attempts to eliminate unions or to at least reduce their power, and a push to deregulate, particularly in occupational and environmental health. In Reagan, they found an even friendlier than usual administration, accelerating the implementation of their agenda. Workers have paid the price with lost wages, benefits, jobs, and prospects. (21)

Fear of job loss has become pervasive in American workplaces as a result of the economic crisis and business's response to it. This makes the mission of an OHCC infinitely harder to accomplish. Workers are more willing to tolerate unhealthy conditions in silence when they feel jobs are threatened, and employers are not reticent about exploiting workers' vulnerability. As a consequence, workers who feel lucky to be employed at all are less likely to seek evaluation at the OHCC, and preventive activities become a luxury.

To make matters worse, the same economic conditions producing fear also increase workplace hazards. With the need to increase exploitation to maintain or increase profits, employers speed production up, skimp on training, decrease maintenance, cut staff, and pursue other measures that can contribute to deteriorating workplace conditions and increase the risk of occupational injuries and illnesses. (22–24) Declining unionization and an administration committed to business compound the situation further. A decreased ability to mobilize workers to act on safety and health concerns in the face of increased risk is a fundamental problem the OHCCs and the occupational safety and health movement face.

OHCCs and the Economy

Every OHCC should be aware of the ways the economy impacts on it and needs to develop a strategy to address these impacts. The ability of an OHCC or an OHCC network to address the problems imposed by the political economic context alone should not be overestimated, however. Publicly funded OHCCs will remain utterly dependent on labor and its allies for funding as well as for the support, resources, and energy necessary to carry out an occupational health and safety agenda. Nonetheless, some of the New York OHCCs have dealt in relatively significant ways with the political economic issues.

One of the major ways the OHCCs can counter some of the political economic trends working against the prevention of occupational illness is by aggressively supporting Workers' Compensation claims. Through Workers' Compensation, the OHCCs may have the opportunity to expand definitions of what is a compensable illness, how the illness should be treated, and how an individual should be rehabilitated. State physicians and MDs employed by insurance carriers may be influenced by the OHCCs' work. In specific workplaces where multiple claims have been filed, preventive efforts may be stimulated.

OHCCs have varied in their concern for people of color as a high-risk group. A coalition in New York City including the OHCCs, NYCOSH, ACTWU and other unions, and the Health Department, has made significant steps, through two conferences, toward bringing together unionized low-income people of color. A COSH and OHCC effort in Syracuse has initiated an embryonic outreach program to church and community groups serving communities of color and is planning a conference. Existing efforts are early and represent a positive step, but the concerns of this group should become a central concern of the OHCC network.The needs of this neglected and high-risk population should be addressed through expansion of resources directed to this effort and the building of strong linkages to church and community groups. Strong efforts also should be made to recruit people of color to OHCC staff.

The other way an OHCC could play a role in the political economic sphere is by recognizing itself as part of a broader social movement aimed at combatting deteriorating living and working conditions. A living wage, job security, and health benefits go hand in hand with workplace safety and health. In its broadest sense, the struggle for occupational safety and health is part of workers' struggle to control their own destinies, a struggle for power and for democracy. Support for union organizing drives where occupational health is an issue, for the OSH Reform Act, and for a National Health Program are current examples of campaigns that OHCCs could support.

FUNDING AND THE RHETORIC OF HARMONY

Public funding is the lifeblood of New York's OHCC network; without it, survival would be impossible. In justifying the need for an OHCC network to the legislature, its independent nature was not emphasized. Instead, the Mt. Sinai report demonstrated the tremendous burden of occupational illness in New York State coupled with the woeful lack of resources to address the problem. This shift in emphasis allowed the OHCC network to be sold to the legislature as a project good for workers, unions, and business, making it difficult for business to mobilize a full-scale assault against the idea.

The rhetoric of harmony, a vision of labor and management cooperatively working on safety and health problems using the OHCC as a technical resource, continues to permeate discussions about the OHCCs. For the COSHes and many in the labor movement, the rhetoric remains rhetoric: a device to be utilized to garner legislative support, expand funding, and mute opposition. For others, both within the OHCC network and without, the rhetoric is a desired and possible reality, reflecting a world view based on the pillars of technocracy and labor/management cooperation. The rhetoric of harmony is a double-edged sword: the one edge a pragmatic and useful device to maintain funding, and the other edge an ideological stance bound to take the OHCC network down the blind alley of negligible effectiveness.

The provision of new services and the creation of services independent of business are both crucial aspects to the original vision of the OHCCs as put forward by the labor-led coalition in New York. While the OHCCs depended absolutely on public funding for their establishment, there are varying currents of thought on the need and the possibility of continued and adequate public funding into the future. The source and amount of funding are obviously critical, not only to survival, but also to the OHCCs' ability to remain independent and to provide services that others do not.

In the OHCC network, there is almost universal unease about future funding. Some believe that public funding will end because of the state's fiscal crisis, or because the state only planned on seed funding that eventually would allow the OHCCs to become self-supporting. Others believe funding will continue, but at an unknown, possibly lower, level.

The OHCCs that are in a position to expand are constrained from doing so by their current funding levels.

Desperately Seeking Revenues

As a consequence of their unease, OHCCs are considering and pursuing other funding sources. These sources are mainly employer contracts or outside grants. Patient revenues, industrial hygiene charges, consultations, or Workers' Compensation testimony have not come close to meeting OHCC funding needs at any OHCC. Union-based activities, including screenings, consultations, or educationals, also have failed, for the most part, to have considerable impact in decreasing the OHCCs' dependency on state funding. If the OHCCs were seriously conceived of to fulfill their mission and to be self-supporting, they were set up to fail. As the Mt. Sinai study demonstrated, the OHCCs were necessary precisely because entrepreneurs recognized that the role envisioned for the OHCCs was not a profit-making market niche. The private-sector occupational health services that exist are employer dependent and employer friendly, marketing strikingly different services than those considered the province of a publicly funded OHCC. (25–27)

OHCC clinical activities are low-volume and labor-intensive. Educationals, industrial hygiene work, and consultations are routinely provided for no payment to individuals, unions, and other groups without substantial resources. OHCC services are meant to be universally accessible. Worker-oriented activities and Workers' Compensation claims do not endear the OHCCs to most employers. Filling this neglected and unprofitable "market niche," the OHCCs will forever require government subsidization.

The idea of shifting niches and competing in the entrepreneurial realm is tempting to OHCCs feeling the funding pressure. Most OHCCs have entered into employer contracts to provide compliance-based services, some more actively and extensively than others. No OHCC provides drug testing yet, but some are considering it. (In addition to employers, some OHCCs have been approached by unions seeking drug-testing services.) Most OHCCs have policies to guide them in their dealings with employers and in maintaining consistency with their mission. Despite these policies, these employer-based activities carry substantial dangers. Of paramount concern is the displacement of scarce occupational safety and health resources away from the diagnostic and preventive work meant to be the core of their activities. Secondly, at a certain point, an OHCC contracting with various employers begins blending in with its entrepreneurial private-sector counterparts, competing to provide similar services to a similar clientele. In reality, or at least in the perception of workers and unions, OHCCs pursuing this course become compromised, working with employers to comply with OSHA, rather than advocating for workers and controlling hazards.

Except at Mt. Sinai, a search for other grant funding is a largely unexplored option. The possibility of grant priorities distorting OHCC activities and leading away from the underlying mission remains a serious concern. More importantly, grant funding cannot provide the long-term financial stability necessary to OHCC survival. It may provide funds for short-term or focused projects not covered by the main funding source.

Medical Allies

Integration of occupational health into the health care system is another important component in building long-term OHCC stability. Recognition by medical schools and health care providers that the OHCCs are important resources as educational centers and as consultants could create a broader constituency supporting the OHCCs politically and financially. Most of the OHCCs have recognized outreach and education to health care providers as an important facet of their work and have pursued these contacts to varying degrees.

The overall importance of this strategy in ensuring OHCC survival is debatable. On the one hand, without a strong labor movement and empowered workers, OHCCs, or individual practitioners, may develop stability but lack effectiveness to significantly improve workplace conditions. On the other hand, in a period of union weakness and worker fear, such a strategy may be important in preserving worker-oriented professionals in positions that will become more significant with a rejuvenated labor movement.

Funding remains a key OHCC concern. The OHCC network was conceived as a publicly funded program, and adequate public subsidization is essential to the survival of OHCCs committed to the original vision of the network. The difficulty of maintaining that vision is directly correlated to the level of funding uncertainty, and the increasing pressure to seek funding from employer-based sources. The OHCCs and their supporters must keep funding in the forefront as a political issue. Only a coalition could push the legislature into originally funding the network, and only a coalition has a chance of maintaining and expanding that funding.

CONCLUSION: WHERE TO FROM HERE?

The existence of a publicly funded OHCC network is by itself a major success of the labor and occupational safety and health movements. The guiding concept animating the network offers a distinct alternative to employer-dependent occupational health services and to the technocratic model of occupational health. (28–30) This provides an alternative that establishes activist OHCCs explicitly advocating for workers in a setting of labor/management conflict over safety and health. The model shares certain similarities with the Worker-Controlled Clinics in Canada and with academic hospital-based clinics in the United States. (31-34)

Existence is obviously only a partial success, and the OHCCs have varied in their ability to put the original vision into practice. As detailed earlier, uneasiness over funding, a difficult political economic climate, the interests of OHCC parent institutions, and the socialization and education of staff all contribute to the barriers keeping the OHCCs from fulfilling their potential. Several OHCCs have partially surmounted these barriers, whereas others have not. The difficulties at some of the OHCCs threaten the integrity of the network, as the needs of significant portions of the state are not being met.

The key to the future fortunes of the OHCC network, if it is to remain loyal to its origins, lies with labor and its allies. A weak labor movement will allow the OHCC network to collapse or to lose its worker-oriented character. A strong labor movement could force solutions to OHCC problems at both the local and state levels, influencing funding, policies, and practices.

With declining union membership and the deteriorating economy, labor will be forced to search for new strategies and allies in the pursuit of its agenda. On the one hand, this sign of weakness does not bode well for the OHCC network or for worker health. On the other hand, it may result in new energized coalitions adapted to changing conditions. What we need to support a worker-oriented OHCC network is clear, but whether those needs will be met is unknown and unknowable. The most we can do is to work toward that end.

ACKNOWLEDGMENTS

The author would like to thank Dom Tuminaro, Matt London, Jim Melius, and especially Lin Nelson and Federica Manetti for reviewing earlier drafts of this chapter. Antoinette Longo also deserves special recognition for her word processing.

NOTES

1. Landrigan, P., and Markowitz, S., *Occupational Disease in New York State*, Report to the New York State Legislature, February 1987.

2. Tuminaro, D., "Organizing for a Statewide Network of Occupational Disease Diagnostic Clinics," *New Solutions* 1 (1990):18–20.

3. Berman, D. *Death on the Job*, New York and London: Monthly Review Press, 1978.

4. Bayer, R., Ed. *The Health and Safety of Workers*, New York and Oxford: Oxford University Press, 1988.

5. Rosner, D. and Markowitz, G. *Deadly Dust*, Princeton: Princeton University Press, 1991.

6. Matt London, New York State Department of Health (personal communication).

7. See Landrigan and Markowitz, op. cit.

8. Yassi, A. "The Development of Worker-Controlled Occupational Health Centers in Canada," *American Journal of Public Health*, 78 (1988): 689–693.

9. *American Academy of Industrial Hygiene Newsletter*, Fall 1991, provides a good example of this mentality.

10. Tarlau, E., "Playing Industrial Hygiene to Win," *New Solutions* 1(1991): 72–82.

11. Castleman, B. I., and Ziem, G. E., "Corporate Influence on TLVs," *American Journal of Industrial Medicine*, 13 (1988): 531–559.

12. Navarro, V. "The Labor Process and Health: A Historical Materialist Interpretation," in *Crisis, Health, and Medicine: A Social Critique*, New York and London: Tavistock Publications, 1986.

13. See Bayer, op. cit.

14. Navarro, V., "Work, Ideology, and Science: The Case of Medicine," in *Crisis, Health, and Medicine: A Social Critique*. New York: Tavistock, 1986, p. 22.

15. Wegman, D.H., et al. "Byssinosis: A Role for Public Health in the Face of Scientific Uncertainty," *American Journal of Public Health*, 23 (1983):188–192.

16. Freire, P., *Pedagogy of the Oppressed*, New York: Continuum Publishing Co., 1987.

17. Wallerstein, N., and Rubenstein, H., Eds., Worker Education Manual for Occupational Health Clinics, Unpublished Manuscript.

19. Cullen, M., et al., "Occupational Medicine," *New England Journal of Medicine*, 322 (1990): 594-601 and 675–683.

20. Weinberg, B., "'Toxic Racism' Dumps on Low-Income Communities," *The Guardian*, September 14, 1991.

21. Harrison, B. and Bluestone, B., *The Great U-Turn: Corporate Restructuring and the Polarizing of America*, New York: Basic Books, 1988.

22. See Navarro, "The Labor Process and Health: A Historical Materialist Interpretation," op. cit.

23. Levenstein, C. and Tuminaro, D., "The Political Economy of Occupational Disease," *New Solutions,* 2 (1991): 25–34.

24. Clayson, Z. and Halpern, J., "Changes in the Workplace: Implications for Occupational Safety and Health," *Journal of Public Health Policy* (1983): 279–297.

25. See Landrigan and Markowitz, op. cit.

26. Zoloth, S., et al., "Asbestos Screening by Non-Specialists: Results of an Evaluation," *American Journal of Public Health*, 76 (1986): 1392–1395.

27. Guidotti, T., and Kuetzing, B., "Competition and Despecialization: An Analytical Study of Occupational Health Services in San Diego, 1974–1984," *American Journal of Industrial Medicine*, 8 (1985): 155–165.

28. McCunney, R., "A Hospital-Based Occupational Health Service," *Journal Of Occupational Medicine* 26 (1984): 375–380.

29. Arling, L., "Industrial Clinics and Occupational Medicine," *Archives of Environmental Health* 12 (1966): 644–646.

30. Kleinman, G., and Morris, S., "The University's Role in Occupational Health: A Model from Washington State," *New Solutions* 1 (1991): 63–71.

31. Rosenstock, L., and Heyer, N., "Emergence of Occupational Medical Services Outside the Workplace," *American Journal of Industrial Medicine*, 3 (1982): 217–223.

32. Rosenstock, L. "Hospital-Based, Academically Affiliated Occupational Medicine Clinics," *American Journal of Industrial Medicine*, 6 (1984) 155–158.

33. Orris, P., et al., "Activities of an Employer-Independent Occupational Medicine Clinic, Cook County Hospital, 1979–1981," *American Journal of Public Health* 72 (1982): 1165–1167.

34. See Yassi, op. cit.

26

Community Occupational Health Clinics in the Free Market

PETER DILLARD

It is common knowledge that the United States has moved toward "deregulation" of its economy since the late 1970s. There can be little doubt that deregulation policy has led to a gradual loss of political power by "nonmarket" forces such as labor unions and other social groups that once had somewhat more ability to bring about regulation of the "free market." Occupational health and safety laws and regulations were born in a time when grassroots political forces and labor unions clearly had more influence in the imposition of regulations on the "free market" than they do now. The continued effectiveness of those laws and regulations requires constant monitoring on the shop floors by health- and safety-conscious workers who have access to information and networks, who can provide stewardship for their work environments, and who have some degree of leverage for making the changes and improvements necessary for safe and healthful workplaces. When the power of these men and women is weakened by market "deregulating" policies (such as the deliberate weakening of unions), market forces are all that remain to determine how our society deals with dangerous and unhealthful workplaces. For example, the problem of dangerous and unhealthful workplaces, as seen by the "free market," is not so much the problem of injuries and illnesses to workers as it is a problem of increased costs to employers. Are the solutions to these two formulations of the problem the same?

This article examines the effects of economic forces acting on community occupational health clinics (COHCs) in the United States in an increasingly "deregulated" market economy, and discusses the questions: Are they forces for improvement of health and safety in their communities' workplaces? And, can community occupational health clinics that arise from these forces provide services equitably to their communities?

ASSUMPTIONS AND DEFINITIONS

One of the goals of medical practice in occupational health clinics or services should be to promote and improve health and safety in their community's workplaces. The words "occupational health" in the name, the special training of occupational physicians, nurses, and other occupational health professionals who work in these clinics, and the thrust of the scientific literature of occupational medicine all suggest that occupational health is a major goal of the practice of occupational medicine. These practices should provide clinical judgments and evaluations that are based on science, ethics, and community standards for confidentiality of health records. Since it is also true that the practice of occupational medicine is deeply affected by economic and social policy, the practice protocols of this health care specialty have important implications for both economic costs and social justice in the community.

In this article, the term COHC means "community-based" occupational health clinic and is meant to exclude from the discussion clinics that are structurally insulated from the marketplace by the fact that they have other bases of financial support, such as those based in an academic setting, clinics that are employee health services for a specific company, or clinics that are funded by government. COHCs could also be called "free-standing" clinics, but since they are not always free-standing, but are often closely related to a community hospital, I have chosen to use the term "community." They may be either for-profit or nonprofit. Whether community-based occupational health clinics are for-profit or nonprofit, as purveyors of services they are almost always expected to compete for revenues in the marketplace at the present time.

Services typically provided by COHCs can include health hazard evaluations of workplaces; consultations regarding the work-relatedness of disease; diagnosis and management of work-related diseases or injuries; "preplacement" (or what used to be "preemployment") exams; legally mandated occupation-related physical exams, such as the physicals required every two years by the Department of Transportation for truck drivers; medical surveillance exams, such as those required by the Occupational Safety and Health Administration (OSHA) for asbestos removal workers, deleading workers, and others, and those examinations that careful employers use to monitor workers who work with hazardous substances; and "fitness-for-duty" exams to ensure physical readiness to work after recent injuries or illnesses. The task of monitoring drug use in the workplace by urine drug testing also has largely devolved to COHCs.

THE PROBLEM

The services technically provided by COHCs, as they are listed above, appear to provide an unbiased menu promoting health and safety in the workplace while helping employers cut the costs of work-related illness and injury. But the relative frequency in which examinations of workers are performed in contrast to the examinations of the workplace, and the actual decision-making processes involved during the performance of many worker examinations can and often do mean that the services, while helping employers reduce costs related to worker injury, actually fail to address workplace health and safety and, furthermore, violate generally accepted civil and privacy rights of workers. Services that emphasize examinations of workers and do not emphasize with equal vigor examinations of workplaces ignore possibly important therapeutic interventions for the individual worker and possibly important measures for the prevention of additional disease for the community in general. What is more, the medical management options that remain, such as medical removal without compensation or the unnecessary labeling of a worker as disabled, often have disastrous effects for the worker economically and psychosocially, and thus may be contraindicated.

The problem is compounded by the fact that many COHCs are staffed by physicians who are unable, by training, to extend their management of work-related injuries beyond the patient and into the workplace. The training required includes work environment analysis, toxicology, epidemiology, biostatistics, and work environment law, and is not part of general medical training. This fact is uncritically substantiated by William Newkirk, who writes:

> The number of board-certified or board-eligible occupational health physicians is grossly inadequate to meet current needs . . . As a result, other physician groups will move in to fill the void. The most likely source for new physicians in occupational health is the emergency department. . . . (1)

I do not mean to describe in this article the behavior of all COHCs, but to describe those forces that I have observed from my experience that pressure COHCs to behave in certain ways. There may be many who do not behave in all the ways that I describe as a result of the market forces acting on them, but that behavior should be seen as resistance to pressure rather than seen as practicing in a place where these forces do not exist.

THE DRIVING FORCES

Since 1970, there have been changes in the costs to industry of workplace health and safety. Some of these have come about as a direct result of political forces, as from labor and environmental interests, with the birth of the Occupational Safety and Health Act (OSH Act) of 1970, which made it the general duty of employers to provide workplaces that were safe and healthful "to the extent feasible." In addition to direct regulation of business, the OSH Act has provided for research, education, and training in the area of occupational health and safety through the National Institute for Occupational Safety and Health (NIOSH). In the last 20 years, at various and uneven rates of growth, a considerable body of new knowledge in the areas of industrial toxicology and industrial safety has come to pass, largely as a result of this governmental intervention, which was a result of erstwhile much stronger political forces in the United States. Much of this knowledge has become the property of the public at large, including people who work in potentially toxic environments.

Other changes have arisen from increases in the cost of health care in general, changes that have been largely market-driven. The cost of workers' compensation insurance has risen along with, and probably largely as a result of, the overall health care costs.

Because of the effects of "deregulating policies" that have included the evisceration of the union movement and other political forces, the author of this chapter views the forces that currently drive and shape COHCs in the United States as primarily market forces. Thus, COHCs, as they are currently growing, are responding (with exceptions) to market rather than political forces, even though the initial impetus for their development came from the OSH Act, a regulating law whose creation was due almost entirely to political forces.

The markets as they now exist have derived, on the one hand, from the needs of U.S. industry to reduce the costs of work-related illness and injury, and the costs created by the government's intervention into workplace health and safety, and on the other hand, from the need of the U.S. public and workers to maintain safe, healthy workplaces and communities. These markets emphasize different solutions and demand different services from occupational health clinics. The relative power of these two markets depends, of course, on their relative abilities to pay for services.

PRINCIPLES OF PRACTICE AND HOW THE MARKET PRESSURES THEM

Occupational health physician practices are governed, or should be governed, by traditional medical ethics and adherence to scientific principles in making clinical judgments, articulately stated by Dr. Patrick Derr. "The fundamental value in medicine is life and health. . . . The corollary of health is that a physician must have a commitment to the truth and logical deductive reasoning to apply this technique." (2)

Physicians should honor the confidentiality of the patient's medical record, and should, above all, as in traditional medicine, strive to do no harm to either the health or the well-being of the patient. These principles are explicitly listed in the "Code of Ethical Conduct" as developed by the American College of Occupational Medicine for physicians providing occupational medical services.

In order to see how market forces affect COHCs on a day-to-day basis, it is not enough to go to the references and textbooks, or simply to read the codes of ethics of practitioners. One also must examine the specific services delivered, see how they are displayed in the marketplace, and weigh the incentive forces at work during the interaction between occupational health provider and worker in the context of the clinic's need to increase its market share in a market economy.

COHCs as they are defined herein do not receive financial help from an umbrella institution, such as a university or government. Since, by far, the major source of the revenues of occupational health services consists of the community's employers, most of the language of occupational health clinic advertising brochures is aimed at employers, promising to save them money, to reduce their occupational health and safety costs, to get their workers back to work in a timely fashion, and so on. Most major "how-to" manuals on the development of COHCs emphasize marketing to employers as opposed to other potential markets from the community at large. (3, 4)

The marketing of services is consciously integrated into the daily service delivery in any service enterprise, and the heart of marketing is in the ability of the enterprise to deliver in high quality the services it has promised. The delivery of services must be followed by continuous reminders to the customers that they are receiving high quality service. Since the primary customers are employers, this means that in a market-sensitive COHC much communication goes on between the clinic providers and the employer-customers, consisting of letters, telephone calls, and reports regarding the outcomes of specific encounters between patients and the provider. Because providers feel eager to please their major customers, there is a constant pressure on COHC service providers to produce what they say their services are intended to produce, namely reduced costs in the areas of occupational health and safety, and in workers' compensation.

These incentive forces have a tendency to undermine the principles of practice as outlined above in several ways. First, they place emphasis on the examination of workers instead of the workplace, putting an undue emphasis on the health of the worker and his or her "fitness-for-duty" instead of the health of the workplace and its fitness as a place in which to work. Not only does this undue emphasis on the fitness of the worker distract from an evaluation of the workplace, but it sometimes puts at unnecessary risk the worker's health, as in the use of lumbosacral X-rays to screen workers for potential back problems. Second, they create a tendency to provide employers with details of a preplacement examination, which violates workers' traditionally held rights to privacy and confidentiality of their health records; and third, they tend to allow excessive influence by employers in the traditional doctor–patient relationship, especially in the case of work-related injuries, and, additionally, to minimize the influence a worker might otherwise have with his or her doctor in getting help to correct workplace hazards.

To see how this works, we will look at specific COHC services.

MEDICAL SCREENING

One of the practices of COHCs involves medical screening. Traditionally, the decision to administer a screening test has been made by the application of scientific principles based on cost-effectiveness, risk, discomfort, invasiveness, and other parameters, all of which are based on concerns for the value of the test to the patient's ultimate well-being and that of the public. These principles have been well thought out and are widely published. (5–7) Under these principles, screening tests are required:

- to be safe;
- to seek diseases that have serious adverse consequences for the patient, and not simply be "fishing expeditions";
- to seek diseases for which practical and judicious remedies exist;
- to have adequate availability and accessibility for confirming positive tests and for follow-up;
- to be generally acceptable to the people being screened; and
- to have greater benefits to the patient if disease is found than the costs and risks to the patient if the test is falsely positive.

However, the exigencies of the marketplace have changed the standard of practice for occupational health clinics in the area of screening. Occupational medical screening is not performed for the benefit of the patient, but for that of the employer. Under this model, the subject of the cost–benefit analysis becomes the employer, not the patient. The worker is frequently a mere bystander in the analysis. The costs to the worker in terms of the health risk of the test and loss of opportunity to get gainful employment are usually not part of the analysis, and the worker may derive no benefit from the screening test. The follow-up or remedy of a suspected condition is often obtained at the worker's expense and initiative, if he or she has enough money to do the follow-up. Often, the condition has no significance except as a risk factor, and there is no "remedy" unless disease actually develops. The worker is caught in the dilemma of either paying money to confirm a mere risk factor, or not paying money and losing an opportunity for employment. Does the worker get due process if he or she is denied employment based on a screening that only suggests disease but does not demonstrate it clearly? Who should be responsible for following up the screening test with more definitive diagnostic studies? In general, employers do not expect to have to follow up positive screening tests that indicate workers might be at risk. There is no requirement for the employer to do this, and most do not. A notorious example of such screening tests includes lumbosacral spine X-rays that are required during preplacement exams by many companies in which a high proportion of the job descriptions include heavy manual labor. These tests are risky to workers, delivering the highest load of ionizing radiation among all the routine X-rays performed at a hospital; and they generally include the gonads in the field of exposure. Yet these X-rays are still frequently performed at many COHCs.

HELPING EMPLOYERS CUT WORKERS' COMPENSATION COSTS

Retooling and redesigning work areas in order to improve workplace safety and health is probably a cost-effective solution, and certainly the most equitable solution to the high costs of workers' compensation. However, U.S. industry has two predominant strategies to cut workers' compensation costs. These are: (a) pre-placement physical examinations, and (b) trying to get their workers back to work sooner. It is interesting to note that the catalogue of research work done to reduce workers' compensation costs by a major industry-funded research group, the Workers' Compensation Research Institute in Cambridge, Massachusetts, lists no research directed at understanding what the effects of intervention in health and safety would be on the costs of workers' compensation. (8)

Both preplacement examinations and "Early Intervention Programs" (EIPs) are services potentially provided by COHCs. The preplacement examination is an examination that, as described above, screens and evaluates the worker's past medical history and physical findings to disclose any incompatibilities between the worker's health condition and the demands of the job description. Early Intervention, sometimes called "Managed Care," for work-related injuries involves care-managing "teams," often third-party vendors, closely following injured workers with phone calls, usually weekly, to determine status; to "encourage" timely doctors' visits and timely diagnostic studies and follow-ups; to reduce "excessive" (in the opinion of the care-managing team) recuperation times; and to develop graduated return-to-work programs with the employer of injury. Early intervention teams try to develop lists of doctors who are willing to cooperate with the goals of this strategy. Occupational health clinics that manage work-related injuries are usually eager to get onto these schedules in order to capture a greater market share of work-related injuries. It is a sort of "preferred provider" system for local employers. Workers are "encouraged" to use physicians on these schedules. Employers often provide transportation directly to these physicians from the place of employment for the initial management of the injury.

PREPLACEMENT EXAMINATIONS

Numerous ethical guidelines exist for conducting preplacement examinations (9), as well as various laws forbidding discrimination against workers who have disabilities. Until recently, the Federal Rehabilitation Act of 1973 was the most protective at the national level. Many states, such as Massachusetts where I practice, have more protective legislation, however. Beginning in 1992, the Americans with Disabilities Act took effect. It provides the most comprehensive legal protection against employment discrimination thus far at a national level.

Under ethical guidelines, there have been three permissible goals for the preplacement examination. One is to identify medical conditions that would put the worker, co-worker, or the public at increased risk of injury if the worker were to carry out his or her job as required in its job description. The second is to determine that there are no medical conditions that would make it impossible for the worker to perform the duties as described in the job description, even if reasonable accommodation were made. The third goal is to determine whether modifications could be made in the job description in order to accommodate a worker who has a disability.

Of course, to make such determinations fairly would reasonably require that the physician making the evaluation should know the job and workplace and be in a position to make suggestions to the employer for accommodation if an apparently incompatible medical condition were found. In addition, I would argue that physicians who perform preplacement examinations should be familiar with the laws concerning employment and worker health. This includes the Federal Rehabilitation Act of 1973, the state statutes, and the Americans with Disabilities Act of 1990. They should be able to communicate to the worker his or her rights if he or she has been unfairly denied a chance to try the job.

The preplacement examination is a process in which the rights of employers to choose the qualifications of their employees and the rights of workers to find employment for which they are qualified hang in the balance. During the preplacement examination, the worker's autonomy must be preserved. The worker must be fully informed about medical conditions the provider has found in taking the history and performing the examination, should be informed as accurately as possible about the increased risks of the job, and, as long as certain conditions are met, should be permitted to make his or her own decision about whether or not to take the job. The only conditions that should interfere with this rightful autonomy should be limited to those instances in which (a) there is significantly increased risk to public health; (b) there is significantly increased risk to coworkers; and (c) the worker's condition creates a significant danger of permanent disability in the opinion of the medical examiner. Even these conditions must be considered in the light of the worker's past work experience and safety record. These conditions should preclude either the provider or the personnel director of the company from making decisions about the employability of a worker based on the probability that the worker will have a future workers' compensation claim. The above principles are not just my own opinion, but are clearly stated as principles of practice in well-accepted literature including textbooks and references on the practice of occupational health.

These judgments would reasonably require a provider trained in occupational medicine, or at least someone familiar with (a) the interface between human biodynamics and job descriptions or work stations, (b) work station evaluations, (c) the physiological effects of medications, (d) work environment law, (e) industrial toxicology, and (f) risk assessment and risk communication. Yet, in Massachusetts, only about half of the clinics that claim to offer occupational health services have either a full-time or part-time board-eligible or -certified occupational physician. (10)

Furthermore, workers rarely come to the clinic with either a job description provided by the employer for the medical evaluator, or even a clear perception of what their job description is going to be. This being the case suggests that the physical characteristics of the worker are often being set against the standard created by a job description rather than the job description being modifiable to meet the characteristics of the otherwise qualified worker. In practice, the physician often gets a job description from the company personnel manager over the phone if a restriction is thought to be suggested by the sound of the job title. Of course, under these circumstances, the worker's impairment is revealed to the employer before the job is revealed to the worker, and before the worker has had an opportunity to determine whether he or she can do the job. Thus the worker's decision-making power is preempted by the employer. In practice, employers do not like to hire workers who are potential liabilities for workers' compensation claims. Preplacement examinations are advertised as a way of helping employers cut their workers' compensation costs; if the occupational health clinic sends a worker to the employer as ready for full duty without restrictions and the worker becomes a workers' compensation claim because of a risk factor that was picked up on the preplacement examination but not communicated to the employer, there is a strong tendency for both the occupational health service and the employer to feel that the service failed. The employer feels that way because he or she paid for the examination, and the occupational health service feels that way because it failed to fulfill a marketing promise.

It is probably true that properly administered preplacement examinations, which preserve workers' autonomy, save employers from compensation liability by identifying workers who have a medical risk factor for an injury. Such workers might often be protected by minor job modifications, sometimes even temporarily, or by minor restrictions. Unfortunately, employers are not yet convinced that they can save money in the long run by spending it in the short run; and, as of now, there are few studies in existence that can prove it to them. With some exceptions, employers in general are not wont to spend money on safety modifications for impaired workers, on the one hand because such modifications are a financial outlay with some risk, and on the other hand, as one employer told me, "If I make changes for one employee, all my employees will want them." (I did not give him the obvious rejoinder that maybe such changes would benefit all the employees and perhaps even save the employer from some unanticipated future compensation liabilities, too.)

It isn't possible to predict what effect such laws as the Americans with Disabilities Act will have on the behavior of occupational health services to employers. However, as I have mentioned already, in the state of Massachusetts there already is a law that forbids employers from withdrawing an offer of employment without making an attempt at reasonable accommodation, and yet there remains tremendous pressure on COHCs to provide employers with the information they would need from the preplacement examination to screen even questionably impaired workers out of the workplace.

Numerous employers circumvent the examining physician's own autonomy, in addition to that of the worker, by having the job applicant bring a physical examination form to be filled out by the examining physician (or nurse practitioner) and then returned to the personnel department of the company. The applicant must sign a release of this medical information to the employer as a condition of employment. This removes the examining physician from the decision-making loop regarding the safe or accommodated employability of the worker. This practice also forces the employee to surrender his or her rights to confidentiality of the medical record. In these cases, the examiner is simply a collector and relayer to the employer of all medical data which he or she can glean from the job candidate.

Finally, it is important to note that laws such as the Americans with Disabilities Act, designed to protect the rights of disabled workers in the preemployment examination process, do not protect the rights of the worker to make his or her own decision regarding risk. In other words, these laws do not protect worker autonomy.

EARLY INTERVENTION PROGRAMS

In some states, including Massachusetts, employers are given an incentive in the form of workers' compensation insurance premium reductions to buy the services of certified EIP vendors. The role of the occupational health clinic physician, if selected as a provider by an EIP vendor, is to cooperate with the EIP vendor in providing feedback to the EIP "counselors" regarding injury status, to make timely decisions in the management of work-related injuries, and to cooperate with the EIP's client employers' return-to-work or restricted duty programs.

Although this may seem like effective management of work-related injuries, whose purpose is to return the injured worker back to productive work life in a timely fashion, such close cooperation with a monitoring third-party permits a significant invasion in the doctor–patient relationship as that relationship is traditionally defined. In Massachusetts, an injured worker still has the right to choose his or her own physician for the primary care of a work-related injury. Nevertheless, many workers either are not aware of this fact, or do not wish to be seen as having suspicious or negative attitudes, and therefore are somewhat passive about going to the physician to whom they are transported by their employer for initial care of the injury. If follow-up care from the same physician who sees the initial injury is available, that course usually seems to make sense and is often assumed by the worker.

In such a relationship, the worker is receiving his or her primary care by a physician who has a tacit relationship with the worker's employer. Since the marketplace has generated this relationship, the physician's financial incentive is to provide results that are satisfactory to the employer, and not necessarily satisfactory to the worker-patient. Numerous potential pitfalls exist for the worker who is receiving his or her care under these conditions. The requirement that the physician confer with the EIP counselor, or employer, gives that third party considerable leverage in determining when the worker is ready to return to work. It can happen, and often does, that a plan for rest or restricted duty made by the physician and worker at the time of the clinic visit can be changed by the employer's intervening phone call before the worker even gets home from the clinic.

As another example, cooperation by the clinician with an employer's return-to-work program can be very degrading for a worker whose employer has an undignified or even punitive program for return-to-work. Under EIPs, the treating physician is under great pressure to surrender control over the worker's return-to-work schedule. These kinds of conflicts over control can have deleterious effects on the normal therapeutic process, which depends very much on a caring and mutually respectful relationship between a worker and the physician.

This is a fact that is perhaps better expressed by the clear apathy toward EIPs of community physicians who are not dependent on employers for their revenue, as occupational health clinics are. Extensive studies by the Workers' Compensation Commissions of Washington and West Virginia have shown that EIPs have had, up to now, no effect on the time of return to work, or on the cost of management of work-related injuries, except for the increase in cost due to the cost of the EIP vendors' services to industry. This appears to be due mainly to a lack of enthusiasm by community doctors (11, 12)—thus the effort to recruit cooperative physicians.

TRYING TO TURN THE TABLES: PUSHING HEALTH AND SAFETY IN COHCs

The potential role of COHCs in promoting health and safety could be performed by providing educational services to workers and employers, or through work site visits and health hazard evaluations. Marketing brochures of occupational health services often list these services to employers.

However, in the case of work site visits, or health hazard evaluations, employers are usually reluctant to engage third parties (such as occupational health clinic physicians or industrial hygienists) in such services. Such services are sometimes provided by workers' compensation insurance carriers at no cost to the employer; but even if they are not provided to a workplace, employers are reluctant to permit third parties to come into work sites without a clear and very narrow purpose. The opportunities for perturbing an often delicately and precariously balanced situation are great, particularly if the employer has ignored the toxic air levels or working conditions around his or her employees. In my experience, employers have usually been made aware of hazards in the workplace by their employees. The employer who has a health hazard usually knows about it and knows that the employees know about it. The absence of a collective action against the employer, or of an OSHA complaint, is the result of a very delicate balance between two uncertainties on the part of employees: uncertainty about the degree of risk of the hazard on the one hand, and the uncertainty about the outcome if they take any action, even a modest one, on the other hand. Employers who have not taken a proactive stance on health and safety may sense that they are operating under this delicate uncertainty level on the shop floor; for pleasantly offered opportunities for work site evaluations (even free ones) by clinic staffers are usually met with cold responses by employers, in my experience. The intervention of third parties in that delicate balance threatens to, and often does, lead to a heightened risk perception on the part of the employees, thus tipping the scales toward action by them.

Again, based on my own experience, employees who are either made sick or are injured by a workplace hazard, or who are suspected of being informed about a workplace hazard by their doctors, are themselves perceived to be a risk to the employer, and are very likely to be terminated under a pretext. Although OSHA's paragraph 11(c) provides theoretical legal protection against employers' terminating workers because of a health or safety issue in the workplace, the track record of OSHA in getting fired workers back into the workplace is poor, and workers know it. At last count, about 17 percent of workers' complaints to OSHA about unfair termination have gone to litigation. (13)

I am reminded of a case I recently had of a Spanish-speaking worker who had been hired to clean some unknown residue from some imported floor tiles. He worked in a corner of the production area in a poorly ventilated cubicle. Soon after starting his job, he became sick with cough and shortness of breath every day at work and got better on the weekends. He developed a rash on his hands soon after starting the job, which got better when he used gloves at work. He didn't know what he was being exposed to and came to our clinic for a consultation. There was no Material Safety Data Sheet (MSDS) available at his work site, and he was afraid to ask for one. He didn't want us to reveal in any way to his employer that he might be sensitive to something at work. His history suggested to us that he might be exposed to an epoxy curing agent. An MSDS that he ultimately found among some containers in another room at work confirmed the fact that a potent epoxy hardener was used at his production site. He eventually agreed that we could ask his employer if we could look at his workstation. We did this. The employer asked us for more information, why we wanted to see the site, what we suspected was the problem. The employer said he would get back to us. The worker was terminated that day. We received a polite letter from the employer saying that the worker was incorrect in attributing his symptoms to the agent in question. We were told he did not work with that agent in that part of the building. It was suggested over the phone and in the letter that the worker had sensitive lungs and his condition was not work-related. We were informed that the worker had been terminated because that particular job had been discontinued anyway.

Thus, work site investigations are very difficult to get permission to perform. Many COHCs have client companies that routinely refuse to permit the medical consultants to make site visits. Such site visits often put the worker in jeopardy if the company feels it is being pressured to make changes to accommodate the worker. I have a number of workers for whom I am caring who are out of work with minor disabilities. Many of these workers could return to work at their regular jobs (and not demeaning or sometimes risky "light duty" jobs) while they were recuperating if their companies would make some minor ergonomic changes in their workstations.

The other way occupational health clinics can enhance health and safety in the workplace is through education of workers about the hazards to which they are likely to be exposed in their workplaces. Such education could take place at the time of routine examinations, such as at pre-placement examinations that the clinic has been hired by the employer to perform. Again, such activity on the part of COHCs who are marketing their services to employers poses some problems. Unless these clinics have been specifically requested to provide such information, undertaking this activity carries some real danger to the COHC–employer relationship. First, there is the problem that is inherent in any effort to communicate risk to anyone, and specifically to workers who are about to start work with an employer whose commitment to safety is completely unknown. The conflict is worse if the employer's safety consciousness is known, and is known to be low. In this situation, the COHC finds its efforts to educate the employee as tantamount to "instigation."

Thus, educational efforts on the part of COHCs must usually be performed gratis and very carefully. These efforts take up clinic and staff time, and are realistically considered threatening by the clinic's marketers to their efforts to advertise clinic services as unqualified benefits to local employers.

In my experience, many COHCs undertake educational activities during routine worker examinations by concentrating on the promotion of such general public health measures as self-breast and self-testicular examinations, the avoidance of smoking, and by encouraging healthy lifestyles. Thus they avoid addressing the specific issues likely to be encountered in the worker's specific workplace.

ENVIRONMENTAL OR OCCUPATIONAL TOXICOLOGICAL CONSULTATIONS

Toxicological consultations are usually worker-, union-, or individual-driven services, as opposed to employer-driven. These consultations often have as their purpose the determination of whether toxicity has occurred or is occurring as a result of work-related exposures.

It is important to note that this kind of consultation often is tedious and time-consuming on the part of the clinic and the patient. The information-gathering interview alone typically takes well over an hour. This is followed up by the physical exam, the lab work (which often involves 24-hour urine collections, stool collections, and other fairly esoteric specimen collection methods), and then the research through the literature by the physician to find often obscure case reports or other references to the chemical(s) and the particular symptoms of the patient. Then there is the generation of a report. The time required means that one consultation takes up valuable clinic time during which more routine revenue-producing examinations could be performed.

Often, patients are forced to pay for these work-ups out of their own pockets. Often, neither their private health insurance nor their employers' workers' compensation carriers will pay, each denying responsibility and trying to burden the other. Therefore, the amount that can reasonably be charged to an individual worker is very limited in comparison to the time spent by the physician, or in comparison to the amount that can be charged to an insurance carrier or a company, entities which more readily pay for services.

As a result, there is a market disincentive to perform this kind of service in COHCs.

SOLUTIONS

There are at least two possible approaches to resolving these problems. The first approach is to remove occupational health clinics from the demands of the marketplace by supporting these clinics with public funds—as through a national health program.

However, given the present environment, another reasonable and practical solution to this problem is to create an accrediting body that evaluates occupational health services for both quality of service and ethical practice.

One way to do this is for the organized forces of labor and other progressive environmental forces in society to set standards of practice and to encourage the use of occupational health clinics that meet these standards. The health and safety problems of the workplace are still very serious. Control of the field in which the battles over who pays for the costs of occupational health and safety are fought, namely the occupational health clinics, should not be abandoned to the community's employers. Unions and other community environmental forces need the consultation services of environmental and occupational physicians as much as employers do. They need fair doctors to interpret risks, to advocate for safer workplaces, and to provide individual consultations to union members who think they may have been injured or made ill in the workplace. In my opinion, many union grievance representatives miss opportunities to make the best cases for their members because they rely too heavily on the opinions of workers' personal physicians and do not use the services of doctors who really understand workplace hazards.

Union officers should seek those clinics that have reputations for both fairness and knowledge about workplace hazards. An occupational physician should be on staff. Occupational health clinics that do not have the capability of providing services to the community in the way of expert consultations on health and safety, including toxicology, ergonomics, and the human factors of workplace design, should be avoided. In this way, unions can put market counterpressures on the employer-driven bias of occupational health services.

I would urge unions and other progressive forces in the community to join efforts to develop a strong accrediting organization, to develop minimum standards of practice, and to require occupational health clinics to abide by principles of ethics, scientifically validated principles of screening, and common standards of fairness in their delivery of occupational health care.

The accrediting of occupational health clinics and the development of standards of practice are ideas that should be applied in any kind of environment, free market health care or universal health care.

In summary, the forces that have brought about the growth of community occupational health clinics in a free market economy could actually contribute to a loss in the momentum to improve the health and safety conditions in their communities' workplaces. Market forces tend to push a menu of services that are biased toward "shortcuts" to cost cutting in health and safety, reducing autonomy and control by workers over both their safety conditions at work and their economic destinies. These forces threaten to keep the newly emerging specialty of occupational medicine as it has been historically, as the squire of industry, and relatively impotent as an independent specialty group to help communities make changes for the better in their environments and workplaces. Furthermore, if these forces succeed in attracting community occupational physicians into market-subservient (and thus industry-subservient) occupational health clinics, industry will have a greater influence than it has ever had in modern times over the practice of community medicine, endangering traditional and culturally held values such as trust between patients and their physicians as a primary aspect of the therapeutic process.

Ultimately, the power employers tend to exert over market-driven occupational health services due to their market power affects the very nature and objectivity of medical fact finding.

The decision whether to hire or terminate a worker is an administrative decision, even if based on medical information. The decisions employers make about workers have immense consequences for the workers involved. Therefore, clinical judgments must be carefully considered, and clinical data must be communicated in such a way that the rights of workers to privacy and autonomy, at least within the boundaries set by considerations of public health and safety, are not infringed.

To state it more strongly, when financially unequal markets, with often competing priorities, are served by any market-driven service, including occupational health services, these services simply become a means whereby the more powerful market exercises its power over the financially weaker market, and the simple act of collecting and conveying medical information promotes, in its own small sphere, the larger inequality between social classes and abets the power that has historically been exerted by the master over the worker in class societies.

Ironically, occupational medicine, which once yearned for "independence" from behind the walls of industry, finds itself in an even more powerless position because of the exigencies of marketplace competition, a competition that usually aims at serving the moneyed interests. Whereas, when behind the walls, occupational physicians had at least some leverage with the managements of their particular companies, now that they are outside those walls, they have little leverage at all. It is slick logic but a self-deception for physicians who simply ride market forces to rationalize their behavior by saying they just collect the data, they don't make the decisions. Such reasoning, it seems to me, is simply a failure to take responsibility for one's role in contributing to social injustice.

It must be remembered that workers' compensation was always meant to be a "no-fault" form of insurance for industrial injuries. It was not a license for employers to interfere in the normal course of therapy. Despite the fact that the economic consequences of workers' compensation are heavy for industry, they also are heavy for workers. The solutions to the high costs of workers' compensation are probably multifactorial, but solutions should not be sought at the expense of workers' basic human rights, or at the expense of important, highly valued social ideals such as the importance of the doctor–patient relationship and medical confidentiality. In fact, workers' compensation costs can be cut both effectively and equitably by industry's efforts to reduce injuries and illness, and by its sincere efforts to accommodate disabled workers by making its jobs as safe as possible for all workers. This is the best solution to the high costs of workers' compensation, and because it is a solution that requires some sophistication and technological expertise, it is most appropriate that occupational health clinics lead the way. However, they must not abandon principles in order to become profit centers.

The goals of the specialty of occupational medicine should include improving community environmental and occupational health and safety. In so doing, it must maintain existing community standards for medicine in general, namely adherence to scientifically developed principles that have always guided proper medical screening; protection of patients' rights to confidentiality of their medical records and health information in general; fairness in making medical judgments that affect patients' lives and general well-being, especially when the judgments are being requested by employers; and finally, remaining guided by that infallible guide to good medical care—what the Greeks called Agape—indiscriminate, unbiased concern for the patient.

ACKNOWLEDGMENTS

The author wishes to thank Howard Hu, M.D., and Glenn Pransky, M.D., for their review and comments on the early drafts.

NOTES

1. Newkirk WL. *Occupational Health Services: A Guide to Program Planning and Management*, Newkirk WL and Jones LD, eds. American Hospital Publishing, 1989, p. 319.

2. Derr PG. "Ethical considerations in fitness and risk evaluations," *Occupational Medicine: State of the Art Reviews* 3(2) (1988):194.

3. Guidotti TL, Cowell JWF, Jamieson AL. *Occupational Health Services: A Practical Approach*. American Medical Association, 1989.

4. Newkirk WL and Jones LD. *Occupational Health Services: A Guide to Program Planning and Management*, American Hospital Publishing, 1989.

5. Himmelstein JS. "Worker fitness and risk evaluations in context," *Occupational Medicine: State of the Art Reviews* 3(2), (1988):169–178

6. Ashford, NA. "Policy Considerations for Human Monitoring in the Workplace," *Journal of Occupational Medicine*, 28(8) (1986): 563–568.

7. Hennekens CH, Buring JE. *Epidemiology in Medicine*. Mayrent SL, ed. Boston: Little, Brown, 1987.

8. WCRI Publication List. Workers Compensation Research Institute, 245 First Street, Cambridge, MA 02141 (1991).

9. Derr PG. "Ethical considerations in fitness and risk evaluations," *Occupational Medicine: State of the Art Reviews* 3(2) (1988):193–208.

10. Survey of Occupational Health Programs in Massachusetts. Unpublished. Division of Health Statistics and Research, Massachusetts Department of Public Health (1991).

11. Greenwood JG, Wolf HJ, Pearson JC, Woon CL, Posey P, Main CF. "Early intervention in low back disability among coal miners in West Virginia: Negative findings," *Journal of Occupational Medicine* 32(10) (1990):1047–1052.

12. Wolfhagen, C. "Comprehensive Intervention Project: Preliminary Findings," Vocational Rehabilitation Section, Industrial Insurance Division, Department of Labor and Industries, Washington (1989).

13. Buckley, T. "Being A Whistleblower," *Job Safety and Health Quarterly* 4(2) Summer 1991:20–22.

SECTION C

PROGRAMS FOR CHANGE

27

Reforming OSHA

MODEST PROPOSALS FOR MAJOR CHANGE

DAVID WEIL

The United States' history of occupational safety and health in the 1900s now has two horrifying milestones. In 1911, nearly 150 women died in the Triangle Shirtwaist Factory fire. The majority of deaths resulted from the fact that most of the fire-escape doors were locked, forcing workers to either jump from the windows of the burning building or be incinerated. (1) Eighty years later, on September 3, 1991, 25 people at the Imperial Food Products chicken processing plant in Hamlet, North Carolina, died. Most of these workers died from smoke inhalation, allegedly because many of the factory's fire escape doors were either locked or blocked. (2)

Despite the creation of the Occupational Safety and Health Administration (OSHA) in 1970, workers in plants across the United States continue to face perils associated with the most mundane and detectable safety problems such as access to adequate emergency exits. Why, after 20 years of OSHA operation, can such tragedies occur? Is the Imperial Food Products disaster (as well as the collapse of the L'Ambiance Plaza project in 1987 that killed 28 construction workers, or the explosion of a Phillips petrochemical plant in 1989 that killed 23 workers) an indictment of the approach undertaken by OSHA?

This chapter explores this public policy question first by comparing the performance of OSHA in theory with OSHA in practice. Such a comparison indicates that OSHA has exerted regulatory pressure over a relatively small subset of all workplaces under its purview. As a result, despite the promulgation of voluminous standards regulating health and safety practices, OSHA's limited ability to *enforce* such standards allows tragedies like Imperial Foods to occur.

As in the past, the recent spate of safety and health disasters has increased pressure to fundamentally reform or even eliminate OSHA as currently structured. I will argue that, rather than scrapping the entire system, major changes in OSHA *strategy* coupled with changes in state workers' compensation laws could create a far more effective regulatory system.

THE REGULATORY SYSTEM: THEORY VERSUS PRACTICE

OSHA in Theory

In theory, OSHA operates on the basis of a simple regulatory model (Figure 27.1): safety and health standards are promulgated to reduce occupational risks in the workplace. OSHA ensures that employers adopt the promulgated standards by conducting workplace inspections. Businesses found out of compliance with standard(s) in the course of inspections face penalties related to the severity of the violation (for example, higher penalties are levied for violations that put a greater number of workers at risk). Those businesses must then bring their workplaces into compliance with violated standards in a short period of time.

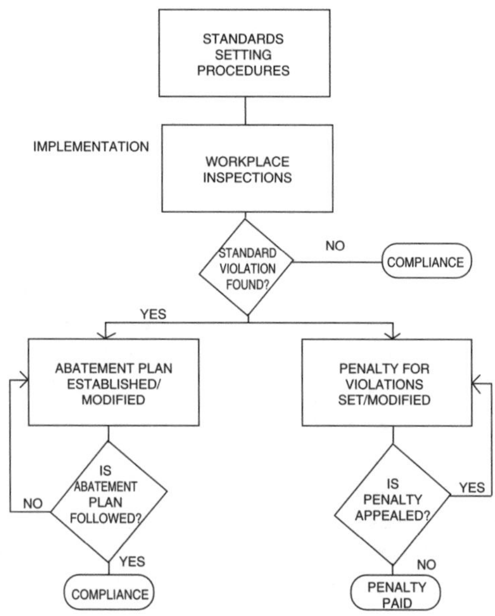

FIGURE 27.1. Schematic structure of OSHA (also of the Mine Safety and Health Act).

This system of enforcement serves two purposes. First, it forces inspected firms to adopt OSHA standards as a result of the "direct" regulatory pressure exerted by inspections. Second, OSHA inspection activity creates a "threat" to all firms covered by the Occupational Safety and Health Act (OSH Act) whether or not they are actually inspected. This threat arises from the combination of the probability of being subjected to an inspection, and the expected penalty costs associated with violations of standards.

Thus, OSHA implementation depends on three critical features of enforcement. First, OSHA must conduct a sufficient number of inspections to enforce standards through the "direct" inspection effect and to create a moderate "threat" of inspection to all firms. Second, inspections must be of sufficient quality and duration to ensure that violations are detected. Third, OSHA must levy significant penalties for violations of standards in order to create sufficient incentives to ensure future employer compliance.

OSHA in Practice

In practice, the simplicity of OSHA's regulatory model quickly breaks down. Problems emerge at each step of the process described above. The standard-setting process has become bogged down, and has failed in recent years to address a large number of workplace chemicals posing potentially serious long-term health risks.

The fundamental problem of OSHA implementation, however, comes down to the often-cited statistics of enforcement. OSHA covers an estimated four million establishments. In 1991, its total inspection work force numbered 1,186. (3) Even if OSHA concentrated solely on the manufacturing sector, the average workplace in that sector would receive an inspection once in 70 years. It is, therefore, not particularly surprising that the Imperial Food Products plant had not received a single inspection by OSHA in more than a decade. (4)

Limited inspection resources have forced OSHA from its inception to focus enforcement selectively. For example, OSHA has always paid greater attention to the manufacturing and construction sectors, given the high safety and health risks associated with those industries relative to other sectors under its mandate. A detailed analysis of OSHA inspection data, however, reveals two critical characteristics of enforcement activity. Across and within different industries, OSHA enforcement is strongly influenced by the size of an establishment and whether the firm is unionized. (5)

Size and Enforcement

Historically, OSHA has pursued a policy that attempts to gain the greatest "bang for its buck" by inspecting larger rather than smaller workplaces.

Similarly, it has tended to inspect workplaces that are part of larger corporate entities rather than single-site employers. This ethos has been strengthened further by OSHA's (and Congress's) continued reluctance to pursue policies that might result in small business failures. As a result, there have been several attempts to exempt small businesses from OSHA's purview entirely. (6)

Union Status and Enforcement

In addition to vesting OSHA with authority to conduct workplace inspections, the OSH Act creates a set of worker rights related to enforcement. These include the right to initiate inspections, to accompany OSHA personnel, and to participate in aspects of postinspection meetings. (7)

Because of the small size of OSHA's inspection force relative to the total number of covered workplaces, employee rights are critical to achieving implementation of the Act. Worker exercise of rights to initiate inspections can appreciably raise OSHA's threat effect. Employee participation in inspections can raise their coverage, duration, and quality. Finally, employee involvement in postinspection meetings can affect the levels of penalties set for cited standard violations.

The exercise of rights is disparate between union and nonunion workplaces for a number of reasons. (8) First, unionized workers are more likely to be informed about the presence of safety and health risks than their nonunion counterparts. Unions in a variety of sectors provide education on these issues to their members. Second, unionized workers are more likely to know about the existence of rights under OSHA, once again as a result of access to resources in the union (for example, shop stewards, international union staff, and safety and health committees). Finally, the Act contains prohibitions against employer discrimination toward workers who have exercised their rights. Unions offer workers a shield against employer discrimination through protection arising under collective bargaining agreements as well as recourse under the Act itself.

As a result of these policies, OSHA enforcement differs markedly across a given industry. Rather than one regulatory environment, there are in fact many, varying on the basis of size and union status. These regulatory environments can be simplified into the following four types of workplaces: large union, large nonunion, small union, and small nonunion. Table 27.1 presents average OSHA enforcement activity for each of these workplace types for the manufacturing sector in 1985. While the discussion below focuses on these results, my research suggests similar patterns for OSHA activity in the construction industry. (9)

TABLE 27.1. Comparative OSHA Enforcement: Manufacturing, 1985

	Workplace type[b]			
Enforcement attribute[a]	Large union	Large nonunion	Small union	Small nonunion
Probability of inspection (%)	95	16	6	14
Employee participation in inspection (%)	70	4	48	3
Inspection duration (hours)	113	89	52	39
Penalty/violation (dollars)	254	98	58	57
Penalty/serious violation (dollars)	495	257	199	191

Source: Based on analysis of data from the U.S. Department of Labor, Occupational Safety and Health Administration, inspection-derived files.

[a]Mean value for selected enforcement attribute.

[b]Large employers (union, nonunion) are workplaces with more than 500 employees; small employers (union, nonunion) are workplaces with fewer than 100 employees.

Large Union Plants

If there is any place where OSHA enforcement over the years has approached that implied by the Act, it is in large unionized establishments. In 1985, 95 percent of these workplaces received OSHA inspections. These inspections typically included employee representatives (70 percent), covered a large portion of facilities, and consequently required considerable investments of OSHA inspectors' time (113 hours per inspection). Finally, employers who violated standards faced significant penalties per violation ($254 on average, and $495 per serious violation).

Large Nonunion Plants

Large nonunion plants are far less likely to receive an inspection than large unionized counterparts. In 1985, only 16 percent of large nonunion plants were inspected. These inspections tended to last far less time than in large union cases (89 versus 113 hours) and virtually never included employees. Finally, average penalties per violation were considerably lower than those assessed in large union plants, even given comparable levels of violation severity.

Small Union Plants

Unionized small workplaces face a low level of scrutiny both as a result of OSHA targeting policy and the difficulties of establishing effective workplace safety and health structures. As a result, in 1985 they did not receive anywhere near the level of scrutiny of large establishments. Where inspections did occur, however, employees were far more likely to participate than in nonunion settings. As a result, these workplaces received long inspections relative to their small size. Unions in small workplaces, however, did not seem to exert much influence on assessed penalties.

Small Nonunion Plants

The small nonunion plant represents the most striking example of the complete breakdown of U.S. occupational safety and health policy. They are seldom inspected. While the probability of inspection in 1985 was about equal to that in large nonunion plants, in nearly one-half of the small workplace cases actual physical inspections were not completed by the OSHA inspector. The few inspections that did occur were short and virtually never included employees. Finally, as in the small union workplace, average penalties were low, not even reaching $60 per violation in 1985.

Clearly, OSHA theory and practice are synonymous *only* in large, unionized workplaces. Because these workplaces represent a small (and declining) percentage of all workplaces covered by OSHA, it is hardly surprising that OSHA performance has been lackluster at best. In a world of such disparate regulatory pressure, what should be done?

A strategy to substantially improve OSHA's performance in its next decade must take into account the different regulatory environments that have resulted from disparate enforcement practices. It must seek to build on areas of strength, while addressing the grave weaknesses of current OSHA policy. Specifically, safety and health policy reform should capitalize on areas of historic OSHA enforcement emphasis (large union workplaces), strengthen private sector safety and health efforts where possible (specifically in small union and large nonunion workplaces) and radically reform enforcement efforts where OSHA has been effectively absent (particularly among small nonunion workplaces).

The remainder of this chapter presents three major areas of policy reform that seek to improve OSHA performance by addressing the enforcement weaknesses described above. However, since these reforms preserve the standards-based system of regulation underlying OSHA, attention must briefly turn to the continuing advantage of a standards-based system.

WHY PRESERVE A STANDARDS-BASED SYSTEM?

The standards-based system created under OSHA has a number of benefits from a regulatory perspective. First, a system of standards implies that there are minimal workplace safety and health practices to which all employers should be held accountable. For example, Imperial Products stands as a reminder of the fundamental need for adequate emergency exits and evacuation procedures in all workplaces. Similarly, provision of adequate scaffolding protection is basic to construction site safety.

Second, standards-based approaches are particularly compelling for occupationally related health problems. Occupational health problems have long latencies (often manifesting themselves after retirement or job changes) and are often multicausal. For example, asbestosis and related lung disorders appear long after initial exposure, and can be aggravated by smoking and other nonwork behavior. As a result, occupational health problems cannot be adequately addressed through reforms such as injury-based taxes that assume more temporal and proximate connections between workplace hazard exposure and regulatory pressure.

Finally, well-crafted standards-based approaches can provide a public good to employers (particularly small- and medium-sized employers). Standards may provide guidance on methods of risk reduction not readily apparent to employers with limited organizational resources. This is particularly advantageous to remediation of health hazards where risk reduction is usually an extremely complex undertaking.

The advantages associated with a standards-based system, however, do not negate the severe problems faced by OSHA in actually promulgating standards, particularly in the area of occupational health. During its history, OSHA has issued standards for only 24 substance-specific health risks. This number is dwarfed by the number of potential substance-related risks discovered over this period. For example, the American Conference of Governmental Industrial Hygienists (ACGIH) has recommended exposure limits on more than 200 chemicals where OSHA has no exposure limits. Reform of standard-setting policies, therefore, requires serious attention, independent of the enforcement-related reforms discussed here. (10)

BUILDING ON STRENGTHS: FOSTER LABOR/MANAGEMENT REGULATION

OSHA's significant presence in large unionized workplaces has provided an important incentive to those employers to work with labor unions in improving internal safety and health activities and practices. In many industries, collective bargaining provisions concerning safety and health increased dramatically following the passage of the OSH Act. In a related vein, many unionized firms formed or strengthened workplace-based safety and health committees. As a result, a "public/private" regulatory system has emerged among many unionized workplaces over the last 20 years.

There is evidence spanning two decades that indicates the potential success of these public/private regulatory systems. Several studies completed during the first decade of OSHA documented the effectiveness of safety and health committees in improving safety and health practices and in some cases reducing injury levels in a number of industries. (11) Labor/management efforts in several industries provide further examples of the benefits of such systems.

In the automobile industry, the United Auto Workers' agreement with the three major automakers in 1973 included for the first time provisions creating safety and health committees and union representatives. Since that time, the scope and sophistication of safety and health provisions have increased. Recently, the provisions expanded to include a "New Technology" section, with the intent of ". . . designing-in health and safety controls on equipment and processes during the early design stage." (12) A recent study by Wokutch found that improvement in injury and illness rates in the auto industry could be associated with the increasing sophistication of labor/management safety and health agreements and practices. (13)

A novel experiment called the Cooperative Compliance Program (CCP) instituted in California between 1979 and 1984 provides another example of effective public/private regulation of safety and health. The CCP was a tripartite agreement between construction unions, contractors, and California OSHA to allow labor/management safety committees to assume many of OSHA's regulatory responsibilities. The program resulted in significant reductions in accident rates compared with comparable construction sites operating outside of the program. (14) Examples of effective systems of public/private safety and health regulation can also be found in petrochemical, steel, rubber, and coal mining industries. (16)

Public policies should actively advance labor/management safety and health systems. The enforcement statistics presented in Table 27.1 indicate pronounced differences in the effectiveness of these efforts between small and large union workplaces. Safety and health performance in the union sector can, therefore, be improved by strengthening labor/management efforts in small workplaces. This can be done in two principal ways: (a) increase resources available to labor/management efforts; and (b) expand worker rights granted under OSHA. (16)

a. *Increase resources.* Limited resources represent a major impediment to instituting effective labor/management efforts in small union workplaces. A vastly expanded program to help create and train safety and health committees could help to build safety and health capabilities at smaller workplaces.

OSHA's "New Directions" program of the pre-Reagan years represents a prototype of such a policy. "New Directions" grants spawned training materials, education programs, and other innovations among a gamut of unions. An expanded program could target resources particularly to smaller workplaces by providing seed money to unions with safety and health programs with such an emphasis, by running regional training conferences and events, and by promoting exchange of information regarding successful programs in small workplaces.

b. *Strengthen and expand employee rights.* Exercise of employee rights granted by OSHA accounts in large part for the significant degree of enforcement in large union workplaces. Improvement of current employee rights can further enhance implementation by strengthening existing rights and expanding them in several respects.

Initiating and Participating in Inspections

Triggering and participating in inspections is critical to implementation and should be improved by expanding rights in two ways. First, OSHA should create a special class of "imminent danger" inspections involving standard violations that pose high levels of risks for workers. Worker-initiated "imminent danger" inspections would require OSHA response within one week of notification. Imminent danger inspections should require that employees accompany OSHA inspectors and the employer as well as require private meetings between affected employees and the OSHA inspector.

Second, OSHA should encourage employee participation in all inspections by requiring that employees receive pay for time spent on such inspections. Collective bargaining agreements in larger workplaces often stipulate that employees (or safety and health committee members) receive such compensation. Mandating pay as a matter of public policy would bring this important incentive to small union (and nonunion) workplaces.

Participating in Postinspection Activities

Many citations by OSHA are appealed by employers in informal and formal settings. Many of these meetings result in significant reductions in penalties and amendments to abatement plans. The end result is a diminution of the costs of noncompliance with safety and health standards. The right of employees or their representatives to participate in such meetings has been contested since 1971. OSHA can strengthen implementation by eliminating these ambiguities in the law.

Unions can act as an important counterweight in informal and formal postinspection proceedings. My research indicates that employers are less able to decrease penalties or contest abatement orders in unionized workplaces. To minimize the dilution of OSHA's regulatory efforts via postinspection proceedings, employees or their representatives should be granted the unconditional right to participate in all such meetings and administrative hearings arising from employer appeals. (17) Further, employee rights to initiate a contest should be extended beyond current restrictions to the length of time provided for abatements. (18)

Refusing Hazardous Work

While there are some precedents to suggest a right to refuse hazardous work under OSHA (19), it does not represent the fundamental right granted miners under the Mine Safety and Health Act. The creation of such an unambiguous right could prove an important right and protection in imminent danger situations. Exercise of such rights would necessitate strengthened discrimination provisions of the Act, which will be discussed below.

ADDRESSING WEAKNESSES: ENHANCING OSHA'S REGULATORY PRESENCE

The vast majority of workplaces in the United States are nonunion. As a result, the second prong of OSHA reform should redress the absence of OSHA enforcement in large and small nonunion workplaces.

It is reasonably clear that government resources at the state and federal level will be constrained for the foreseeable future. As a result, improving enforcement in the nonunion sector will require targeting limited OSHA resources more effectively, while enhancing the incentives for voluntary compliance to the greatest extent possible. This can be achieved in a number of ways.

Target OSHA Resources

The risk of inspection for large and small nonunion workplaces is extremely low. Thus, OSHA must find means of increasing its potential threat to these firms. In this regard, OSHA should base enforcement targeting on *substantive and strategic* criteria. Substantive criteria would entail choosing industries and industry segments facing the most critical safety and health problems. Strategic criteria involve focusing inspection resources to maximize the "threat" impact of OSHA efforts on noninspected employers.

On the basis of substantive targeting, OSHA should enforce a subset of key standards, judged of particular importance to the specific industry in question. By focusing on a few standards, OSHA inspections would be more intensive, yet require less time per inspection. Activity in this model would therefore be double-targeted: first, to a subset of firms and, second, to a subset of critical standards. In a related vein, penalties for the subset of violations should be raised, appeals procedures delimited, and criminal penalties considered (particularly in the case of repeat or "willful" violators).

Strategic targeting would attempt to maximize the effect of OSHA inspections on all relevant firms, whether or not they actually received inspections. OSHA would concentrate inspection resources in a manner that substantially raised OSHA's threat effect for all firms in a targeted group (for example, by focusing on a subindustry in a limited geographic area). It would choose to inspect those firms within the targeted group which sent the loudest "message" to the remaining firms. Finally, inspection efforts would be carefully publicized to give them as high a profile as possible.

For example, assume that OSHA discovered an increasing incidence of health problems associated with chemical inhalation among workers in electroplating firms. In response, OSHA would create a list of standards related to the specific problem of airborne chemical exposure in electroplating. It would then institute an intensive inspection program concentrating solely on the subset of relevant standards for electroplating firms.

In order to ensure that a maximum number of electroplating firms were affected by these efforts, the program would focus on firms in a specific geographic area where a large percentage of the industry operated (for example, New England). Early results of the inspection efforts would be publicized, with particular attention paid to willful or repeat violators. Finally, after a significant percentage of New England firms had received "blitz" inspections, OSHA would publicly announce the next regional target for intensive enforcement.

The notion of targeting is not new. During the Carter administration, OSHA concentrated inspections in high-hazard manufacturing industries, defined as those with a lost-time injury rate above the average for the entire manufacturing sector. The Reagan administration instituted a procedure that subjected only those firms with an injury rate exceeding the national industry average to inspections, while entirely exempting firms that fell below the average rate. (20)

Previous schemes, however, did not treat OSHA's regulatory threat as an explicit targeting criteria. Further, by focusing on a narrow range of standards, the productivity of inspections could be increased and the total number of inspections expanded, even if few new inspectors are hired. Finally, increasing penalties dramatically for a more limited number of violations would further increase the threat effect posed by OSHA.

Increase Employee Exercise of Rights

OSHA rights are far less likely to be exercised in nonunion settings. OSHA should foster nonunion worker exercise of rights by reducing fears of discrimination or employer reprisal.

OSHA could send a powerful message about the importance of employee rights by instituting a high-profile effort to penalize employers who discriminated against employees in a manner similar to OSHA's recent efforts on record-keeping violations. Similarly, OSHA inspectors could encourage employee participation, and particularly private meetings with employees, as a major part of their inspection efforts.

Even given these actions, however, the effectiveness of OSHA's current antidiscrimination clause is limited. First, many workers know little about their rights or recourse under OSHA. Second, workers who do attempt to find recourse against discrimination have very limited success. A study of 249 workers in Wisconsin who were fired for job-safety protests between 1981 and 1986 found that only 6 percent eventually won reinstatement. (22)

As a result, OSHA's provisions on discrimination must be improved in two ways. First, the review process for discrimination cases should be expedited. OSHA should be required to initiate investigations within a brief (and stipulated) period of time. If initial investigations found that the employee claim was not frivolous, OSHA should require temporary reinstatement pending resolution of the claim. Second, the costs of discrimination should be increased. Employers who retaliate against workers who exercise OSHA rights should pay stiff penalties in addition to being required to reinstate employees with back-pay. (23)

Policies for Large Employers

Large employers' human resources practices in general and safety and health policies in particular are considerably better than those of small employers. (24) Many studies cite the existence of corporate cultures that foster safer workplace conditions. These attitudes may arise from a combination of factors, including more enlightened corporate leadership, more abundant resources and human resource planning capacities, or fear of exposure to adverse publicity in the event of accidents.

OSHA should examine the sophisticated safety and health policies of progressive large firms to better understand the reasons for their voluntary adoption. In so doing, it could create regulatory carrots to foster corporate cultures that promote safety and health. (25) However, given that major problems in recent years also have occurred at plants run by large, well-known corporations, OSHA also should not fail to capitalize on the aversion of large firms to negative publicity as a regulatory tool for noncompliers.

Finally, large nonunion employers seem the most conducive audience for OSHA efforts to promote exercise of employee rights. Several large corporations have attempted to encourage "whistleblower" behavior in other policy areas (for example, environmental laws). The identity of a potential whistleblower is far more concealable in a large operation than in a small workplace where such a person may feel far more vulnerable. OSHA's efforts to inform workers of the existence of rights and protection against reprisal might therefore most productively be focused on larger operations.

Policies for Small Workplaces

While small workplaces are the least subjected to OSHA enforcement, they also tend to be the most dangerous. Many studies indicate that the rate of serious injuries or fatalities is highest for establishments with fewer than 50 employees. (26)

It is unlikely that workers unprotected by labor unions can ever be expected to exercise their OSHA rights to a level sufficient to provide enforcement in these workplaces. It seems equally unlikely to expect to foster corporate cultures conducive to safer practices in most small firms. Finally, adverse publicity seems an unlikely lever (or credible stick) to improve performance. Thus, safety and health policies for the small workplace must ultimately depend on concerted governmental enforcement. This can come only from explicit targeting strategies like those discussed above.

The huge number of small workplaces and the small size of the inspection workforce make intensive surveillance difficult. OSHA should focus on those industries where workers in small establishments face the greatest risks (for example, the poultry industry). These efforts should focus on a very limited number of critical standards, allowing inspectors to quickly and efficiently inspect a large number of employers. Once again, penalties for these standards would be significantly higher than currently prevailing levels. As described above, these industry "blitzes" could be publicized to enhance their spillover impact on uninspected firms. (27)

Over the years, small businesses have complained that compliance with OSHA (as well as other regulatory policies) would put them out of business. These pleas have been taken as a justification for explicit exemption of small businesses from any number of public policies. Yet one must ask whether a firm that cannot meet its baseline safety and health responsibilities should be allowed to remain in operation. In fact, it might be an entirely desirable policy outcome to force small, marginal firms out of business for this reason. (28)

IMPROVE OVERALL INCENTIVES

The injury tax is usually considered a polar alternative to OSHA's standards-based system. This need not be the case. By increasing the benefits to employers of decreasing their injury rates (while refocusing OSHA in the manner described above), overall safety and health performance could be enhanced.

Rather than requiring new legislation, injury taxes exist in an imperfect form under the present workers' compensation system. Current workers' compensation premiums are based in part on the injury experience of employers. Unfortunately, the system is only imperfectly experience-based. As a result, the most safe employers (usually larger firms) subsidize the least safe employers (usually smaller firms) in a given industry. By both improving experience rating in workers' compensation systems and expanding the basis of experience rating beyond injury outcomes, the present workers' compensation system could supplement the policies discussed above.

Improving Experience Rating

Analysts long have cited the need for workers' compensation systems to link premiums more closely to a firm's injury experience. (29) Several recent studies indicate the positive role that improved experience rating can bring. For example, Krueger has found that firms that have opted out of imperfectly rated systems and become self-insured improve safety and health performance. (30) Thus, moving state systems closer to the "self-insurance" case would improve the incentives for employers to reduce injury levels further.

Broadening the Basis of Experience Rating

Expanding experience rating to consider safety and health practice as well as injury rates would allow the workers' compensation system to support the OSHA reforms discussed above. For example, workplaces that instituted safety and health committees could receive benefits through rate reductions. Firms that invested in committee efforts (for example, provided pay for committee work or trained committee members) could be similarly rewarded. Finally, companies that demonstrated concerted efforts to create better corporate safety and health practices or improve protection for whistleblowers could similarly benefit through premium reductions.

A standards-based and injury tax system could therefore productively function together. Under this integrated regime, certain firms would face a double regulatory presence where aggressive enforcement of baseline standards would occur side by side with experience rating. In other lower-risk industries, employers might only face the incentives provided by experience rating.

CONCLUSION: A TRACTABLE REGULATORY STRATEGY

The strategy described here is based on explicit consideration of issues of enforcement and implementation. It seeks to recognize the virtues of the present system and expand upon them while paying equal attention to the vice that afflicts most labor policies: lack of implementation among unorganized workers. The three-pronged strategy would represent a marked departure from the present system of safety and health regulation. Yet, it is feasible without the complete abandonment of the present OSHA system.

The Comprehensive Occupational Safety and Health Reform Act (S. 1622) introduced by Sen. Edward Kennedy (D.-Mass.) in August 1991 embodies many of the proposals discussed in this chapter. Among other provisions, it would create stronger rights to refuse hazardous work, strengthen antidiscrimination provisions, and enhance worker rights to initiate inspections. (31)

The feature of this legislation, however, that has received greatest attention is that it would mandate that all employers with 11 or more workers set up management/employee safety and health committees. The committees would be vested with the authority to review employer safety and health programs, investigate fatalities, injuries, and illnesses, conduct inspections once every three months, and make recommendations to employers. In effect, it would create a "right to act" for workers in all workplaces.

At first glance, these proposals would seem to constitute a fundamental change in safety and health enforcement. By mandating safety and health committees in all workplaces, the Act would seek to create the productive private/public regulatory systems currently operating in many large unionized workplaces. By empowering workplace-based safety and health committees, risks could be addressed in an ongoing fashion, even absent OSHA intervention.

The stumbling block of this proposal, however, is the problem of implementation. As this discussion of OSHA enforcement in large versus small union workplaces has suggested, creating effective safety and health committees is not easy. If they have not taken root in many unionized settings, they certainly will not emerge by merely legislating them into existence in general. Further, nonunion workers are reluctant to exercise their rights under the circumscribed set of employee rights currently granted them by OSHA. The probability that these workers will exercise more extensive rights in committee settings seems highly questionable, particularly in small nonunion workplaces. (32)

If enacted, the committee-related provisions may strengthen existing labor/management efforts by unions and in some large nonunion workplaces. However, in the short to medium term, the proposed legislation's most immediate effect would arise from the other provisions that strengthen features of the existing system.

Major disasters have moved occupational safety and health policy forward throughout this century. The Triangle Shirtwaist fire precipitated the enactment of New York's earliest protective labor legislation. Rank-and-file coal miner reaction to the epidemic of black-lung cases in the 1960s precipitated the passage of the Mine Safety and Health Act of 1969. (33) The explosion of a mining operation in Farmington, West Virginia, gave final impetus to passage of OSHA in 1970. (34) The L'Ambience, Amoco, and Imperial disasters hopefully may provide a similar impetus to draw on all we have learned from 20 years of experience under OSHA in order to build a truly effective system of health and safety regulation.

NOTES

1. For one account of the Triangle fire, see Alice Kessler-Harris, "Organizing the Unorganizable: Three Jewish Women and Their Union," *Labor History*, v. 17, no. 1 (Winter 1976).

2. Ronald Smothers, "25 Die, Many Reported Trapped, As Blaze Engulfs Carolina Plant," *The New York Times*, September, 4, 1991, pp. Al, B7; Ronald Smothers, "North Carolina Examines Inspection Lapses in Fire," *The New York Times*, September 5, 1991, p. D25.

3. U.S. Department of Labor, Occupational Safety and Health Administration, 1991.

4. In 1990, North Carolina had 27 safety and health inspectors (including trainees) to oversee more than 180,000 employers. See Susan B. Garland, "What a Way to Watch Out for Workers," *Business Week*, September 23, 1991, p. 42.

5. The following section arises largely from my studies of OSHA enforcement. For specific results, see David Weil, "Enforcing OSHA: The Role of Labor Unions," *Industrial Relations*, v. 30, n. 1 (Winter 1991), pp. 20–36; and "Building Safety: The Role of Labor Unions in the Construction Industry," *Journal of Labor Research*, v. 13, n. 1 (Winter 1992), pp. 121–132.

6. Three bills introduced in the 1970s sought to exempt small businesses from OSHA entirely.

7. Benjamin W. Mintz, *OSHA: History Law, and Policy* (Washington, DC: Bureau of National Affairs, Inc., 1984), Chapter 13 .

8. I examine this issue more generally in "Employee Rights and the Implementation of Labor Laws," *Proceedings of the Council on Employee Rights and Responsibilities*, October 1991.

9. Table 27.1 is based on the complete OSHA inspection records for 1985 for federally run programs, encompassing 28 states. It does not include inspections administered by state-run OSHA programs.

10. For several views, see Mark Rothstein, "OSHA After Ten Years: A Review and Some Proposed Reforms," *Vanderbilt Law Review*, v. 34 (1981), pp. 71-139; John Mendeloff, "Regulatory Reform and OSHA Policy," *Journal of Policy Analysis and Management*, v. 5 (1986), pp. 440-460; and Sidney Shapiro and Thomas McGarity, "Reorienting OSHA: Regulatory Alternatives and Legislative Reform," *Yale Journal on Regulation*, v. 6, no. 1 (Winter 1989), pp. 1-63.

11. These studies include: Thomas Kochan, Lee Dyer, and David Lipsky, *The Effectiveness of Union-Management Safety Committees* (Kalamazoo, MI: W. E. Upjohn Institute, 1977); William Cooke and Frederick Gautschi, "OSHA, Plant Safety Programs, and Injury Reduction," *Industrial Relations*, v. 20, no. 3 (Fall 1981), pp. 245-257; and Leslie Boden, Judith Hall, Charles Levenstein, and Laura Punnett, "The Impact of Health and Safety Committees," *Journal of Occupational Medicine*, v. 26, n. ll (November 1984), pp. 829-834.

12. For the text of this agreement, see "Health and Safety Agreement Between General Motors and the UAW," *New Solutions*, v. l, no. 4 (Spring 1991).

13. See Richard Wokutch, *Cooperation and Conflict in Occupational Safety and Health: A Multination Study of the Automotive Industry*, (New York: Praeger, 1990).

14. Joseph V. Rees, *Reforming the Workplace: A Study of Self Regulation in Occupational Safety*, (Philadelphia: University of Pennsylvania Press, 1988).

15. The Mine Safety and Health Act (MSHA) regulates safety and health for metallic and nonmetallic mining including underground coal mining. Similar to the findings presented above, underground coal mines organized by the United Mine Workers of America (UMWA) receive far greater regulatory scrutiny than their nonunion counterparts. The UMWA has created a sophisticated system of mine committees, supported by a regional and national union program, to monitor mine conditions on a daily basis. While mine committees are encouraged to work out disputes voluntarily with mine operators, committees also use their right to involve MSHA as a further incentive to reach such agreements. For a more complete discussion, see David Weil, "Guarding the Mine Face: The UMWA and the Implementation of MSHA," unpublished manuscript, 1990.

16. Even absent changes in the existing law, union policies to build their capacities to address safety and health can improve safety and health performance. Unions can provide critical training for workplace committees (particularly in small workplaces), and through collective bargaining attempt to build the strength of those committees. This includes bargaining for inspection rights, pay for time on committee work, and winning the right to refuse hazardous work as matters of collective bargaining. Safety and health remains a powerful organizing tool for unions. Such efforts serve both the union movement and occupational safety and health goals.

17. See Stephen A. Bokat and Horace A. Thompson III, *Occupational Safety and Health Law* (Washington, DC: Bureau of National Affairs, 1988), chapter 17, for a full discussion of employee and union participation in litigation under OSHA.

18. In contrast to OSHA, MSHA provides for an employee right to contest penalties.

19. See Nicholas Ashford and Judith Katz, "Unsafe Working Conditions: Employee Rights Under the Labor Management Relations Act and the Occupational Safety and Health Act," *Notre Dame Lawyer*, v. 52, (June 1977), pp. 802-837.

20. *See* Wokutch (1990), pp. 40-46, for a discussion of these targeting policies.

21. Given the collapse of the cold war, OSHA might turn to former Defense Department experts on deterrence theory to help design targeting strategies.

22. Based on a study conducted by Joan McManus, director of Wisconsin Committee on Occupational Safety and Health, and reported in David Moberg, "Weak Workplace Safety Law Needs Strength to Do Its Job," *In These Times*, May 23-June 5, 1990.

23. The Mine Safety and Health Act provides far more extensive discrimination provisions that could be adopted by OSHA. See Federal Mine Safety and Health Act of 1977, 30 United States Code, Sections 801 through 962, Section 105(c).

24. For a study of this issue generally, see Fred K. Foulkes, *Personnel Policies in Large Nonunion Companies* (Englewood Cliffs, NJ: Prentice-Hall, 1980); Charles Brown, James Hamilton, and James Medoff, *Employers Large and Small* (Cambridge, MA: Harvard University Press, 1990).

25. For example, the 3M Corporation has created a comprehensive safety and health and environmental program, primarily through its own initiation. Other major nonunion corporations have similarly developed state-of-the-art safety and health policies (often actively including their employees in these processes). OSHA should gain greater insight into the internal cultures that fostered these policies and find means to encourage similar efforts among other employers. For a description of the 3M program, see A. Ron Zigman, "Minimizing the Impact of Technology on People and the Environment," in Association of Trial Lawyers of America, *Safe Work: Preventing Injury and Disease at the Workplace*, Conference Report (Washington, DC: ATLA, 1991). For a description of other programs among large nonunion employers, see Roger Kasperson, Jeanne Kasperson, Christoph Hohenemser, and Robert Kates, *Corporate Management of Health and Safety Hazards: A Comparison of Current Practices* (Boulder, CO: Westview Press, 1988).

26. Mendeloff and Kagey find that the probability of a fatal injury is inversely related to firm size. See John Mendeloff and Betsy Kagey, "Using Occupational Safety and Health Administration Accident Investigations to Study Patterns in Work Fatalities," *Journal of Occupational Medicine*, v. 32, no. 11 (November 1990), pp. 1117-1123. See also J. Paul Leigh, "No Evidence of Compensating Wages for Occupational Fatalities," *Industrial Relations*, v. 30, n. 3 (1991), pp. 382-395.

27. Thus, while the overall probability of an inspection at small workplaces might remain relatively small, the probability of inspection in hazardous target industries would be boosted considerably during the course of a particular "blitz" program.

28. The most disturbing example of this problem can be found in underground coal mining. Many small mine operations persist at the edge of profitabil-

ity and would be closed if MSHA standards were fully enforced. Yet 80 percent of all mining fatalities occur in these operations, despite the fact that they account for a small percentage of total coal output. Thus, forcing these operations out of business arguably serves larger policy goals. For recent discussions, see Albert Karr, "Labor Department Plan Could Close Unsafe Mines," *Wall Street Journal*, June 1, 1989; Peter Kilborn, "Scraping by, Illegally, Mining Kentucky Coal," *The New York Times*, March 3, 1991, pp. 1, 32.

29. For a concise discussion of this issue, see James R. Chelius, "Role of Workers' Compensation in Developing Safer Workplaces," *Monthly Labor Review*, September 1991, pp. 22-25.

30. See Alan Krueger, "Incentive Effects of Workers' Compensation Insurance," *Journal of Public Economics*.

31. For a text of the proposed Act, see *Congressional Record*—Senate, pp. 511833-511845.

32. The limited success of Canadian provincial government efforts to mandate safety and health committees provides another cautionary note for the U.S. See Bob DeMatteo, "Health and Safety Committees: The Canadian Experience," *New Solutions*, v. 1, no. 4 (Spring 1991), pp. 11-15.

33. See Barbara E. Smith, *Digging Our Own Graves: Coal Miners and the Struggle Over Black Lung Disease* (Philadelphia: Temple University Press, 1987).

34. For a more complete discussion of the origins of OSHA, see Robert Asher, "Organized Labor and the Occupational Safety and Health Act," *Labor's Heritage*, January 1991, pp. 56-76.

28

A Women's Occupational Health Agenda for the 1990s

SUSAN KLITZMAN
BARBARA SILVERSTEIN
LAURA PUNNETT
AMY MOCK

The occupational health problems that are specific to women workers primarily result from patterns of employment rather than from physiological sex differences. (1) Women's position in the labor force and in society limits their ability to obtain protection from and compensation for work-related illness and injury.

The difficulties encountered by women workers in getting their occupational health and safety concerns addressed result from several interrelated problems. Relatively little is known about many of the hazards to which women are exposed. Related to this, there are commonly held misperceptions about women's position in the labor force, such as "women's work is safe" (compared to male-dominated industries such as mining and construction) or that "women don't really work at all." Yet hazards do exist in many female-dominated industries such as health care, garment, and microelectronics. Moreover, the risk of certain types of accidents and illnesses may be high in some female-dominated workplaces, but is often unrecognized and therefore not compensated.

In addition to direct occupational health risks, many barriers to dealing with these problems exist. The current employment system is not structured to accommodate the needs of employees of either sex who have child-care, elder-care, and other home-related responsibilities. By and large, women continue to bear a disproportionate burden for these obligations. The dual obligations of women at work and at home often conflict, especially for the increasing number of women who head single-parent families. This double burden limits many women's ability to participate in educational, union, and political activities that could empower them to improve health and safety conditions. Finally, women of color, lesbians, older, and physically handicapped women often face additional educational and employment discrimination that may lead to their being placed in more hazardous and stressful working conditions.

THE CONTEXT OF OCCUPATIONAL HEALTH IN THE UNITED STATES

The 1980s have witnessed a general decline in government support of occupational health activities. The overall impact of budget cuts on regulation, enforcement, and research on occupational safety and health hazards and subsequent work-related illness and injury has been discussed in detail elsewhere. (2, 3) These cutbacks have had a particular impact on women. OSHA inspections are rare in most of the nonmanufacturing industries, where more than two-thirds of the female workforce is employed. In addition, many hazards common in female-dominated workplaces—such as ergonomic hazards (for example, repetitive manual work and static postures), video display terminals (VDTs), waste anesthetic gases, and indoor air pollution—continue to be unregulated by specific OSHA standards.

At the same time, OSHA has made some progress in developing standards for certain hazards that affect many women workers, such as ethylene oxide, formaldehyde, and blood-borne infectious diseases. These actions have come about largely in response to pressure exerted by labor unions, environmental, public interest, and women's organizations, and by congressional oversight activities. For example, the Service Employees International Union (SEIU) and the American Federation of State, County and Municipal Employees (AFSCME) recently led a successful campaign that resulted in OSHA's proposed rulemaking on blood-borne diseases. (4) In addition, since 1980, 14 unions worked to get OSHA to promulgate an Emergency Temporary Standard on formaldehyde. A standard was finally issued in 1988. (It should be noted, however, that both of these standards have been the subject of some union criticism.) (5, 6)

These efforts are particularly noteworthy, since unions generally have been under political and economic attack in recent years, both by hostile employers and by the government. (7, 8) As a result, many unions have only limited resources that can be directed toward organizing women, tackling health and safety issues, and taking direct political action. In addition, corporate interests generally have organized strong opposition to the limited government attempts at regulating health and safety hazards. For example, of particular relevance to women, the major U.S. computer manufacturers testified against comprehensive VDT legislation in Suffolk County, New York. (9) Once the law was enacted industry raised legal challenges that resulted in a New York State Supreme Court ruling striking down the law. (10)

WOMEN'S EMPLOYMENT STATUS

Occupational Segregation

Approximately 45 million women in the United States today are employed outside the home. (11) Despite significant affirmative action gains, the vast majority are still concentrated in jobs characterized by low wages, low rates of unionization, and limited opportunities for career development and advancement. Eighty-four percent of all employed women work in the service (i.e., nonmanufacturing) sector; within this sector, women are overrepresented in selected job classifications. In 1985, women made up 95 percent of registered nurses, but only 17 percent of physicians; 98 percent of secretaries, but only 7 percent of engineers; and 73 percent of primary and secondary school teachers, compared with 17 percent of college and university teachers. (12) Many female-intensive jobs, such as domestic child care and cleaning and industrial and electronic homework, also are socially isolated.

Inequality between men and women is dramatically illustrated by the difference in their respective earning power. In 1986, the average earnings for women who worked full-time were 70 percent of those for men. Several studies have concluded that a significant portion of this difference is due to outright sex discrimination. (13–15) Also, substantial numbers of women in the labor force work part-time (41 percent in 1984). Although some people choose to work part-time because of family responsibilities, school, or transition to retirement, many workers do so involuntarily, because full-time work is not available. Women constitute an estimated 60 percent of this part-time involuntary workforce, where wages and benefits are disproportionately lower than those of full-time workers (16, 17).

In addition to gender segregation, women of color, lesbian, elder, and handicapped women often face additional forms of discrimination that may result in their being placed in lower-paying, dirtier, and heavier jobs. Black workers have a 37 percent greater chance of suffering an occupational injury or illness; a 50 percent greater likelihood of becoming severely disabled from job injuries and illnesses; and a 20 percent greater chance of dying from job-related injuries and illnesses than white workers. (18)

Unionization

It is reasonable to expect that unionized workers have relatively more resources and opportunities to improve health and safety conditions. Workers who lack union representation may be less likely to report or take action to reduce health and safety hazards for fear of retaliation by the employer. Nonunionized workers are therefore more likely to work under conditions that endanger their health and safety. Women disproportionately lack union representation and its respective benefits.

While only a small and declining minority of U.S. workers are unionized (18 percent in 1985), women are even less likely to be represented by unions than men. Only about 30 percent of union members are women, although 44 percent of the civilian labor force is female. In general, the proportion of workers represented by unions is higher among male- than among female-dominated occupations. For example, 86 percent of motor vehicle and equipment operators (90 percent male) and 57 percent of machinists (96 percent males) are union members, compared with only 14 percent of service workers (39 percent male). (11, 12) Women are also underrepresented among union officers and staff, especially at the national level. One study of 15 major U.S. unions found that women made up only 9.7 percent of their national governing boards, even though the membership was 45 percent female. (19)

Despite the overall low rates of unionization among women workers, some unions (for example, AFSCME, SEIU, NUHHCE, ILGWU, UFT) have traditionally organized women in female-dominated industries, such as light manufacturing, educational and health services, and government. Other unions facing declines in membership due to a loss of manufacturing jobs, such as the UAW, have begun organizing efforts in these sectors of the labor force. The successes may be attributed, in part, to using women organizers in union campaigns and to addressing issues of particular interest to women, such as job stress and maternity benefits. (19, 20)

Unionization has a direct impact on women's wages, benefits, and working conditions. In 1986, women in unions earned an average of nearly one-third more than their nonunionized counterparts ($368 compared with $274 per week, respectively). (11) In 1980, the earnings of service workers represented by unions were 47 percent higher than the earnings of unorganized workers. (21) Collective bargaining has achieved a variety of other improvements affecting women, such as comparable worth clauses

(AFSCME); job training and retraining (UAW); parental leaves with guaranteed retention of wages and seniority (ILGWU, AFT); and child-care benefits (ILGWU, SEIU, AFGE, AFSCME).

In the area of health and safety, many unions with large female memberships (for example, CWA, AFSCME, PEF) have become involved in issues relevant to women, such as office work, ergonomics, indoor air pollution, and job stress. Additionally, local joint health and safety committees have become a focus of some women's labor groups, such as CLUW, 9 to 5, NAWW, and COSHs (Coalitions for Occupational Safety and Health).

Expectations Regarding Women's Social Role

Despite a prevailing social attitude that wage labor and family obligations are in conflict, it is estimated that 85 percent of working women in the United States will become pregnant at least once during their working lives. (22) The cycle of caretaking—first for children, then for aging parents and other family members—continues throughout many women's lives. Current employment practices do not support the multiple roles of working, birthing, parenting, and elder care, however. Employment policies that would allow employees of both genders to combine working and dependent care on a long-term basis are limited in the United States, in contrast to Canada, Japan, and virtually all Western European nations. (23)

The unmet need for child- and other dependent-care services is difficult to estimate. One recent survey found that only 2,500 out of 6 million employers in the United States offer any child-care assistance. (24) Yet 52 percent of women with children under six years old were employed outside the home in 1986. (25) Furthermore, part-time and temporary employment, which could better accommodate dual roles for some workers, generally do not provide job security, adequate salary benefits, or career development.

THE OCCUPATIONAL HEALTH SIGNIFICANCE OF WOMEN'S EMPLOYMENT STATUS

Women's employment status is characterized by occupational segregation, lack of unionization, and societal expectations of multiple roles in the home and in the workplace. As a result of these conditions, women often face differential exposure to workplace hazards and often discriminatory responses to genuine health and safety problems. Four health and safety issues that illustrate this are: (a) exposure to mixed and poorly characterized hazards; (b) disproportionate focus on occupational reproductive hazards, excluding similar hazards to men; (c) inadequate attention to ergonomic hazards, such as highly repetitive manual work; and (d) work-related sources of stress, particularly high workload demands with little ability to control one's own work. Below we summarize each of these.

Mixed and Poorly Characterized Exposures

Traditionally, occupational epidemiology has focused on assessing exposures to relatively high levels of toxic materials that are found largely in male-dominated industries, such as asbestos and coal dust. Quantitative risk assessments generally use males as the model for workplace exposures and children as the model for environmental exposures.

Many workers—both male and female—are employed in workplaces that are characterized by multiple exposures, often at unknown levels, such as in offices, semiconductor plants, and health care facilities. This mixed-exposure picture is exacerbated for women who hold more than one part-time job. Moreover, these workplaces are often perceived to be "clean," thus obscuring the very real hazards that do exist. (26) The health problems associated with these work environments often are difficult to link with specific workplace exposures, given the limitations of epidemiologic and industrial hygiene methods. As a result, the potential hazards of female-intensive occupations are relatively ignored and the health risks seriously underestimated.

Reproductive Hazards and Employment Discrimination

Work-related damage to the reproductive system encompasses a wide range of adverse health effects, not limited to pregnancy, that may occur in adult workers of either sex. (27) The relative paucity of research on reproductive toxicity makes it difficult to estimate both the full range of effects and the extent of exposure to reproductive hazards in the workplace. Relatively few chemical and physical agents have been thoroughly studied for reproductive toxicity in both women and men. (28) There is little evidence that any reproductive toxin affects only women. (29) Furthermore, all known reproductive toxins have other adverse effects on the health of adult workers.

Because women are the ones to physically bear children, reproductive hazards have often been considered a "woman's problem," as if men's role in reproduction was not important. Employer policies excluding either fertile or pregnant women from certain jobs are becoming increasingly common. (30) These exclusionary policies often have the stated purpose of protecting the unborn fetus. There is evidence, however, that they are prompted by economic rather than by health concerns, because they only address one type of reproductive effect and only via maternal exposure. Men are rarely excluded from employment, even when exposed to known reproductive toxins, such as lead or DBCP (Dibromo-3-chloropropane). These policies are largely applied to women in higher-paying, male-dominated occupations. When reproductive hazards have been identified in predominantly female jobs, such as waste anesthetic gas exposure among nurses, solutions for controlling the hazard have been found, rather than forcing women out. (31) Hence, although exclusionary policies promote a higher regard for the health of the fetus than for the health of adult workers, they fail to fully protect either. It may be argued that exclusionary policies are used by employers to provide another rationale for continued sex segregation in nontraditional jobs. (29)

Because many exclusionary policies are unwritten, they may be difficult for individual workers to challenge. Such policies are often implemented in a subtle manner, for example, through regular pregnancy testing of workers, or "medical counseling" that steers women away from areas with known reproductive hazards. Many women who are denied employment opportunities may not be aware of court decisions protecting their rights and do not have the resources to challenge employment discrimination. (27)

Although the federal Pregnancy Discrimination Act (PDA) of 1978 prohibits an employer from treating pregnancy differently than other medical conditions, the scope of this law is weak. Coverage is not provided where there is no preexisting medical disability program. In addition, the law limits leave to six weeks for a "normal" pregnancy, excludes businesses with fewer than 50 employees, and does not cover the needs of male employees. (32) To overcome these limitations, some states have enacted broader coverage, such as mandated disability insurance.

Perhaps more importantly, though, the PDA represents a compromise between the need for parental leave and concern about a potential backlash against working women, as has occurred with other protective labor and health laws for women. (29) The treatment of pregnancy as a "medical disability"—akin to ulcers and heart attacks—is in direct conflict with a more positive view of childbirth and parenting as fulfilling a unique societal need for continuity.

The issues surrounding employment and reproduction take on particular importance in the current legal and political climate. In a U.S. Court of Appeals ruling in *UAW v. Johnson Controls* (33), the employer's policy of excluding women of childbearing capacity from working in its battery division was upheld. This decision is in clear conflict with earlier rulings that established a substantial burden of proof on employers to justify exclusionary policies (27) and represents an obvious backlash against women. The union appealed the decision and the U.S. Supreme Court later decided against exclusionary policies.

Ergonomics

According to recent data from the Bureau of Labor Statistics, musculoskeletal injuries are one of the largest and rapidly increasing types of occupational disease. (34) (Whether this represents an increase in the occurrence of musculoskeletal disorders, or simply increased recognition, is difficult to say.) While it appears that men and women employed in jobs with identical ergonomic demands suffer similar rates of musculoskeletal problems (35, 1), the specific problems they face vary according to the type of employment and extent of sex segregation. Women are more likely to be assigned to highly repetitive, precise manual tasks (for example, word processing, scanning, and packaging), where significant numbers of workers may develop hand and wrist disorders. (36) Substantial numbers of women whose jobs require prolonged sitting or standing, heavy lifting, or infrequent lifting or stretching after prolonged sitting (nursing, health care aides, office workers, and machine operators) are at risk for back injuries. (37)

Another ergonomic problem affecting many women is improperly protective clothing and equipment. Often, such equipment does not reflect anatomical differences between men and women, and as a result may hamper agility and increase the risk of injury. (37)

Ergonomic issues may be used to justify discriminatory employment policies, as in the case of preplacement strength testing. Since women, on average, have less upper-body strength compared to men, women may be denied well-paying employment in jobs where such preemployment testing is used. This emphasis on strength is not based on scientific evidence, however, since the actual predictive value of strength tests for future back pain or carpal tunnel syndrome has not been demonstrated. Moreover, strength testing is often used in lieu of proper ergonomic job design.

Psychosocial Working Conditions

"Women's work" is often characterized by a high level of demands (such as performing clerical duties for several superiors under deadline or caring for several sick people) with little control over the nature and content of that work. Many jobs in the service and light manufacturing sectors combine a substantial mental workload with highly routinized tasks. (38) The combination of these conditions may, over time, result in stress-related symptoms and illness, especially when combined with structural problems, such as lack of child- and elder-care provisions, and the "double duty" of home responsibilities. (39, 40) Conversely, women whose work is challenging and who have high levels of social support and job status have been shown to experience better mental and physical health. (41, 42)

Threats and fears of violence and sexual harassment against women also contribute to occupational stress. NIOSH found that homicide accounted for 42 percent of the traumatic occupational fatalities among women, compared to only 11 percent among men. (43) Regarding sexual harassment, in 1986, the U.S. Supreme Court ruled in *Meritor Savings Bank v. Vinson* that sexual harassment "severe or pervasive enough" to create a "hostile or abusive work environment" constitutes an unlawful form of employment discrimination based on gender. Many employers have adopted "antiharassment" policies, at least partially in an effort to avoid potential liability. (44) Some women have been successful in receiving workers' compensation for stress resulting from sexual harassment. (29) Most women, however, never receive compensation or even recognition of the problem, and are left to bear the economic and emotional burdens alone.

In summary, women's employment status is characterized by occupational segregation, low levels of unionization, and social expectations of dual roles in the home and workplace. This has resulted in differential occupational exposures and adverse health effects that have been inadequately studied and regulated. These include poorly characterized multiple exposures, discrimination based on women's reproductive role, and placement in highly repetitive jobs resulting in physical and psychological strain. Women's position in the labor force also has limited their access to health and safety resources available to protect them from exposure, employer retaliation, and discrimination at work. Women's attempts to organize for health and safe working conditions on their own behalf have been hindered by: (a) location in isolated jobs, (b) lack of effective organizing approaches by unions, and (c) a concerted employer offensive against independent organization.

RECOMMENDATIONS

The recommendations we propose are aimed at addressing the political and economic context of women's employment as well as the particular occupational health and safety problems faced by women workers, and are prioritized by the scope of their impact. They are intended to supplement related proposals made by other bodies, such as resolutions by the American Public Health Association (APHA) on: enforcement of occupational health and safety standards; revision of OSHA regulations and occupational health standards; ending Office of Management and Budget (OMB) abuse of the Paperwork Reduction Act (45); strengthening worker/community right to know; reproductive health and rights of workers; the need for accurate data; and reform of workers' compensation. (46) We also support various legislative proposals, such as the High Risk Occupational Disease Notification and Prevention Act and expanding workers' compensation coverage to include illnesses prevalent in female-intensive occupations, such as musculoskeletal disorders, provisions for a national medical care system (47, 48), and efforts to secure workers' rights to act to protect themselves from occupational hazards.

The following principles have served as a guide to the remedies proposed here:

- Primary prevention, that is, reducing the usage of hazardous materials, is always preferable to other control measures;
- Equity for women, in all aspects of social, economic, and political life, should not and need not be sacrificed in order to attain protection for workers' safety and health;
- Workers have the right to be fully informed about all occupational health and safety risks, and to act on the basis of that information;
- Workers have the right to play an active role in decisions about what products are used, in establishing priorities for research, legislation, regulation, and workplace policies;
- Workers should not bear the burden of scientific uncertainty. Actions to protect health and safety should be taken when there is reasonable doubt, until more conclusive evidence is available;
- Violent and aggressive behaviors, which disproportionately affect women, are unacceptable.

Legislation

Congress should undertake the following initiatives:

1. Amend Title VII of the Civil Rights Act to prohibit occupational health policies that target pregnant or fertile women, even if employers can claim a health justification for such policies.
2. Adopt legislation guaranteeing to all workers—in both the public and private sectors, regardless of income level—access to and control over high-quality child- and dependent-care services at low cost.
3. Adopt legislation guaranteeing to all workers parental leave for both parents after the birth or adoption of a child, or when a child is seriously ill.
4. Adopt legislation establishing a minimum benefit package to all full- and part-time workers on a proportional basis.
5. Adopt legislation to ban the use of electronic surveillance incorporated in computer hardware and software to monitor productivity of individual workers. Workers covered by this legislation

should include, but are not limited to, operators of video display terminals, telephone systems, assembly lines, robots, electronic letter sorters, supermarket laser scanners, and other machinery.

Regulation

OSHA should undertake the following actions:

1. Explicitly define the terms "serious physical harm" and "material impairment of health or functional capacity"—as used in the Occupational Safety and Health Act of 1970—to include harm or impairment of the reproductive, procreative, and sexual functions of any employee. This would allow occupational safety and health standards to be set at levels so as to protect the most vulnerable employees and to protect the fetus. (27, 49)
2. Promulgate a generic reproductive hazards standard, to contain the following provisions:

   - Evidence that workers of either sex are at risk or can expose a fetus to a hazard should trigger the assumption that all workers are at risk until proven otherwise.
   - Reproductive hazards should be controlled in a manner similar to other hazards; that is, first, by eliminating the hazard, second, by using engineering and work practice controls, and not discriminatory personnel practices.
   - Voluntary light duty and medical removal protection with retention of pay, seniority, and benefits should be guaranteed to all current potential parents and pregnant workers. Medical removal policies should be written so as to protect the most vulnerable employees, pregnant or otherwise, from chemical or physical stressors.

3. Promulgate a standard addressing ergonomic stressors arising from seated work with static body posture, high visual demands, and repetitive manual tasks. Specific requirements should address: workstation and equipment design, lighting, rest breaks, eye examinations and corrective lenses, ionizing and non-ionizing radiation exposure, maximum number of hours of continuous work; and also prohibit electronic surveillance of individual workers.
4. Adopt more stringent standards governing indoor air quality in nonindustrial settings, such as office buildings, retail establishments, and airplanes. Specific requirements should address: the quality, quantity, and distribution of fresh air, special ventilation where toxic materials are generated, temperature and humidity, and

remediation of low levels of common contaminants, such as radon and formaldehyde.

Enforcement

Regulatory agencies should increase efforts at enforcing existing employment and occupational health regulations, as follows:

1. The Equal Employment Opportunities Commission (EEOC) should more stringently enforce Title VII of the Civil Rights Act to prevent discrimination in hiring, setting wages and benefits, job advancement, and other employment practices in all workplaces. Adequate staff and resources should be made available to resolve all cases on a timely basis. Bias-free job evaluations should be used uniformly. A strategy to empower the agency to enforce penalties in a reasonable time frame should be developed.
2. The Office of Civil Rights (OCR) should more strictly enforce Title IX of the Civil Rights Act, which prohibits sex discrimination in education. The OCR should ensure that, as a mimimum, educational institutions establish grievance procedures.

TABLE 28.1. Glossary of Acronyms

AFGE	American Federation of Government Employees
AFSCME	American Federation of State, County and Municipal Employees
APHA	American Public Health Association
BLS	Bureau of Labor Statistics
CLUW	Coalition of Labor Union Women
CWA	Communications Workers of America
DBCP	Dibromo-3-Chloropropane
EEOC	Equal Employment Opportunities Commission
ILGWU	International Ladies Garment Workers Union
NAWW	National Association for Working Women
NIH	National Institutes of Health
NIOSH	National Institute for Occupational Safety and Health
NUHHCE	National Union of Hospital and Health Care Employees
OCR	Office of Civil Rights
OMB	Office of Management and Budget
OSHA	Occupational Safety and Health Administration
PDA	Pregnancy Discrimination Act
PEF	Public Employees Federation
SEIU	Service Employees International Union
SSA	Social Security Administration
UAW	United Auto Workers

3. OSHA should undertake the following:

   - Amend the present criteria for targeting workplaces for compliance inspections, so that female-dominated industries are routinely inspected.
   - Strictly enforce and support with adequate resources recent field directives in areas that especially affect women workers, such as ergonomic hazards and indoor air quality (50, 51).
   - Develop additional field directives covering other hazards, such as chemotherapeutic drugs, until such time as relevant standards have been promulgated.

Education and Training

Another important strategy for improving women's occupational health and safety is through education and training. Training should be directed toward present workers, future workers, and health professionals and government leaders whose jobs involve education, regulation, and enforcement. The following mechanisms are proposed to address the training needs of each of these groups:

1. Federal and state funding should be made available for developing and expanding educational programs aimed at women workers (especially those in segregated, isolated, and unorganized jobs), unions, other labor-supported groups, and community-based organizations with large female memberships. These programs should include both technical information on health and safety issues and strategies for improving working conditions, such as through collective bargaining.
2. Federal and state funding should be provided for training primary and secondary school teachers to develop curricula and special programs (for example, "Career Days") on issues related to female students' future employment, including: occupational health and safety, pay equity, sex discrimination, and the role of unions.
3. Employers and unions should hire more women in key positions and provide ongoing training programs for their managers, plant health and safety specialists, and union leadership, respectively, that emphasize all of the issues of relevance to women workers, as described above.
4. Schools of public health and other professional educational institutions should systematically incorporate women's employment and health and safety issues into their curricula.
5. Relevant state and local government agencies (such as, health, social service, and labor departments) should develop the capacity to provide consultation, materials, and support for training on hazards in female-dominated workplaces.

Organization

Mechanisms are proposed for: (a) enhancing union effectiveness in addressing the concerns of female members; (b) enhancing organizing efforts toward the majority of women workers who are currently not in unions; and (c) strengthening the relationship between women workers and public health professionals.

1. Labor unions should improve compliance with contract rights affecting women by filing unfair labor practice charges to get gender and wage data from employers, and by grieving sexual harassment under expedited contract procedures. (19)
2. Efforts to organize women workers should be improved by employing female organizers, thus facilitating further development of their organizing styles and leadership roles; and by focusing organizing campaigns on issues of concern to women, such as pay equity, ergonomics, job stress, reproductive hazards, and child care.
3. Public health workers should provide technical support to unions on health and safety issues of concern to their female members.
4. Public health workers should support and participate in the formation of coalitions such as COSH groups, and other issue-specific coalitions (for example, the AdHoc Coalition of the Semiconductor Industry Association) that promote women's occupational health and safety issues.
5. Public health and local government workers should play a role in reaching unorganized women workers through community organizations (churches, community associations, grassroots toxics organizations) and networks for working women (9 to 5, National Association for Working Women, Women in the Building Trades, Tradeswomen Inc., and Rainbow Tradeswomen). Many women are already involved in health issues—such as environmental toxins and reproductive health—that are related to occupational health, and this link should be emphasized in organizing efforts.
6. Finally, many public health workers—50 percent of whom are women—are themselves employed in nonunionized workplaces and may be directly affected by the issues described above. Thus, public health workers should be involved in organizing and health and safety issues in their own workplaces. These activities may include forming or participating in union or joint occupational health and safety committees.

Research

A key problem in evaluating and controlling occupational hazards for women workers is a lack of information about health consequences of women's working conditions. Several recommendations for analyzing existing data, collecting additional data, and conducting research are proposed.

1. NIOSH should establish, as a funding priority, a meta-analysis of existing databases in which data on women were collected but not sufficiently analyzed. These analyses should be evaluated conservatively, in light of the often poor information on actual working conditions in many female-dominated jobs; they should be used to inform the development of new methodological approaches for epidemiologic studies of women workers.
2. The National Center for Health Statistics should create and maintain a national database that links employment and health records by unique personal identifiers. Cooperation among the various government agencies involved in data collection (for example, NIOSH, BLS, SSA, NIH) is necessary to develop and maintain such a database. Access to the data must be guaranteed to labor unions, public health workers, educators, and employers in such a way so as to protect the confidentiality of individual subjects. This proposal complements, rather than supplants, the concept of a comprehensive occupational disease surveillance system based at NIOSH. (52)
3. Public and private sector initiatives should be developed to fund research on issues relevant to women's employment and health. This includes research on the relationship between broad employment factors (such as availability of child care, parental leave, union representation, and similar benefits) and health. In addition, it includes research on occupational hazards of concern to women, such as: (a) evaluation and control methods for mixed exposures; (b) reproductive hazards to men and women; (c) design and effectiveness of tools and personal protective equipment used by women; and (d) stress-related health effects of machine pacing, repetitive work, job schedules, and psychosocial demands.
4. Workers should be involved in all stages of epidemiologic research concerning them, including: priority setting, study design, interpretation of findings, and recommendations for control measures (53) through such mechanisms as joint labor/management-administered programs. Disagreements on research protocols and related issues should be subject to mediation by mutually agreed upon third parties.

Litigation

Individual workers who can escape the exclusive provision of the workers' compensation system may bring tort liability suits for occupational injury or illness involving, for example, reproductive hazards or poor tool design. These suits are individual, rather than collective, remedies and only indirectly prevent future injury and illness. Therefore, this method is less preferable than regulation, legislation, organizing, or education for achieving public health goals. Moreover, it is generally time-consuming and expensive, with employers having disproportionate access to financial resources, compared to employees. For those individuals who already suffer from occupational injuries and illness, however, few other remedies may be available, and rulings from such cases may set important precedents for the future.

Other forms of litigation, such as class-action suits on sex discrimination, or OSHA hearings on emergency temporary standards, have a broader impact and support the types of actions discussed above, such as stronger enforcement of existing laws or promulgation of new regulations. Public health professionals should vigorously assist unions and others who are involved in legal actions affecting women's occupational health and safety.

CONCLUSIONS

We have attempted to show how the occupational health of women workers is affected by their social roles and employment patterns. The specific remedies we propose are aimed at chipping away at some of the social and economic barriers to women's full and active participation in the labor force and the occupational hazards they face in very concrete and tangible ways. At the same time, no single proposal outlined here—or even all of them combined—is sufficient to secure healthy and safe work environments for all women. We recognize that there are certain problems that directly affect some but not all women. These include racism, heterosexism, agism, and discrimination against handicapped women. They have not been fully addressed in this chapter.

More importantly, a gender-specific approach is limited because it ignores the occupational health problems that are common to all workers. They include inadequate health and safety standards, poor and uneven enforcement, absence of right-to-act provisions, lack of available information about hazards and safer alternatives, and related rights and protections. As the recent court decision in *UAW v. Johnson* Controls shows, policies that are aimed solely at protecting women workers may actually serve to discriminate against them, while leaving male workers unprotected. To fully address women's occupational health and safety needs will require ongoing dialogue and coalition building between labor, feminist, government, public interest, and other health and safety organizations and activists. We hope that raising these issues will help to increase the discussion around women's health and safety issues and help to forge such a coalition.

ACKNOWLEDGMENTS

Contributions to earlier drafts were made by Lise Anderson, Barbara Boylan, Winifred L. Boal, Madeline Caporale, Diana Echeverria, Rebecca Head, and Michelle Van Ryn. The authors thank the following people for their thoughtful comments on earlier drafts: Joan E. Bertin, Wendy Chavkin, Fran Conrad, Mary Sue Henifin, Donna Mergler, Sharon Morris, Elise Pechter Morse, Rafael Moure-Eraso, Cathy Schwartz, Clare Sullivan, Eileen Tarlau, and also several anonymous reviewers.

NOTES

1. Mergler, D., Brabant, C., Vezina, N., and Messing, K., "The weaker sex? Men in women's working conditions report similar health symptoms," *Journal of Occupational Medicine*, 29(5): 417–421, 1987.

2. U.S. Congress, Committee on Government Operations, "Occupational health hazard surveillance, 72 years behind and counting," Washington, DC: U.S. Government Printing Office, 1986.

3. Council on Occupational and Environmental Health, National Association for Public Health Policy, "Occupational safety and health legislative agenda, 1989," *Journal of Public Health Policy*, 9(4): 544–555, 1988.

4. U.S. Department of Labor, Occupational Safety and Health Administration, Proposed OSHA rule governing exposure to blood-borne pathogens, 54 *Federal Register* 23042, May 30, 1989.

5. Bureau of National Affairs (BNA), *Occupational Safety and Health Reporter*, "OSHA failed to adopt most stringent rule on formaldehyde, unions say in court brief," 18(30): 1398, December 21, 1988.

6. Bureau of National Affairs (BNA), *Occupational Safety and Health Reporter*, "OSHA's proposed blood-borne disease rule welcomed reservedly by health care unions," 18(52): 2147, May 31, 1989.

7. Goldfield, M., *The decline of organized labor in the United States*: 123–148, Chicago: University of Chicago Press, 1987.

8. Prosten, R., "The longest season: Union organizing in the last decade:" 240- 249, *Proceedings of the 31st Annual Meeting of the Industrial Relations Research Association*, 1978.

9. *VDT News*, 4(4), July/August 1987.

10. *ILC Data Device Corp. v. Suffolk County*, NY Sup Ct, Suffolk Cty, No.12149-1988, 12/29/89.

11. U.S. Department of Labor, Bureau of Labor Statistics, "1986 Employment and Earnings," Washington, DC: Government Printing Office, 1987.

12. U.S. Department of Labor, Bureau of Labor Statistics, "1985 Household Annual Averages," Washington, DC: Government Printing Office, 1986.

13. *The New York Times*, "Women reduce lag in earning but disparities with men remain," September 4, 1987: 1.

14. Conference on Alternative State and Local Policies, "A women's rights agenda for the states," 1981.

15. Treiman, D., and Hartmann, H. I., *Women, Work and Wages: Equal Pay for Jobs of Equal Value*, Washington, DC: National Academy Press, 1981.

16. National Association for Working Women, "Working at the margins: Part-time and temporary workers in the United States," September 1986.

17. Meurs, M., and San Juan, K. A., *Beyond growth: Women for economic justice*, Boston, 1987.

18. "Women of Color and Occupational Health," *Proceedings of the Conference on Women in the Workplace*, Boston, November 8-9, 1985, Volume 1, Daniels C.R., ed.

19. Baden, N., "Developing an agenda: Expanding the role of women in unions," *Labor Studies Journal*, 10(3): 229-249, 1986.

20. Kistler, A., "Union organizing: New challenges and prospects," *Annals of the American Academy of Political and Social Science*, 473: 96-107, 1984.

21. U.S. Department of Labor, Bureau of Labor Statistics, "Earnings and other characteristics of organized workers, 1980," Washington, DC: Government Printing Office, 1981.

22. Kamerman, S. B., Kahn, A. J., and Kingston, P., *Maternity Policies and Working Women*, New York: Columbia University Press, 1983.

23. Wagner, M. G., "Infant mortality in Europe: Implications for the United States." Statement to the National Commission to Prevent Infant Mortality, *Journal of Public Health Policy*, 9(4): 473-484, 1988.

24. Friedman, D. E., "Prevalence of child care initiatives," The Conference Board, 1985.

25. Logan, M., "Statement on work and family," presented to the AFL-CIO Executive Council, February 21, 1986.

26. Chavkin, W., "Double Exposure: Women's Health Hazards on the Job and at Home," New York: Monthly Review Press, 1983.

27. Office of Technology Assessment (OTA), U.S. Congress, "Reproductive Health Hazards in the Workplace," Washington DC: Government Printing Office, 1985.

28. Rosenberg, M. J., Feldblum, P. J., and Marshall, E., "Occupational influences on reproduction: A review of recent literature," *Journal of Occupational Medicine*, 29(7): 584-591, 1987.

29. Bertin, J., and Henifin, M. S., "Legal issues in women's occupational health," *Women and Work: An Annual Review 2*, Stromberg, A. H., Larwood, L., and Gutek, B. A., eds.: 93-115, Newbury Park, CA: Sage Publications, Inc., 1987.

30. Paul, M., Daniels, C., and Rosofsky, R., "Corporate responses to reproductive hazards in the workplace: Results of the family, work and health survey," *American Journal of Industrial Medicine*, 16(3): 267-280, 1989.

31. Quinn, M., and Woskie, S., in Levy, B. and Wegman, D., eds., *Occupational Health: Recognizing and preventing work-related disease*, Boston: Little Brown, 1988.

32. Kelemin, Gardin, S., and Richwald, G. A., "Pregnancy and employment leave: Legal precedents and future policy," *Journal of Public Health Policy* 7(4): 458–469, 1986.

33. Bureau of National Affairs, *Auto Workers v. Johnson Controls, Inc.*, *Occupational Safety and Health Cases*, 14(8): 1217–1256, October 11, 1989.

34. U.S. Department of Labor, Bureau of Labor Statistics, 1988, "Injuries and Illnesses," Washington, DC: Government Printing Office, 1989.

35. Silverstein, B. A., Fine, L. J., and Armstrong, T. J., "Hand wrist cumulative trauma disorders in industry," *British Journal of Industrial Medicine*, 43: 779- 784, 1986.

36. Armstrong, T. J., and Silverstein, B. A., "Upper-extremity pain in the workplace—role of usage in causality," in 333–354 in *Clinical Concepts in Regional Musculoskeletal Illness*, Hadler, N. M., ed. Orlando: Grune and Stratton, Inc., 1987.

37. Bureau of National Affairs, "Working Women's Health Concerns: A Gender at Risk?" Washington, DC: 1989.

38. Teiger, C., "Les constraintes du travail dans les travaux repetitifs de masse et leurs consequences sur les travailleuses:" 38–66, *Les effets des conditions de travail sur la sante des travaileuses*, CSN, ed. Montreal, 1984.

39. Haynes, S. G., and Feinlieb, M., "Women, work and coronary heart disease: Prospective findings from the Framingham Heart Study," *American Journal of Public Health*, 70: 133–141, 1980.

40. Haynes, S. G., "The effect of high job demands and low control on the health of employed women," *Work Stress and Health Care*, Quick, J. C., Bhagat, R. S., Dalton, J., and Quick, J. P., eds. New York: Praeger Scientific Publishers, 1986.

41. Muller, C. M., "Health and health care of employed adults: Occupation and gender," *Women and Health*, 11(1): 27–45, 1986.

42. Verbrugge, L. M., "Role burdens and physical health of women and men," *Women and Health*, 11(1): 47–77, 1986.

43. CDC. "Traumatic Occupational Fatalities—United States, 1980–1984." MMWR 1987; 36(28): 461–470.

44. *The New York Times*, "A grueling struggle for equality," 12–13, November 9, 1986.

45. "The Nation's Health," American Public Health Association, September 1987.

46. American Public Health Association Public Policy Statements, 1948–Present, Cumulative, American Public Health Association, Washington, DC, 1987.

47. Terris, M., "Toward a national medical care system, I." (editorial), *Journal of Public Health Policy* 7(2): 152–155, 1986a.

48. Terris, M., "Toward a national medical care system, II." (editorial), *Journal of Public Health Policy* 7(3): 290–292, 1986b.

49. Bertin, J., "Reproductive laws for the 1990's: Proposed legislation regarding reproductive health hazards in the workplace," Women's Rights Project, American Civil Liberties Union, New York, June 16, 1985 (draft).

50. OSHA Compliance Directives, CPL 2, May 1986.

51. Bureau of National Affairs (BNA), *Occupational Safety and Health Reporter*, OSHA preparing to issue compliance officers technical guide on indoor air investigations," 18(50): 2032, May 17, 1989.

52. Pollack, E. S., and Keining, D. G., *Counting Injuries and illnesses in the workplace: Proposals for a better system*, Washington, DC: National Academy Press, 1987.

53. Mergler, D., "Worker participation in occupational health research: Theory and practices," *International Journal of Health Services*, 17(1): 151–167, 1987.

29

Pollution Prevention and Income Protection

FIGHTING WITH EMPTY HANDS— A CHALLENGE TO LABOR

LIN KAATZ CHARY

In the summer of 1988, while acid rain legislation was being debated in Congress and in the media, the local newspaper in Gary, Indiana, ran a front-page picture of officers of downstate United Mine Workers (UMW) locals on a podium alongside the chiefs of the downstate coal operators. All were smiling, hands clasped in a victory sign, as they celebrated joining forces to work to defeat legislation addressing acid rain. The issue was jobs, they agreed, and the continued prosperity of Indiana's high sulfur coal-related industries. It was a moment to make John L. Lewis turn over, and a low point for labor and environment.

In the course of that same attempt to pass the Clean Air Act, amendments proposed by Senator Byrd to offer some protection to coal miner jobs against the significant changes mandated by the legislation were soundly defeated. Not a peep of support had been heard from the environmental community, and not much more—at least to the public ear—from the labor movement either. At a Congressional field hearing on the acid rain legislation that summer in Indianapolis, the frequently repeated thrust of overwhelming negative testimony from unions and industry alike was jobs, jobs, jobs. Environmentalists had little to offer in response, even though Indiana consistently ranks among the worst states in the union for air pollution.

By the time the Clean Air Act Amendments were passed in 1990, the situation had improved to the extent that at least the need to address the plight of workers displaced due to environmental protection measures was recognized in law by the "Clean Air Employment Transition Assistance Program" (CAETA). But this law provided an extremely flawed model, as Oil, Chemical, and Atomic Workers (OCAW) President Bob Wages pointed out in recent testimony to a Senate Committee regarding the "Circle of Poison Prevention Act of 1991." Designed in such a way as to "explicitly direct workers into poverty," CAETA failed resoundingly to offer any real solution to the problem of income protection and job replacement. (1)

A better answer is clearly necessary. Job displacement, and even, unavoidably, job elimination are going to occur at some level as the increasing demands for pollution prevention begin to be implemented. This is the reality, regardless of some environmentalists' well-meaning but ill-directed efforts to sugarcoat the situation by insisting that new jobs will be created. It is also the reality regardless of some attempts by labor to block environmental efforts because of the issue of jobs. A new solution to the dilemma of protecting the environment while also protecting and improving the standard of living of workers is the challenge before both movements as the decade progresses.

This challenge has begun to be addressed in several places, including in the pages of *New Solutions*, where discussion has ranged from Michael Merrill's "No Pollution Prevention Without Income Protection: A Challenge to Environmentalists," to the United Steel Workers of America (USWA)'s landmark document, "Our Children's World: Steelworkers and the Environment," and others. (2-4). In this chapter, I want to respond to Merrill in the current context of this discussion, from the perspective of one environmental activist who, as a former Steelworker and coke plant laborer, comes directly out of the trade union movement.

WHOSE RESPONSIBILITY?

In his "Challenge," Merrill begins by taking the environmental community to task for being "on a collision course with a declining labor movement and its increasingly desperate membership." (2) He bases this on the acknowledged failure of some environmentalists to appreciate the significant impact on jobs from the changes they are calling for, and the historical failure of the environmental community overall to incorporate the demand for income protection in the fight for toxics use reduction. He also puts responsibility on environmentalists to "convince workers that toxics use reduction equals more jobs and not just more jobs in general, but more jobs for them in particular." (2)

But this is a serious misunderstanding of the problem. The environmental movement has neither the obligation, the objective, nor most significantly of all, the ability to persuade workers or the labor movement about anything relating to jobs. It is the responsibility of environmentalists to educate people about the nature of *environmental* problems and threats, to promote alternatives, and to persuade various constituencies to join together in working to promote environmental health for the planet and all the creatures who live on it. The obligation is to do this in a *responsible* way, which includes fully appreciating the social, ethical, and certainly the economic implications of these actions. The obligation is to understand that demands for changes as drastic as zero discharge and toxics use reduction will have drastic impacts that are both desirable and undesirable, and that what we are looking at is a realignment of the entire social order. Toxics use and production have become central to productivity in virtually all the world's economies, intertwined irrevocably with both industrialized and agricultural systems alike. There can be no such thing as environmentalism, no matter how "deep," that does not acknowledge this reality and face up to the responsibilities inherent in it.

Even so, it is still not the environmental community's responsibility to convince labor that environmental changes are going to create jobs. For one thing, while new jobs definitely will be created, they will be different jobs than those that have been lost, and there is little likelihood that they will directly affect the workers whose jobs have been eliminated. In addition, given the continuing trend away from labor-intensive operations (which was obviously not initiated by environmental considerations however much it may serve prevention efforts), the number of heavy industry production jobs is going to continue to shrink.

More importantly, to frame the issue in this way is a serious misassessment of where the real work lies. The task before us is to build a coalition that is powerful enough to achieve the common objective of a society in which no one is forced to choose between a healthy economy and a healthy planet. The USWA seems to understand this very clearly, stating unequivocably: "The environment is an essential union issue. . . . We cannot expect the company or the government, or for that matter, *the environmental community to defend our interests for us . . . In the long run, the real choice is not jobs or environment. It's both or neither.*" (3, emphasis added)

Over the last two years, the solution suggested by both Merrill and the USWA, along with a growing number of trade unionists, follows the lead of the OCAW in promoting the need for what has come to be known as the Superfund for Workers: guaranteed income protection coupled with educational advancement/job retraining for workers who are displaced by changes due to environmental protection. This idea has found significant support among many environmentalists who have been searching for a way to integrate their understanding of the need to protect jobs while protecting the environment.

THE CHALLENGE TO LABOR

By the terms of Merrill's challenge, proponents of toxics use reduction—which must be extended to also include the ratcheting down of toxic emissions towards zero discharge if it is going to be meaningful in the broad scope of what is happening in the real world—"should pledge to support specific pollution prevention measures if and only if they include an adequately funded, broadly conceived Superfund for Workers." (2) The question is, how does this translate into the real world?

Let's assume the challenge is accepted; the pledge taken. Now the task is to apply it to specific situations. This is the environmentalists' challenge to labor. Merrill's point that had the pledge been taken prior to the passage of the Clean Air Act (CAA), the outcome is likely to have been quite different (the environmental community would not have been able to support the final result as it occurred) is exactly on the mark. Surely, however, he does not imply that there was adequate leadership and organization from labor or adequate proposals to work with on the question of income protection to have changed the outcome this time around?

Regulatory and policy problems now on the table cannot be shelved until the next time the CAA comes up for authorization. Should environmentalists now eschew a role in setting coke oven rules and withhold support for proposals for stricter standards because the CAA failed to provide for worker protection as well? Examples of this vacuum—between what we want, what we need, what we have, and what is being demanded—and how it plays out at the grassroots are abundant.

THE CLEAN AIR ACT REVISITED

Under exemptions granted by the Clean Air Act, regulations governing coke oven emissions—among the worst in a steel mill—are being negotiated separately from the rest of the act, with companies being given up to *30 years* to come into full compliance. This has raised significant questions among environmentalists with a labor perspective over how to deal with the many issues on the table.

In Northwest Indiana, a struggling steel industry still operates several coke batteries—large, lumbering facilities that still belch out coke and pollution in an ancient symbiotic relationship. The decisions over how long these batteries will be able to keep operating will undoubtedly be influenced by the regulations that come out of current Clean Air Act negotiations in which environmentalists are involved. If some batteries are forced to go down sooner rather than later because of the new regulations (which is probably not likely, but also not impossible), several hundred jobs will be on the line.

This may be good news for the health of the surrounding communities, and in fact for the health of the coke plant workers themselves, who suffer highly disproportionate rates of lung and other cancers from long exposures to coke plant carcinogens. But it's terrible news not only for the workers who will lose their jobs, but for the economic health of the whole community, which has never recovered from the disastrous steel industry disintegrations of the 1980s.

On the other hand, there is also the possibility that by the time the 30 years is up, due to the pressures of economics and technology having nothing to do with environmental regulations, there won't even *be* any more of these old coke ovens around, so this whole negotiation is discussing the wrong questions altogether. Perhaps it *should* be discussing what is going to happen to the coking industry itself in the United States in the next 30 years, and how are the interests of workers going to be protected regardless of the decisions of the industry? And how do environmental regulations fit in to *that*? On the other hand, because the batteries *will* continue to operate for several more years, it is as important as ever to minimize emissions to the greatest possible extent now, and how does that affect jobs?

A familiar story. What position should environmental activists, many of whom are also union members, or have been union activists before they lost their steel jobs, take? What do we have to offer? The *idea* of a Superfund for Workers is frequently met with skepticism here, to be perfectly frank. "Retraining," even as a college education, has little meaning in an area where there are virtually no jobs to be retrained for; where the economy has been stagnant for a decade; where retraining also means relocation for many people—a tremendous financial burden, often impossible; and where "retraining" brings back memories of the useless though well-meaning resume-writing sessions that were set up when the mills went down about 15 years ago, even though the current Superfund for Workers plan has nothing to do with that at all. And what environmentalists also see here, on top of the economic misery, is a situation ripe for the siting of a new hazardous waste incinerator, new toxic industries, new schemes to exploit the community's problems with toxic solutions. This can mean whole new arenas for the "jobs vs. environment" blackmail issue to spring up, with environmentalists' pleas for clean economic development no answer to major unemployment.

THE CLEAN WATER ACT REAUTHORIZATION

A second example, which demonstrates the depth of the need for an organized labor/environment campaign, is the reauthorization of the Clean Water Act (CWA), the outcome of which is pending at this writing. In the Great Lakes, where the fight for zero discharge will be taken straight to the heart of the reauthorization, the issue of jobs comes up in different ways. If environmentalists are successful in forcing the paper and pulp industry to stop using chlorine and to substitute other, cleaner technologies, the greatest impact will not be on paperworkers, who support this campaign, but on the workers who make chlorine. The same will be true regarding changes in the auto industry, where chlorine-alternative processes would not affect autoworkers as significantly as the chemical workers.

Clearly, income protection language will be crucial to gaining the support of some sectors of labor in the fight to ban chlorine in Lake Superior and to promote tighter ratcheting down throughout the Great Lakes. Where will this language come from? Should it be language in the CWA itself, or should it be separate, independent legislation that takes a more global, more universal approach to the issue? These are not decisions that environmentalists can make alone; they are also not decisions that the fight for the CWA will wait for. The fight for zero discharge will proceed and there will be significant labor support from unions such as the Paperworkers who see very clearly how the health and safety issues in the plant are connected to the larger movement. (5, 6) Whether or not the mandate for income protection will be a driving part of this fight is still to be determined.

It is not a profound observation that U.S. EPA is far from recognizing that labor's agenda should be an integral aspect of the environmental agenda. *But the failure of labor to develop an aggressive, comprehensive, and broad-based campaign on the issue of jobs and the environment and jobs and pollution prevention in the form of a Superfund for Workers or anything else has left environmentalists with empty hands at the bargaining table. It is a key barrier to the ability of the environmental movement to put that issue in the forefront of its efforts*. Environmentalists, even the most committed to income protection, cannot bring to the table a labor agenda that does not exist, cannot promote a strategy that exists only in the conceptual realm; environmentalists cannot bargain for labor.

Obviously, it is only through the debate and discussions that an organized movement for income protection will provide that the answers to some of these problems will begin to emerge. The outline of the Superfund for Workers concept as developed in greater detail by Wykle, Morehouse, and Dembo in their book *Worker Empowerment in a Changing Economy* and discussed by Wages in his testimony to the Senate Committee on Agriculture, Nutrition, and Forestry has provided a starting point. (7, 1) Now the work is to take it forward to the next level where it will be able to be used in the real world of environmental policy negotiation.

WHAT WE NEED NOW

We need a national, labor-wide organization to mobilize the fight for income protection now. We need to be organizing for specific legislation starting now; we need concrete models for different industries and different situations that can be taken into all negotiations for standards and regulations. We need concrete program definitions on the table as we develop environmental policy. We need a labor-wide, environmental-wide movement that is working together to develop the strategies, communication, and support to win pollution prevention with income protection in this decade, in this century.

A LEGACY OF SOCIAL JUSTICE

The development of a labor agenda for income protection and a labor-wide strategy to back it up is what Merrill begins to touch on but does not go far enough with when he outlines what the pledge of "no pollution prevention without income protection" could do.

Such a pledge could also be the focus of a comprehensive educational and organizing campaign among rank-and-file members of both movements. Among trade union activists, such a campaign would be designed to encourage in them the notion that they have a right to continued income and tuition support if their job is eliminated, and that such support provides an acceptable alternative to their continuing dependence on the toxic economy; whereas, among environmentalists, it would be intended to win support for a Superfund for Workers as an inseparable and nonnegotiable part of toxics use reduction. (2)

The troublesome part of this pledge is that it carries within it the potential for the differences between labor and the environment to be exploited; inherently dialectical, it can easily be turned into its opposite, that is, no income protection and no toxics use reduction at all. There is unarguable truth in many assertions by labor that the concerns of workers are sacrificed in the name of environmental progress or necessity—that, as the OCAW points out, we spend more money on dirt than on people. (1) So what else is new? This is the history of the capitalist economic system in the western world. The job here is to figure out a way to escape the old dichotomies, the old "not-this-without-that" thinking, the ultimatums that alienate people rather than bring them closer together. Taking the pledge may, in fact, be a dynamite way to get people thinking about the issue and started on the road to action. But putting this fight for income protection into its correct historical perspective is what will ultimately get the motor going at a speed that will not be stopped.

At the heart of the struggle for economic and social justice, the movement for income protection must not only be *recognized* as, but *organized* as, quite literally, the successor to the great social movements of the past that are the honored legacy of the labor movement: the fight for the eight-hour day, the fight for unemployment insurance, and the fight for the very right to unionize at all. These were ultimately great social struggles that engaged the entire spectrum of both activists and the rank-and-file public because they represented, over many decades of effort, the most fundamental needs and demands of working people.

The current fight for income protection must be seen as the struggle to hold on to, to rebuild, and to safeguard these historical gains; it is the struggle to take the union movement into the twenty-first century as a viable and rejuvenated force among working people and all people who believe that social progress is the motor of history. In this context, the difficulties projected in sustaining an "unwavering commitment to a Superfund for Workers by the environmental movement," the difficult choices to be made on the sides of both labor and the environment, would be greatly mitigated.

The environmental movement, and most especially at the grassroots where the identification with the movement for social and economic justice is clearest and strongest, also carries a powerful legacy for social change. The conservationists of the past have been joined by a new generation of activists at the grassroots, many of whom grew up in the context of earlier struggles such as the civil rights movement and the women's movement. Many others, newer to the political arena, are understanding more and more that the power and enthusiasm of the environmental movement joined in solidarity with the power and experience of the labor movement will represent a social force not seen in this country for a very long time.

Labor and the environmental community have the opportunity to create social history, to define an entirely new relationship between pride of work and pride of place. The environmental realities that engender the need for toxics use reduction, zero discharge, process changes and realignments as social and economic imperatives are not going to go away; they are going to get more intense. The labor movement can choose to lead the way in dealing with these inevitable transitions with progressive and aggressive strategies such as the Superfund for Workers, or it can fight them every inch of the way into oblivion. These are the realities that all people who are committed to social and economic justice must address, and the sooner the better.

ACKNOWLEDGMENTS

The author thanks Babette Neuberger and Richard Miller for their thoughtful consideration and helpful comments in the development of this article.

NOTES

1. Wages, Robert, "Statement of Robert E. Wages, President, Oil, Chemical, and Atomic Workers International Union before the U.S. Senate Committee on Agriculture, Nutrition, and Forestry Concerning the Employment Impact on Pesticide Manufacturing Workers from the Circle of Poison Prevention Act of 1991 (5. 898)." Communication from Richard Miller.
2. Merrill, M. "No Pollution Prevention Without Income Protection: A Challenge to Environmentalists," *New Solutions* I:3 (1991):9–11.
3. USWA Environmental Task Force. "Our Children's World: Steelworkers and the Environment," *New Solutions* 2:2 (1991):75-87.
4. Lewis, S. "Author Sanford Lewis Replies," *New Solutions* I:4 (1991):17–19.
5. "U.S. Paper Industry's Slow Response to Chlorine Problem," *The Paperworker*, United Paperworkers International Union, 19:10(1991):8–9.
6. "Why the UPIU is Concerned with IP's Disregard for the Environment," *The Paperworker*, United Paperworkers International Union, 19:6(1991):8–9.
7. Wykle, L., Morehouse, W., Dembo, D. *Worker Empowerment in a Changing Economy*. New York: The Apex Press, 1991.

THE GREAT LAKES UNITED LABOR/ENVIRONMENT TASK FORCE

The Great Lakes United (GLU) Labor/Environment Task Force was formed in May 1990 at the GLU annual meeting. Great Lakes United is a binational coalition dedicated to preserving the Great Lakes/St. Lawrence ecosystem that includes several local unions around the Great Lakes basin. Historically, the role that labor had played in GLU and similar organizations was to lend support to environmental issues in an effort to diffuse attempts at job blackmail over improvements in areas such as water quality standards. In addition, many locals around the basin, in particular the United Auto Workers (UAW) in Michigan and the Canadian Auto Workers (CAW) in Windsor, recognized very early the relationship between health and safety issues and environmental issues. The UAW's Solidarity House, for example, supported GLU's efforts from the beginning, with the UAW holding at least one seat, and often more, on GLU's board of directors.

Much of the traditional relationship between environmental groups and labor had been based on soliciting, and often gaining, support from labor for an environmental agenda. With the growing momentum of the campaign for zero discharge (of persistent toxics into the Great Lakes) and toxics use reduction, however, and especially with the growing focus on efforts to stop chlorine discharges into Lake Superior, it was increasingly clear to several activists in GLU that environmentalists had to take a more aggressive and visible interest in what the implications of this campaign were going to be for jobs.

By the spring of 1990, several GLU members who were activists in both the labor and environmental movements had recognized that this was an issue that was going to require significantly more attention in the future, and that environmentalists were going to have to develop significantly greater sensitivity to the question of jobs if their overall goals were going to be successful. For many environmentalists, especially those whose roots were in the conservation and more traditional land-use movements, the appreciation of how important the labor/ environment connection was going to be was somewhat slow in coming. For those whose relationship with labor was based on recruiting labor's support of environmental issues, the idea that environmentalists were going to have to become much more sensitive to labor issues represented a significant challenge.

Nonetheless, the GLU Task Force was proposed and passed at the annual meeting in May 1990 with surprisingly little opposition or discussion. From the beginning, it was understood that its main objective would be to work with the labor movement, not in the old way of looking to labor for its support for zero discharge, but in supporting labor efforts to explore and address solutions to the problem of job dislocation brought about by environmental initiatives. At the top of the discussion list was the Superfund for Workers and Right to Act.

Although the task force was slow getting off the ground, due primarily to organizational and financial restraints, by one year later, in May 1991, it had adopted a revised mission statement that clearly enunciated its goals. Later, at the same meeting, this time after significant discussion, Great Lakes United became one of the first environmental organizations to adopt in spirit, if not in exact language, the pledge that Michael Merrill demanded from environmentalists in his article: no pollution prevention without income protection.

From the beginning, one of the greatest concerns of many of the GLU board members was that, in general, many of them had little experience in understanding labor issues, and less understanding of the practical implications of what incorporating worker protection and compensation into

RESOLUTION FROM GREAT LAKES UNITED TASK FORCE ON LABOR AND THE ENVIRONMENT

WHEREAS, the labor movement in the U.S. and Canada has supported Great Lakes United and other environmental programs for the protection and restoration of the Great Lakes ecosystem; AND

WHEREAS, the issue of environmental protection and quality jobs in the Great Lakes Basin is imperative, AND

WHEREAS, toxics reduction and zero discharge could impact on the stability and quality of present and future jobs; AND

WHEREAS, the ability to achieve zero discharge through pollution prevention, toxics use reduction, and other changes in production processes and production processes and production choices will be integrally related to the mutual cooperation and efforts of the labor movement in the affected industries.

THEREFORE BE IT RESOLVED, that GREAT LAKES UNITED place a high and immediate priority on obtaining new funding to support the work of the Labor/Environment Task Force, with the objective of supporting paid staff time and providing other necessary resources; AND

BE IT FURTHER RESOLVED, that GREAT LAKES UNITED will introduce and promote the principle of worker compensation and economic protection as well as other options in all its pollution prevention policies and initiatives; AND

BE IT FURTHER RESOLVED, that GREAT LAKES UNITED'S Task Force on Labor and the Environment will take responsibility for reviewing and evaluating current worker protection and compensation programs as they exist in the Great Lakes Basin with the future task of providing further policy recommendations to the Board as required.

pollution prevention policy would mean on a day-to-day basis. Recognizing this, the task force committed itself to the job of providing concrete direction when needed, which would be based on its own efforts to further explore and become involved with already existing worker protection strategies.

In the early years after the task force was formed, progress was slow and frustrating. While a major part of this was due to internal development issues, it is also clear that little leadership was forthcoming from the labor movement *as a whole* on the question of a Superfund for Workers and other significant aspects of the jobs and toxics use reduction questions. The GLU task force persevered, however, and the organization began to devote dedicated staff time to building the task force and developing the issue. Major grants to fund GLU and Task Force work on linking pollution prevention strategies to job retention and worker protection strategies are being developed and submitted, even as Canadian initiatives also are being developed and pursued. Free trade, its impact on the environment, and ways to link up with the labor movement to fight it, also are emerging with greater strength as issues for the task force.

MISSION STATEMENT FOR THE LABOR/ENVIRONMENT TASK FORCE OF GREAT LAKES UNITED, SPRING 1991

With the commitment of Great Lakes United to zero discharge as a political goal, the Labor/Environment Task Force has been formed with the mission of building the alliance between the labor and environmental movements to the advantage of both.

It will be our task to look boldly at the question of jobs, what the impact of the struggle for source reduction and zero discharge will be on them, and how the environmental movement can support and contribute to the efforts of labor to fight job blackmail, job displacement, and job loss. We will also work for policies and programs that protect and compensate workers whose jobs may be affected by the fight for source reduction, process changes, product terminations, and zero discharge.

We will fight against the efforts of employers and governments to present the issue of environmental protection and environmental clean-up as a choice between earning a living and living in an environmentally healthy community. Following in the tradition of "an injury to one is an injury to all," the mission of this task force is to break down the barriers which have been used to separate environmentalists from workers, welding together the respective strengths of the labor and environmental movements into a unified force to fight in a common cause.

30

Accepting the Challenge

A RESPONSE TO LIN KAATZ CHARY

MICHAEL MERRILL

First, I want to thank Lin Kaatz Chary for her stimulating intervention. While I disagree with much of what she has to say [in Chapter 29], I admire the forthrightness with which she has stated her position.

According to Chary, the environmental movement has "neither the obligation, the objective, nor most significantly of all, the ability" to convince labor that environmental changes are going to create jobs. On the contrary, protecting jobs is the labor movement's responsibility. Environmentalists cannot bargain for workers. If income protection is going to be an indispensable part of the environmental agenda, it is labor's responsibility to put it there. The labor movement needs to develop its own agenda for income protection, and "an aggressive, comprehensive, broad-based campaign on the issue of jobs and the environment."

I would respond that the labor movement already has a program for income protection—economic growth. Whatever helps the economy grow, helps to protect the income of workers. For the most part, however, this income protection program is opposed to the environmentalist agenda. According to the conventional wisdom, growth requires deregulation and environmental destruction. Some day the labor movement might take the lead to develop an alternative program—should it become desperate enough, or should economic growth turn out to provide less income protection than environmentalism. But workers are not going to abandon a tried-and-true path easily. Economic growth has provided a greater measure of income protection over a longer period in the recent history of humankind than has environmentalism—at least I think it seems that way to most wage earners. If it seemed otherwise, more wage earners would be environmentalists.

NOT OFF THE HOOK

The environmental movement cannot be let off the labor hook that easily. If the movement is going to grow, it must address the real questions of what the world will look like if it wins and environmentalists are close enough to winning for that to be a real concern for many people. What place, working people are asking themselves, are we going to have in the brave new, green world of tomorrow? The environmental movement's failure to provide a convincing answer to this question is the principal obstacle to a meaningful labor-environmental coalition.

At the same time, if the labor movement is to grow, it too must address the real concerns of the millions of people who currently don't identify with it. The current income protection program of the labor movement is inadequate. It is based on environmental destruction, not environmental protection. And there are many reasons to think that nature is going to close off this alternative.

Which brings me to the other issue raised in her chapter: the Superfund for Workers. According to Chary, the labor movement needs to provide environmentalists with guidance about how to apply the demand in specific situations. But the problem goes deeper than simply a failure to be specific. Most efforts to translate the demand for a Superfund for Workers into concrete proposals don't get very far because people haven't yet agreed on what it is.

A "GREEN PARACHUTE"

To some, the Superfund for Workers is simply a bigger and better job training program. I agree with Chary that this is an impoverished conception. Job retraining is to the Superfund for Workers what litter patrols are to environmentalism—not a serious assault on the problem. If there is to be an environmentally sound income protection program that makes sense to working people, then it is necessary to be bolder.

The Superfund for Workers, in effect, ought to be seen as a "green parachute" for workers who lose their jobs because of environmental regulation—or any other reason for that matter. The Superfund for Workers is not about retraining workers. It is about redefining work. The only way to stop people from getting paid for destroying the environment is to begin to pay them for not destroying it. If you want loggers to stop cutting down trees, then pay them not to cut down trees. If you want chemical workers to stop making hazardous chemicals, and nuclear workers to stop making plutonium, then pay them not to make hazardous chemicals and plutonium. That is what the Superfund for Workers ought to be about.

The way things are now, people are paid to destroy the environment. If the environmental movement wants workers to support an alternative program, they are going to have to come up with a way of paying people not to destroy the environment—even if that means they are paid to do nothing. Or paid to do what bosses do—which is to sit around all day and think and talk and figure out how to do things differently. It is precisely these kinds of unthinkable things that the original proposal for a Superfund for Workers imagined. Until people accept the general notion, concrete proposals will only be of limited utility. As the Situationists used to say, the most realistic thing for us to do is demand the impossible. Only the impossible can make the unthinkable real.

31

Toxics Use Reduction and Pollution Prevention

KEN GEISER

The mid-1980s marks an important turning point in state and federal hazardous waste management policy. During this period, a new policy of reducing and preventing the generation of hazardous waste was grafted on to conventional policies that for 10 years had focused primarily on regulating and controlling hazardous waste treatment and disposal.

Although, at first glance, this new orientation may not appear as a bold departure, it has laid the groundwork for an increasingly fundamental reconsideration of waste policy and opened the possibility of a new approach to toxic chemical risk management.

THE LEGACY OF POLLUTION CONTROL POLICIES

In 1976, the United States Environmental Protection Agency (EPA) issued a policy statement outlining its preferred options for managing hazardous wastes. (1) With that policy statement, the EPA officially set the government's highest priority on reducing the generation of hazardous waste at the source. Yet, over the past 40 years, the volume of hazardous waste has continued to increase unabated. The EPA estimates that at the close of World War II the nation generated about a billion pounds of hazardous waste a year. In 1987, more than 22 billion pounds of hazardous wastes were discharged to the air, water, and land. (2) Further, the government has all but ignored its stated highest priority. During the 1980s, federal and state spending on all pollution programs ran at about $16 billion a year.

The United States Congressional Office of Technology Assessment (OTA) found that in 1984 the total amount spent on reducing the generation of hazardous waste was about $4 million, or less than one percent of all government spending. (3)

In reviewing this cautious history, former EPA Administrator Lee Thomas referred to the conventional regulatory approach as the "strategy of the cork": It consists of putting a regulatory cork in every pollution source you can find as quickly as you can. . . . The idea is that if you get enough corks, and put enough pressure behind them, pollution will eventually be eliminated. (4)

By the mid-1980s, this regulatory strategy began to come under increasing criticism. A diverse assortment of environmental advocates and grassroots community activists began seeking a shift in the conventional policy. What they demanded as their objective was very similar to the original government commitment stated nearly a decade before: the prevention of pollution at the source. (5)

This policy shift involves not simply a different program direction; it requires a shift in the conventional paradigm that has guided waste management policy for nearly 10 years. The new prevention paradigm differs from the conventional regulatory approach in terms of the problem definition, the domain of policy intervention, and the policy instruments it espouses.

Problem Definition

The new pollution prevention approach includes a different definition of the hazardous waste problem that focuses attention less on hazardous waste and more on the toxic chemicals used in production and maintenance. Waste management becomes a subset of toxic chemical management.

Domain of Intervention

The domain of conventional policy implementation has been the end of the privately owned discharge pipe where waste is emitted into the public media. Under the pollution prevention approach, the point of intervention is shifted up into the production process where decisions traditionally have been management prerogatives well hidden from public review.

Policy Instruments

The conventional waste management approach relied on enforceable "command and control" regulations that set standards and monitored compliance. The pollution prevention approach relies more on government-mandated planning, goal setting, and performance standards coupled with government technical assistance and financial incentives.

This new policy paradigm has not emerged all at once or found its expression in any one single published document. Rather, the pollution prevention approach has developed out of ongoing social and intellectual struggles among environmental advocates, grassroots activists, government officials, and business managers.

THE EMERGENCE OF POLLUTION-PREVENTION POLICIES

The stage was set in 1984 when Congress passed the "Hazardous and Solid Waste Amendments" to the Resource Conservation and Recovery Act. In the amendments, Congress recognized the emerging possibilities of hazardous waste reduction. At the time, waste reduction was often called "source reduction," a term that had been borrowed from the solid waste policy discussion of the mid-1970s. Source reduction had become an increasingly popular demand of citizen groups that were protesting the siting of new waste management facilities. Instead of building new landfills, incinerators, and treatment facilities to dispose of the ever growing volume of waste, these activists advocated that the government enact programs to reduce the growth in the volume of waste. Waste reduction or "source reduction" meant processes or technologies that reduced the amount and toxicity of hazardous waste before it left the production process (that is, at the "source").

The idea has a good history. Michael Royston, a French engineer, had written a book called *Pollution Prevention Pays* in 1979. (6) In this book, Royston had documented and praised the "3P" ("Pollution Prevention Pays") Program at the Minneapolis-based 3M Corporation. The 3M Corporation first began its 3P Program in 1975. By 1982, the corporation reported more than 600 projects worldwide eliminating 125,000 tons of industrial sludge. (7) State officials in North Carolina had been so impressed with the "3P" concept that in 1983 they launched a small state program to encourage North Carolina firms to save money by initiating practices that reduced the amount of waste for which waste treatment services would be required. (8) Within the following year, state administrative staffs in Minnesota, Illinois, and Massachusetts began experimenting with modest assistance programs as well. (9)

The Congressional hearings of 1984 took all of this into account. (10) The majority of the technical testimony pointed at the large number of hazardous waste landfills that had ruptured and were leaking contaminants into groundwater. The cost of treating and disposing of hazardous waste was rising rapidly. The strict "joint and several" liability standards written into the 1980 federal Superfund law had made all waste generators liable for waste handling no matter how long after the waste had left the plant site. Finally, the growing grassroots resistance to the siting of treatment facilities had created a virtual moratorium on new hazardous waste incinerator construction.

With these growing pressures, Congressional leaders returned to the earlier federal priority on waste reduction. The 1984 amendments stated, "The Congress hereby declares it to be the national policy of the United States that, whenever feasible, the generation of hazardous waste is to be reduced or eliminated as expeditiously as possible." (11)

Congress wrote into the 1984 amendments requirements that firms manifesting waste transport and filing biennial waste-treatment reports must report on their efforts to reduce the volume and toxicity of waste prior to treatment. In addition, the amendments included a charge to the EPA to conduct a two-year study and report back to Congress on the need for and feasibility of additional federal initiatives to encourage "waste minimization." (12)

This new term, "waste minimization," which Congress left undefined, introduced a significant debate among waste-reduction advocates. At root, this debate centered on how waste reduction was to be measured. The EPA defined waste minimization with broad language to include all processes and technologies that reduced the volume of hazardous waste that a generator sent to landfills for disposal. Thus, if a waste generator burned a waste on-site or sent the waste to an off-site materials recycler, this was considered a form of waste minimization. Source reduction advocates sought to reduce all environmental exposure. They argued, instead, that the definition of waste reduction should include only on-site processes and technologies that reduced the volume and toxicity of the waste.

In 1986, the OTA released a landmark report that helped to clarify the language and the policy shift.(13) Entitled *Serious Reduction of Hazardous Waste: For Pollution Prevention and Industrial Efficiency*, the OTA report was useful in documenting in-plant technologies and processes for reducing hazardous wastes. Yet, the most significant contribution of the OTA report was in clarifying and advocating the emerging shift in government policy paradigms. Prior to 1986, the OTA found, most federal and state activity was directed at regulating the discharge, treatment, and disposal of hazardous waste. This, the report called "pollution control" because it was not directed at reducing waste, only managing it properly. In contrast, the report identified and recommended a new paradigm that it called "pollution prevention," under which firms are encouraged, aided, and pressed to reduce or eliminate the generation of hazardous waste.

The OTA argued that pollution prevention is superior to pollution control for several reasons. First, it benefits government because it means less waste entering the environment. Thus, there is reduced risk of mismanagement and lower costs of environmental treatment and remediation in cases where proper management fails. Second, it benefits companies because it reduces the expenses of waste treatment and the total costs of raw materials that can now be recycled and reused in production. Third, it benefits society because it conserves materials and slows the depletion of nature's virgin resource base.

The OTA report promoted and encouraged state government efforts to deepen investments in this new pollution-prevention direction. In the 1988 budget, Congress authorized $3 million in funds to support state hazardous waste-reduction programs. By 1989, 34 states had set up separate administrative units or added new functions to existing state divisions to encourage some form of hazardous waste reduction. In addition, the EPA set up a new Office of Pollution Prevention to work with divisions inside the agency to develop and coordinate waste-minimization policies.

THE MOVE TO TOXICS USE REDUCTION

Although waste reduction has emerged or reemerged on the frontier of federal policy, it has been heavily burdened by the legacy of the past waste-management culture. Advocates of source reduction had always included in the definition changes in feedstock chemicals and the redesign of industrial production in order to reduce the volume and toxicity of the chemicals in use. Yet, even where the EPA's waste-minimization program promotes chemical substitutions, the primary goal is to reduce waste generation. The focus remains on waste. In promoting source reduction, advocates came to see waste as merely one of the forms of toxic chemicals used in production. Some environmental activists realized that the prevention of pollution could be more effectively achieved by reducing the volume and toxicity of the production chemicals themselves. (14) This focus on chemical use was dubbed "toxics use reduction" by its advocates.

The objective of toxics use reduction is the reduction or elimination of toxic chemicals in production whether the chemicals appear as wastes, by-products, intermediaries, feedstocks, or constituents of finished consumer products. By expanding the focus from waste reduction to chemical use reduction, several new opportunities emerge.

First, there is the opportunity to form new alliances and broaden the emerging grassroots "toxics movement." With toxics use reduction, the professional and legal gulf between occupational and environmental advocacy can be bridged. Reducing the use of toxic chemicals in production reduces both occupational and environmental risk and offers common ground between environmentalists and the occupational safety and health movement. Health and safety leaders in both the Oil, Chemical, and Atomic Workers Union and the International Chemical Workers Union have begun internal discussions on toxics use reduction and the possible effects such policies may have on labor. To date, organized labor has had little participation in the crafting of the toxics use reduction concept, yet labor remains a critical determinant of the successful implementation of the policy. Changes in chemical use patterns in production will have effects on workers ranging from simple substance changes in cleaning equipment to the required learning of new skills and, in some cases, changes in the conditions of employment. Workers and their representatives need to be better integrated into the designing of these new pollution prevention programs both to guarantee meaningful protection for workers and to add new ideas that may only arise from those with real shop-floor experience.

Toxics use reduction also provides a new convergence between environmentalists and consumer advocates concerned about the safety of food, drugs, and products on the commercial market. The term has been recently adopted in California campaigns around pesticide use and food safety.

Second, there is the potential to realign the public and private domains of industrial decision making. Decisions about the use of toxic chemicals in production traditionally have been proprietary. Occupational health and safety laws provide regulations on the conditions under which employees can use chemicals. The federal Toxic Substances Control Act regulates the production of new chemicals. Except for federal bans on the use of asbestos, poly-chlorinated byphenyls and some pesticides, there are few direct prohibitions on the use of toxic chemicals in manufacturing. The selection and use of conventional chemicals in industrial production has generally proceeded unchallenged by a demand for public accountability. The concept of toxics use reduction opens up a new opportunity for the public (or at least the government) to penetrate into this traditional domain of management rights.

Finally, there is the opportunity to introduce a new private–public dialogue based on industrial production planning. It is generally agreed that regulations are inappropriate policy instruments for encouraging toxics use reduction. American industry is so varied in size, technology, market, and product that it would be impossible to set general regulations. Austere government budgets provided no hope for the hundreds of government agents it would take to monitor such process-specific regulations. Besides, there had been clear reluctance among state and federal officials toward the introduction of more business regulations. The result has been policies that rely on management-oriented and market-oriented instruments to encourage toxic chemical use reduction planning, corporate goal setting, and government assistance. Negotiated plan making offers new opportunities for public participation and local accountability.

The emerging horizon for toxics use reduction involves preparing plans now for converting industrial production and rethinking materials use, technology design, and social need for the future. Pollution prevention relies on a new policy paradigm. As we have learned from the more studied paradigm shifts in the history of science, the outline and consequences of this policy shift can only vaguely be predicted while the shift is in process.

THE MASSACHUSETTS TOXICS USE REDUCTION CAMPAIGN

Recognizing the slow motion of the federal shift toward pollution prevention, environmental activists turned to state governments. In 1987, toxics use reduction bills initially drafted by organizers in the National Toxics Campaign and the state Public Interest Research Groups (PIRGs) were submitted in four state legislatures. The Massachusetts campaign took the lead with a bill that was the product of six months of work by representatives of environmental, health, and consumer organizations. The state's business leaders at first remained hostile and skeptical.

During the legislative struggle in 1988, it became clear that the state PIRG would attempt to pass the toxics use reduction bill through a popular ballot initiative if the bill made no progress in the legislature. Fearful of the consequences of a ballot campaign, the state business association wrote and submitted its own "Hazardous Materials Waste Elimination and Management Act." Faced with two competing bills in the 1989 legislative session, the legislative leaders forced the two sides to the bargaining table to produce a "consensus" bill.

The negotiations continued through four months. The struggle involved both power and conceptual issues. The business representatives persisted in seeing the bill as focused on waste reduction, while the environmental representatives appealed for a broader chemical use definition. For instance, the business negotiators wanted to include a waste-treatment facility siting provision in the bill, whereas the environmentalists argued successfully that waste treatment was irrelevant to chemical use reduction. The environmentalists, reflecting the earlier national debate about definitions, wanted firms to report an annual toxics use reduction index, while the business representatives wanted the index to be a broader waste-reduction rating. A compromise resulted in both indices being required.

The issue of public penetration into traditional managerial areas emerged around the planning requirement and the trade secret provision. The original toxics use reduction language mandated that firms prepare toxics use reduction plans and submit them to the state. These plans were to be reviewed by a representative of the firms' employees and signed off by a certified professional. The business representatives accepted the mandatory planning requirement, but successfully resisted the state submittal element and the worker review provision. Instead, the plan, once produced and approved by a certified planner, will remain at the facility available for state, but not public, inspection. A matrix summary of the plan will be sent to the state and this and the two indices will be made public. The trade secret provision was expanded to include all proprietary information where a firm could prove a loss of competitive position if the information was revealed.

In June, a consensus bill was achieved, and the next month the Massachusetts Toxics Use Reduction Act was passed by the legislature and signed by the governor. (15) As enacted, the act requires four new initiatives: administrative reorganization, toxics use reduction planning, creation of a research and training institute, and state-mandated performance standards for targeted industrial sectors. The administrative reorganization provision requires the state to establish a new Administrative Council on Toxics Use Reduction to coordinate the shift in policy directions within the various state agencies.

Under the law, Massachusetts firms that manufacture or use any substance on a special toxic chemical substance list (16) are required to file annual toxic chemical inventories by production unit as well as both a toxics use reduction index and a waste reduction index. Further, beginning in 1994, each of these firms must prepare "toxics use reduction plans" that are updated every two years, certified by a planner, and kept at each facility site. The plans must include past and projected changes in toxic chemical use, assessments of available technologies or chemical substitutes that would reduce toxic chemical use, and schedules for introducing economically feasible reduction technologies or practices.

In order to encourage the implementation of the plans, the law calls for establishment of a university-based Toxics Use Reduction Institute to be established at the University of Lowell to provide training and conduct research on new "clean technologies." The institute must prepare curricula for training the nongovernment professionals who are delegated to certify the plans and must provide public and professional education in toxic chemical reduction. Further, the institute is to sponsor and conduct research on toxics use reduction methods and impacts with special emphasis on the social and economic consequences of the state's program.

Further, the law grants authority to the state to target specific industry segments where it appears that new technologies or practices could achieve substantial toxic chemical use reductions. After a study of the plans from the firms in the sector, the state can set performance standards to pressure all firms to achieve high levels of reduction.

THE FUTURE STRUGGLE FOR POLLUTION PREVENTION

The passage of the innovative law in Massachusetts does not guarantee clear and effective implementation or a high standard of business compliance. While the resources for administration are protected through the establishment of an earmarked Toxics Use Reduction Fund fed by fees on chemical users, the state has little experience with such a system, and the projected revenues are a matter of guesswork. The effectiveness of the required use reduction plans has yet to be tested. Corporate plans are based on good-faith efforts accountable only to professional review and periodic state inspections. Also, there remains an unresolved issue over substitutions. Reducing toxic chemical use suggests that equally effective alternative substances or processes can be found or developed. Fundamental to this issue are the definitions for safe or safer materials and clean or cleaner technologies. New materials, new technologies, new practices mean changes in products and markets. Little of this had been clearly discussed or planned. There remains much work to do. The Massachusetts experiment is probably best seen as a laboratory to explore these more concrete problems of the pollution-prevention approach.

Elsewhere, other pollution-prevention bills have been introduced in state legislatures. On the same day that the Massachusetts law was signed, the governor of Oregon signed a toxics use reduction law that in many ways paralleled the Massachusetts bill. (17) Similar bills passed during 1990 in Washington, Indiana, and Maine.

Yet, pollution prevention, for all its attractiveness, remains the lesser policy approach. In October of 1989, the EPA required that all states submit "capacity-assurance plans" demonstrating state capacity for managing all hazardous waste generated internally. While the agency indicated that it would allow consideration of waste-reduction goals as part of such state plans, the large majority of states are conservatively relying on conventional recycling, treatment, and disposal policies in order to satisfy the capacity-assurance requirements. (18)

Conventional treatment and disposal strategies rely on off-site recycling, incineration, landfilling, and underground injection. The EPA reported that there were, in 1987, at least 100 hazardous waste landfills across the country, of which 33 were commercial facilities; and there were 79 active deep well injection sites. Today, there are 82 permanently licensed hazardous waste incinerators in operation, and there are another 140 proposals seeking permits. (19) These technologies are products of the traditional pollution-control paradigm. While these technologies are advancing in sophistication and expense, they can do little to change the problems of the past policies.

For now, the national agenda appears to be composed of two different policy directions. Proponents of both approaches vie for dominance. History provides strong lessons about relying on treatment and disposal technologies, even those that look most promising during construction. Toxics use reduction breaks sharply with that past. It introduces a new policy paradigm. It shifts attention to the chemicals of production. But pollution prevention will require changing old programs, processes, and attitudes. For well over a decade, the nation relied on a regulatory—"cork"—approach. The time has come for more sophisticated and effective instruments and for a more far-sighted and preventative policy approach.

NOTES

1. U.S. Government, 41 *Federal Register* 35050, August 18, 1976.

2. U.S. Environmental Protection Agency, *The Hazardous Waste System*, Washington, DC, June 1987.

3. U.S. Congress, Office of Technology Assessment, *From Pollution to Prevention*, Washington, DC, June 1987, p. 39.

4. Lee M. Thomas, "A Systems Approach: Challenge for EPA," *EPA Journal*, September, 1985, p. 22.

5. Roland Alsop, "Local Citizen Groups Take a Growing Role Fighting Toxic Dumps," *Wall Street Journal*, April 18, 1983.

6. Michael Royston, *Pollution Prevention Pays*, New York: Pergamon Press, 1979.

7. Russell Susag, "Pollution Prevention Pays: The 3M Corporate Experience," in Donald Huising and Vicki Bailey, eds., *Making Pollution Prevention Pay*, New York: Pergamon Press, 1982, p. 20.

8. North Carolina Department of Natural Resources and Community Development, "The North Carolina Pollution Prevention Pays Program: 1985 Program Summary and Status," Raleigh, NC, 1986.

9. New York was the first state to establish a recognizable source reduction program in 1981. For a review of state programs, see U.S. Congress, Office of Technology Assessment, *Serious Reduction of Hazardous Waste: For Pollution Prevention and Industrial Efficiency*, Washington, DC, September, 1986, pp. 197–223.

10. U.S. House of Representatives, Committee on Energy and Commerce, *Legislative History of the Hazardous and Solid Waste Amendments of 1984*, 98th Congress, Washington, DC, June 1983.

11. U.S. Congress, *Hazardous and Solid Waste Amendments of 1984*. P.L. 98-616 (98 Stat. 3221). November 8, 1984.

12. U.S. Congress, P.L. 98-616, 1984.

13. U.S. Congress, Office of Technology Assessment, *Serious Reduction of Hazardous Waste: For Pollution Prevention and Industrial Efficiency*, Washington, DC, September 1986.

14. The term was first coined and the concept developed by David Allen in unpublished papers prepared for the National Toxics Campaign. See National Campaign Against Toxic Hazards, *The Toxics Prevention Act: Model State Legislation*, Boston, MA, January 7, 1987.

15. Masssachusetts General Court, "An Act to Promote Reduced Use of Toxic and Hazardous Substances in the Commonwealth," House Bill No. 6161, Boston, MA, June 26, 1989.

16. The base list is the Toxic Chemical List defined in Section 313 of the U.S. Emergency Planning and Right-to-Know Act, P.L. 99-499, 1986. Over four years this list will expand to include the additional substances listed in Section 101 (14) and 102 of the U.S. Comprehensive Environment Response and Compensation Liability Act, P.L. 92-500, 1980.

17. Christopher Daly, "New Laws to Reduce Industrial Poisons," *Washington Post*, July 25, 1989.

18. Sanford Lewis and Marco Kaltofen, *From Poison to Prevention*, National Toxics Campaign Fund, Boston, MA, August 17, 1989.

19. Lewis and Kaltofen, *From Poison to Prevention*.

PART IV

WORK ENVIRONMENT, HEALTH, AND DEMOCRACY

Throughout this volume we have consistently presented material that defines the work environment broadly, with each chapter addressing in one way or another the problems that arise for the health of workers and the environment in relation to production. Given the fundamental role that the production of goods and services (the economy itself) plays in any society, we argue that what happens on the factory floor, in the office, or in the store is the real subject matter of environmental health and regulation. But the workplace is only one locus of political and social relations in society. A multiplicity of institutions, social organizations, laws, and even belief systems affect and are affected by what goes on at the point of production. In short—as we noted in Part I in setting out our primary organizing theme—there exists a *political economy of the work environment.*

We began by discussing the ways in which economic structures determine the health of workers, communities, and the environment. In Part II we saw how these features interact with regulatory policy and political institutions, and we provided case studies illustrating the various relationships among regulating authorities, science, and the politics of risk assessment. In Part III we examined how social conflict may arise from work environment issues, further illustrating with numerous examples the ways in which such conflicts become manifest and how they might conceivably be resolved. In this final section, Part IV, we provide three perspectives that sum up many of the issues elaborated earlier.

Part IV takes a broad view of the problems and the potential for work environment reform. In Chapter 32 Grossman and Adams present an innovative and controversial strategy for dealing with the problems raised throughout this book, in effect arguing that we should resurrect the original conception of the "corporate charter." Originally, in the eighteenth century, such charters represented an agreement between corporations and the community that defined the limits of corporations' actions. Charters typically specified that corporations could conduct business only if they obeyed all laws, served the common good, and caused no harm. Unfortunately, this original conception of charters of incorporation was subverted in the nineteenth century by big business, which eventually led to our current plight. Grossman and Adams call for a reexamination of the original concept of charters, suggesting that we can utilize the nation's legal establishment to hold corporations responsible, through their charters, for the damage they do to workers, communities, and the environment.

Terris, in Chapter 33, raises many of the same questions that we began this volume with, concluding in general that public health (that is, the general health of all citizens) is *socially* determined. For Terris, the health problems facing workers and the community are primarily a function of inequalities in income and opportunity, reflecting the absence of decent jobs, adequate housing, and good nutrition. Until we begin to remedy these profound inequalities, he argues, we can never really have a healthy community.

Quinn and Buiatti, in Chapter 34, discuss, and present in detail, an Italian proposal that shows how we might conceivably regard the issues of work, quality of life, and sustainable economic development very differently, were we to give the matters some really deep thought.

The Italian proposal breaks new ground by suggesting that work should not necessarily be the all-consuming central priority of people's lives. It calls for integrating work into the general life of each individual, a life that might put greater emphasis on family, friends, personal development, recreation, and in general on greater flexibility in the organization of time. Work has been central to people's lives since the dawn of the industrial revolution. Technology and automation have not necessarily freed us from the grind of work itself—quite the contrary, in fact! The Italian proposal suggests an innovative approach to how we should think about work. This proposal addresses head-on one of the central health problems of contemporary society, namely, how can we go about diminishing the stress caused by our best efforts to maintain and improve our personal standard of living. Two-wage-earner families, after all, are fast becoming the norm. The overarching question remains: can we meet the demands of family and careers, and yet leave time for personal development and a life we find personally satisfying?

32

Taking Care of Business

CITIZENSHIP AND THE CHARTER OF INCORPORATION

RICHARD L. GROSSMAN
FRANK T. ADAMS

Corporations cause harm every day. Why do their harms go unchecked? How can they dictate what we produce, how we work, what we eat, drink, and breathe? How did a self-governing people let this come to pass?

Corporations were not supposed to reign in the United States.

When we look at the history of our states, we learn that citizens intentionally defined corporations through charters—the certificates of incorporation.

In exchange for the charter, a corporation was obligated to obey all laws, to serve the common good, and to cause no harm. Early state legislators wrote charter laws and actual charters to limit corporate authority, and to ensure that when a corporation caused harm, they could revoke its charter.

During the late nineteenth century, corporations subverted state governments, taking our power to put charters of incorporation to the uses originally intended.

Corporations may have taken our political power but they have not taken our Constitutional sovereignty. Citizens are guaranteed sovereign authority over government officeholders. Every state still has legal authority to grant and to revoke corporate charters. Corporations, large or small, still must obey all laws, serve the common good, and cause no harm.

To exercise our sovereign authority over corporations, we must take back our political authority over our state governments.

CLAIMING OUR LEGACY

Today, in our names, state legislators give charters to individuals who want to organize businesses. Our legislators are also supposed to oversee how every corporation behaves. Corporations cannot operate—own property, borrow money, hire and fire, manufacture or trade, sign contracts, sell stock, sue and be sued, accumulate assets or debts—without the continued permission of state officeholders.

Our right to charter corporations is as crucial to self-government as our right to vote. Both are basic franchises, essential tools of liberty.

At first only white men who owned property could vote, and gaining the vote for every person has taken years. But as we were winning that struggle, corporate promoters were taking away our right to have a democratic say in our economic lives.

Corporate owners claim special protections under the U.S. Constitution. They assert the legal authority over what to make and how to make it, to move money and mountains, to influence elections and to bend governments to their will.

They insist that once formed, corporations may operate forever. Corporate managers say they must enjoy limited liability, and be free from community or worker interference with business judgments.

The lord proprietors of England's colonial trading corporations said the same things, even boasting that their authority came not from a constitution, but from God. Since the colonists used guns to take land from the Indians, they could easily see the source of that corporate authority was the king's militia.

The colonists did not make a revolution over a tax on tea. They fought for many reasons, but chiefly to create a nation where citizens were the government and ruled corporations.

So even as Americans were routing the king's armies, they vowed to put corporations under democratic command. As one revolutionary, Thomas Allen, said: "It concerned the People to see to it that whilst we are fighting against oppression from the King and Parliament that we did not suffer it to rise up in our Bowels . . . [and to have] Usurpers rising up amongst ourselves."

The victors entrusted the chartering process to each state legislature. Legislators still have this public trust.

A HOSTILE TAKEOVER

The U.S. Constitution makes no mention of corporations. Yet the history of constitutional law is, as former Supreme Court Justice Felix Frankfurter said, "the history of the impact of the modern corporation upon the American scene."

Today's business corporation is an artificial creation, shielding owners and managers while preserving corporate privilege and existence. Artificial or not, corporations have won more rights under law than people have—rights that government has protected with armed force.

Investment and production decisions that shape our communities and rule our lives are made in boardrooms, regulatory agencies, and courtrooms. Judges and legislators have made it possible for business to keep decisions about money, production, work, and ownership beyond the reach of democracy. They have created a corporate system under law.

This is not what many early Americans had in mind.

People were determined to keep investment and production decisions local and democratic. They believed corporations were neither inevitable nor always appropriate. Our history is filled with successful worker-owned enterprises, cooperatives and neighborhood shops, efficient businesses owned by cities and towns. For a long time, even chartered corporations functioned well under sovereign citizen control.

But while they were weakening charter laws, corporate leaders also were manipulating the legal system to take our property rights. "Corporations confronted the law at every point. They hired lawyers and created whole law firms," according to law professor Lawrence M. Friedman. "They bought and sold governments."

In law, property is not merely a piece of land, a house, a bicycle. Property is a bundle of rights; property law determines who uses those rights. As legal scholar Morris Raphael Cohen said, property is "what each of us shall receive from our work, and from the natural resources of the earth . . . the ownership of land and machinery, with the rights of drawing rent, interest, etc., [which] determine the future distribution of the goods . . ."

Under pressure from industrialists and bankers, a handful of nineteenth-century judges gave corporations more rights in property than human beings enjoyed in their persons. Reverend Reverdy Ransom, himself once a slave treated as property, was among the many to object, declaring "that the rights of men are more sacred than the rights of property."

PROFITS AS PROPERTY

Undeterred by such common sense, judges redefined corporate profits as property. Corporations got courts to assume that huge wealthy corporations competed on equal terms with neighborhood businesses or with individuals. The courts declared corporate contracts, and the rate of return on investment, were property that could not be meddled with by citizens or by their elected representatives.

Within a few decades, judges redefined the common good to mean corporate use of humans and the earth for maximum production and profit. Workers, cities and towns, states and nature were left with fewer and fewer rights corporations were bound to respect.

Wielding property rights through laws backed by government became an effective, reliable strategy to build and to sustain corporate mastery.

Some citizens reacted to this hostile takeover by organizing to maintain their rights over corporations. Mobilizing their cities and towns, citizens pressured legislators to protect states' economic rights for many decades.

Others turned to the federal government to guarantee worker and consumer justice, to standardize finance and stock issues, to prevent trusts and monopolies, to protect public health and the environment.

The major laws that resulted, creating regulatory and managing agencies, actually give corporations great advantages over citizens. Some, like the National Labor Relations Act and the National Labor Relations Board, intended that the government aid citizens against the corporation.

But these laws and agencies were shaped by corporate leaders, then diminished by judges. They neither prevent harms, nor correct wrongs, nor restore people and places. These regulatory laws were—and remain—reporting and permitting laws, laws to limit competition and to manage destruction.

Congress, betraying its obligation to preserve, protect, and defend the U.S. Constitution, has been giving away citizen sovereignty to the EPAs, OSHAs, NLRBs, FTCs, NRCs, SECs, BLMs, RTCs.

Agency administrators act under the assumption that corporations have prerogatives over labor, investment, and production. They regard land, air, and water as corporations' raw materials, and as lawful places to dump corporate poisons. Business leaders and politicians are given license to equate corporations' private goals with the public interest.

Regulators and regulatory laws treat labor as a cost and employees as disposable. They equate efficiency and freedom with maximum resource extraction, maximum production, and maximum profits. They shift what had been the corporate burden to prove no harm onto the citizen, who must prove harm.

Corporations chartered by our states are the cause of political, economic, and ecological injury around the globe. Little wonder so many citizens lament today, as Thomas Paine did 200 years ago: "Beneath the shade of our own vines are we attacked; in our own house, and on our own lands, is the violence committed against us."

A HIDDEN HISTORY

For 100 years after the American Revolution, citizens and legislators fashioned the nation's economy by directing the chartering process.

The laborers, small farmers, traders, artisans, seamstresses, mechanics, and landed gentry who sent King George III packing feared corporations. As pamphleteer Thomas Earle wrote: "Chartered privileges are a burthen, under which the people of Britain, and other European nations, groan in misery."

They knew that English kings chartered the East India Company, the Hudson's Bay Company, and many American colonies in order to control property and commerce. Kings appointed governors and judges, dispatched soldiers, dictated taxes, investments, production, labor, and markets. The royal charter creating Maryland, for example, required that the colony's exports be shipped to or through England.

Having thrown off English rule, the revolutionaries did not give governors, judges, or generals the authority to charter corporations. Citizens made certain that legislators issued charters, one at a time and for a limited number of years. They kept a tight hold on corporations by spelling out rules each business had to follow, by holding business owners liable for harms or injuries, and by revoking charters.

Side by side with these legislative controls, they experimented with various forms of enterprise and finance. Artisans and mechanics owned and managed diverse businesses. Farmers and millers organized profitable cooperatives, shoemakers created unincorporated business associations. Joint-stock companies were formed.

The idea of limited partnerships was imported from France. Land companies used various and complex arrangements, and were not incorporated. None of these enterprises had the powers of today's corporations.

Towns routinely promoted agriculture and manufacture. They subsidized farmers, public warehouses and municipal markets, protected watersheds and discouraged overplanting. State legislatures issued not-for-profit charters to establish universities, libraries, firehouses, churches, charitable associations, along with new towns.

Legislatures also chartered profit-making corporations to build turnpikes, canals, and bridges. By the beginning of the 1800s, only some 200 such charters had been granted. Even this handful issued for necessary public works raised many fears.

Some citizens argued that under the Constitution no business could be granted special privileges. Others worried that once incorporators amassed wealth, they would control jobs and production, buy the newspapers, dominate elections and the courts. Craft and industrial workers feared absentee corporate owners would turn them into "a commodity being as much an article of commerce as woolens, cotton, or yarn."

WIDESPREAD EARLY OPPOSITION

Because of widespread public opposition, early legislators granted very few charters, and only after long, hard debate. Legislators usually denied charters to would-be incorporators when communities opposed their prospective business project.

Citizens shared the belief that granting charters was their exclusive right. Moreover, as the Supreme Court of Virginia reasoned in 1809: if the applicants' "object is merely private or selfish; if it is detrimental to, or not promotive of, the public good, they have no adequate claim upon the legislature for the privileges."

Citizens governed corporations by detailing rules and operating conditions not just in the charters but also in state constitutions and in state laws. Incorporated businesses were prohibited from taking any action that legislators did not specifically allow.

States limited corporate charters to a set number of years. Maryland legislators restricted manufacturing charters to 40 years, mining charters to 50, and most others to 30 years. Pennsylvania limited manufacturing charters to 20 years. Unless a legislature renewed an expiring charter, the corporation was dissolved and its assets were divided among shareholders.

Citizen authority clauses dictated rules for issuing stock, for shareholder voting, for obtaining corporate information, for paying dividends and keeping records. They limited capitalization, debts, landholdings, and sometimes profits. They required a company's accounting books to be turned over to a legislature upon request.

LARGE SHAREHOLDERS' POWER LIMITED

The power of large shareholders was limited by scaled voting, so that large and small investors had equal voting rights. Interlocking directorates were outlawed. Shareholders had the right to remove directors at will.

Sometimes the rates that railroad, turnpike, and bridge corporations could charge were set by legislators. Some legislatures required incorporators to be state citizens. Other legislatures bought corporate stock in order to stay closely engaged in a firm's operations.

Early in the nineteenth century, the New Jersey legislature declared its right to take over ownership and control of corporate properties. Pennsylvania established a fund from corporate profits that was used to buy private utilities to make them public. Many states followed suit.

Turnpike charters frequently exempted the poor, farmers, or worshippers from paying tolls. In Massachusetts, the Turnpike Corporation Act of 1805 authorized the legislature to dissolve turnpike corporations when their receipts equaled the cost of construction plus 12 percent. Then the road became public. In New York, turnpike gates were "subject to be thrown open, and the company indicted and fined, if the road is not made and kept easy and safe for public use."

Citizens kept banks on particularly short leashes. Their charters were limited from 3 to 10 years. Banks had to get legislative approval to increase their capital stock or to merge. Some state laws required banks to make loans for local manufacturing, fishing, agriculture enterprises, and to the states themselves. Banks were forbidden to engage in trade.

Private banking corporations were banned altogether by the Indiana constitution in 1816, and by the Illinois constitution in 1818.

People did not want business owners hidden behind legal shields, but in clear sight. That is what they got. As the Pennsylvania legislature stated in 1834: "A corporation in law is just what the incorporating act makes it. It is the creature of the law and may be moulded to any shape or for any purpose that the Legislature may deem most conducive for the general good."

In Europe, charters protected directors and stockholders from liability for debts and harms caused by their corporations.

American legislators rejected this corporate shield. Led by Massachusetts, most states refused to grant such protection. Bay State law in 1822 read: "Every person who shall become a member of any manufacturing company . . . shall be liable, in his individual capacity, for all debts contracted during the time of his continuing a member of such corporation."

SHAREHOLDERS MADE LIABLE

The first constitution in California made each shareholder "individually and personally liable for his proportion of all [corporate] debts and liabilities." Ohio, Missouri, and Arkansas made stockholders liable over and above the stock they actually owned. In 1861, Kansas made stockholders individually liable "to an additional amount equal to the stock owned by each stockholder."

Prior to the 1840s, courts generally supported the concept that incorporators were responsible for corporate debts. Through the 1870s, seven state constitutions made bank shareholders doubly liable. Shareholders in manufacturing and utility companies were often liable for employees' wages.

Liability laws sometimes reflected the dominance of one political party or another. In Maine, for example, liability laws changed nine times from no liability to full liability between 1823 and 1857, depending on whether the Whigs or the Democrats controlled the legislature.

Until the Civil War, most states enacted laws holding corporate investors and officials liable. As New Hampshire Governor Henry Hubbard argued in 1842: "There is no good reason against this principle. In transactions which occur between man and man there exists a direct responsibility—and when capital is concentrated . . . beyond the means of single individuals, the liability is continued."

The penalty for abuse or misuse of the charter was not a fine or slap on the corporate wrist, but revocation of the charter and dissolution of the corporation. Citizens believed it was society's inalienable right to abolish an evil.

Revocation clauses were written into Pennsylvania charters as early as 1784. The first revocation clauses were added to insurance charters in 1809, and to banking charters in 1814. Even when corporations met charter requirements, legislatures sometimes decided not to renew those charters.

States often revoked charters by using *quo warranto*—by what authority—proceedings. In 1815, Massachusetts Justice Joseph Story ruled in *Terrett v. Taylor*: "A private corporation created by the legislature may lose its franchises by a misuser or nonuser of them. . . . This is the common law of the land, and is a tacit condition annexed to the creation of every such corporation."

ATTACK ON SOVEREIGNTY

Four years later, the U.S. Supreme Court tried to strip states of this sovereign right. Overruling a lower court, Chief Justice John Marshall wrote in *Dartmouth College v. Woodward* that the U.S. Constitution prohibited New Hampshire from revoking a charter granted to the college in 1769 by King George III. That charter contained no reservation or revocation clauses, Marshall said.

The court's attack on state sovereignty outraged citizens. Protest pamphlets rolled off the presses. Thomas Earle wrote: "It is aristocracy and despotism, to have a body of officers, whose decisions are, for a long time, beyond the control of the people. The freemen of America ought not to rest contented, so long as their Supreme Court is a body of that character."

Said Massachusetts legislator David Henshaw: "Sure I am that, if the American people acquiesce in the principles laid down in this case, the Supreme Court will have effected what the whole power of the British Empire, after eight years of bloody conflict, failed to achieve against our fathers. "

Opponents of Marshall's decision believed the ruling cut out the heart of state sovereignty. They argued that a corporation's basic right to exist

—and to wield property rights—came from a grant that only the state had the power to make. Therefore, the court exceeded its authority by declaring the corporation beyond the reach of the legislature that created it in the first place.

People also challenged the Supreme Court's decision by distinguishing between a corporation and an individual's private property. The corporation existed at the pleasure of the legislature to serve the common good, and was of a public nature. New Hampshire legislators and any other elected state legislators had the absolute legal right to dictate a corporation's property use by amending or repealing its charter.

State legislators were stung by citizen outrage. They were forced to write amending and revoking clauses into new charters, state laws, and constitutions, along with detailed procedures for revocation.

In 1825, Pennsylvania legislators adopted broad powers to "revoke, alter or annul the charter . . ." at any time they thought proper.

New York state's 1828 corporation law specified that every charter was subject to alteration or repeal. Section 320 declared that corporate acts not authorized by law were *ultra vires*, or beyond the rights of corporations and grounds for charter revocation. The law gave the state authority to secure a temporary injunction to prevent corporations from resisting while legal action to dissolve them was under way.

TIME LIMITS ON CHARTERS

Delaware voters passed a constitutional amendment in 1831 limiting all corporate charters to 20 years. Other states, including Louisiana and Michigan, passed constitutional amendments to place precise time limits on corporate charters.

President Andrew Jackson enjoyed wide popular support when he vetoed a law extending the charter of the Second Bank of the United States in 1832. That same year, Pennsylvania revoked the charters of 10 banks.

During the 1840s, citizens in New York, Delaware, Michigan, and Florida required a two-thirds vote of their state legislatures to create, continue, alter, or renew charters. The New York legislature in 1849 instructed the attorney general to annul any charter whose applicants had concealed material facts, and to sue to revoke a charter on behalf of the people whenever he believed necessary.

Voters in Wisconsin and four other states rewrote constitutions so that popular votes had to be taken on every bank charter recommended by their legislatures. Rhode Island voters said charters for corporations in banking, mining, manufacturing, and transportation had to be approved by the next elected state legislature before being granted.

Over several decades starting in 1844, 19 states amended their constitutions to make corporate charters subject to alteration or revocation by legislatures. Rhode Island declared in 1857: "the charter or acts of association of every corporation hereafter created may be amendable or repealed at the will of the general assembly."

Pennsylvanians adopted a constitutional amendment in 1857 instructing legislators to "alter, revoke, or annul any charter of a corporation hereafter conferred . . . whenever in their opinion it may be injurious to citizens of the community . . ."

As late as 1855, citizens had support from the U.S. Supreme Court. In *Dodge v. Woolsey*, the court ruled the people of the states have not "released their powers over the artificial bodies which originate under the legislation of their representatives. . . . Combinations of classes in society . . . united by the bond of a corporate spirit . . . unquestionably desire limitations upon the sovereignty of the people. . . . But the framers of the Constitution were imbued with no desire to call into existence such combinations."

STRUGGLES FOR CONTROL

Massachusetts mechanics who opposed a charter request by the men who wanted to start the Amherst Carriage Company in 1838 told the legislature: "We . . . do look forward with anticipation to a time when we shall be able to conduct the business upon our own responsibility and receive the proffits of our labor. . . . We believe that incorporated bodies tend to crush all feable enterprise and compel us to Work out our days in the Service of others."

Contests over charters and the chartering process were not abstractions. They were battles to control labor, resources, community rights, and political sovereignty. This was a major reason why members of the disbanded Working Men's Party formed the Equal Rights Party of New York State. The party's 1836 convention resolved that lawmakers "legislate for the whole people and not for favored portions of our fellow-citizens. . . . It is by such partial and unjust legislation that the productive classes of society are compelled by necessity to form unions for mutual preservation. . . . [Lawmakers should reinstate us] in our equal and constitutional rights according to the fundamental truths in the Declaration of Independence, and as sanctioned by the Constitution of the United States . . ."

This political agenda had widespread support in the press. A New Jersey newspaper wrote in an editorial typical of the 1830s: "the Legislature ought cautiously to refrain from increasing the irresponsible power of any existing corporations, or from chartering new ones," else people would become "mere hewers of wood and drawers of water to jobbers, banks and stockbrokers."

With these and other prophetic warnings still ringing in their ears, citizens began to feel control over their futures slipping out of their communities and out of their hands. Corporations were abusing their charters to become conglomerates and trusts. They were converting the nation's treasures into private fortunes, creating factory systems and company towns. Political power began flowing to absentee owners intent upon dominating people and nature.

As the nation moved closer to civil war, farmers were forced to become wage earners. Increasingly separated from their neighbors, farms and families, they became fearful of unemployment—a new fear that corporations quickly learned to exploit.

In factory towns, corporations set wages, hours, production processes, and machine speeds. They kept blacklists of labor organizers and workers who spoke up for their rights. Corporate officials forced employees to accept humiliating conditions, while the corporations agreed to nothing.

Julianna, a Lowell, Massachusetts, factory worker, wrote: "Incarcerated within the walls of a factory, while as yet mere children—drilled there from five till seven o'clock, year after year. . . . What, we would ask, are we to expect, the same system of labor prevailing, will be the mental and intellectual character of the future generations. . . . A race fit only for corporation tools and time-serving slaves? . . . Shall we not hear the response from every hill and vale, 'EQUAL RIGHTS, or death to the corporations'?"

Recognizing that workers were building a social movement, industrialists and bankers pressed on, hiring private armies to keep workers in line. They bought newspapers and painted politicians as villains and businessmen as heroes. Bribing state legislators, they then announced legislators were corrupt, that they used too much of the public's resources and time to scrutinize every charter application and corporate operation.

Corporate advocates campaigned to replace existing chartering laws with general incorporation laws that set up simple administrative procedures, claiming this would be more efficient. What they really wanted was the end of legislative authority over charters.

Cynically adopting the language of early charter opponents, corporate owners and their lawyers attacked existing legislative charters as special privileges. They called for equal opportunity for all entrepreneurs, making it seem as if they were asking that everyone have the same chance to compete.

ACCUMULATIONS OF CAPITAL

But these corporations were not just ordinary individual entrepreneurs. They were large accumulations of capital, and getting larger. By 1860, thousands of corporations had been chartered—mostly factories, mines, railroads, and banks.

Government spending during the Civil War brought these corporations fantastic wealth. Corporate managers developed the techniques and the ability to organize production on an ever grander scale. Many corporations used their wealth to take advantage of war and Reconstruction years to get the tariff, banking, railroad, labor, and public lands legislation they wanted.

Flaunting new wealth and power, corporate executives paid "borers" to infest Congress and state capitals, bribing elected and appointed officials alike. They pried loose from the public trust more and more land, minerals, timber, and water. Railroad corporations alone obtained more than 180 million free acres of public lands by the 1870s, along with many millions of dollars in direct subsidies.

Little by little, legislators gave corporations limited liability, decreased citizen authority over corporate structure, governance, production, and labor, and ever longer terms for the charters themselves.

Corporations rewrote the laws governing their own creation. They "left few stones unturned to control those who made and interpreted the laws . . ."

CITIZENS HOLD ON

Even as businesses secured general incorporation laws for mining, agriculture, transportation, banking, and manufacturing businesses, citizens held on to the authority to charter. Specifying company size, shareholder terms, and corporate undertakings remained a major citizen strategy.

During the 1840s and 1850s, states revoked charters routinely. In Ohio, Pennsylvania, and Mississippi, banks lost charters for frequently "committing serious violations . . . which were likely to leave them in an insolvent or financially unsound condition." In Massachusetts and New York, turnpike corporations lost charters for "not keeping their roads in repair."

"No constitutional convention met, between 1860–1900, without considering the problems of the corporations," according to Friedman.

In 1876, New York's constitutional convention authorized the attorney general to bring an action to "vacate the charters" of any corporation which violated the state chartering law or abused their rights and privileges. Eighteen years later, the Central Labor Union of New York City asked the attorney general to request the state supreme court to revoke the charter of the Standard Oil Company of New York. He did.

CHARTERS REVOKED

New York, Ohio, Michigan, and Nebraska revoked the charters of oil, match, sugar, and whiskey trusts. Courts in each state declared these trusts illegal because the corporations—in creating the trusts—had exceeded the powers granted by their charters. "Roaming and piratical corporations" like Standard Oil of Ohio, then the most powerful corporation in the world, refused to comply and started searching for "a Snug Harbor" in another state.

Rhode Island enacted a law requiring corporate dissolution for "fraud, negligence, misconduct . . ." Language was added to the Virginia constitution enabling "all charters and amendments of charters to be repealed at any time by special act."

Farmers and rural communities, groaning in misery at the hands of railroad, grain, and banking corporations, ran candidates for office who supported states' authority "to reverse or annul at any time any chartered privilege . . ."

The Farmers' Anti-Monopoly Convention, meeting in Des Moines in 1873, resolved that "all corporations are subject to legislative control; [such control] should be at all times so used as to prevent moneyed corporations from becoming engines of oppression."

That same year, Minnesota Grangers resolved: "We, the farmers, mechanics, and laborers of Minnesota, deem the triumph of the people in this contest with monopolies essential to the perpetuation of our free institutions and the promotion of our private and national prosperity."

Because these and other powerful resistance movements directly challenged the harmful corporations of their times, and because they kept pressure on state representatives, revocation and amendment clauses can be found in state charter laws today.

JUDGE-MADE LAW

But keeping strong charter laws in place was ineffective once courts started aggressively applying legal doctrines that made protection of corporations and corporate property the center of constitutional law. As corporations got stronger, government became easier prey; communities became more vulnerable to intimidation.

Following the Civil War, and well into the twentieth century, appointed judges gave privilege after privilege to corporations. They freely reinterpreted the U.S. Constitution and transformed common law doctrines.

Judges gave certain corporations the power of eminent domain—the right to take private property with minimal compensation to be determined by the courts. They eliminated jury trials to determine corporation-caused harm and to assess damages. Judges created the right to contract, a doctrine that, according to law professor Arthur Selwyn Miller, was put forward as a "principle of eternal truth" in "one of the most remarkable feats of judicial law-making this nation has seen."

By concocting the doctrine that contracts originated in the courts, judges then took the right to oversee corporate rates of return and prices, a right entrusted to legislators by the U.S. Constitution. They laid the legal foundation for regulatory agencies to be primarily accountable to the courts—not to Congress.

Workers, the courts also ruled, were responsible for causing their own injuries on the job. The Kentucky Court of Appeals prefigured this doctrine in 1839: "Private injury and personal damage . . . must be expected" when one goes to work for a corporation bringing "progressive improvements." This came to be called the assumption of risk, what Professor Cohen dismissed as "a judicial invention."

Traditionally under common law, the burden of damage had been on the business causing harms. Courts had not permitted trespass or nuisance to be excused by the alleged good works a corporation might claim. Nor could a corporation's lack of intent to cause harm decrease its legal liability for injuries it caused to persons or the land.

Large corporations—especially railroad and steamship companies—pressured judges to reverse this tradition, too. Attentive to lawyers and growing commercial interests, judges creatively interpreted the commerce and due process clauses of the U.S. Constitution. Inventing a new concept that they called substantive due process, they declared one state law after another unconstitutional. Wages and hours laws, along with rate laws for grain elevators and railroads, were tossed out.

Judges also established the managerial prerogative and business judgment doctrines, giving corporations legal justification to arrest civil rights at factory gates and to blockade democracy at boardroom doors.

Corporations were enriched further when judges construed the common good to mean maximum production—no matter what was manufactured, who was hurt, or what was destroyed. Unfettered corporate competition without citizen interference became enshrined under law.

Another blow to citizen constitutional authority came in 1886. The Supreme Court ruled in *Santa Clara County v. Southern Pacific Railroad* that a private corporation was a natural person under the U.S. Constitution, sheltered by the Bill of Rights and the 14th Amendment.

"There was no history, logic or reason given to support that view," Supreme Court Justice William O. Douglas was to write 60 years later.

But the Supreme Court had spoken. Using the 14th Amendment, which had been added to the constitution to protect freed slaves, the justices struck down hundreds more local, state, and federal laws enacted to protect people from corporate harms. The high court ruled that elected legislators had been taking corporate property "without due process of law."

Emboldened, some judges went further, declaring unions were civil and criminal conspiracies, and enjoining workers from striking. Governors and presidents backed judges up with police and armies.

SABOTAGING SOCIAL MOVEMENTS

By establishing "new trends in legal doctrine and political-economic theory" permitting "the corporate reorganization of the property production system," the Supreme Court effectively sabotaged blossoming social protest movements against incorporated wealth. Judges positioned the corporation to become "America's representative social institution," "an institutional expression of our way of life."

Legislative "chartermongering" attracted as many corporations as possible to their states. In exchange for taxes, fees, and whatever else they could get their hands on, some state governments happily provided new homes to Standard Oil and other corporations.

Led by New Jersey and Delaware, legislators watered down or removed citizen authority clauses. They limited the liability of corporate owners and managers, then started handing out charters that literally lasted forever. Corporations were given the right to operate in any fashion not explicitly prohibited by law.

After such losses of citizen sovereignty, 26 corporate trusts ended up controlling 80 percent or more of production in their markets by the early 1900s. There were trusts for almost everything—matches, whiskey, cotton, alcohol, corks, cement, stoves, ribbons, bread, beef.

During the Progressive Era, corporations operated as ruthlessly as any colonial trading monopoly in the 1700s. Blood was often spilled resisting these legal fictions.

Jo Battley, a West Virginia miner, was beaten severely and stabbed trying to organize a union at the Consolidated Coal Company. Mother Jones, one of his rescuers, said, "We tried to get a warrant out for the arrest of the gunmen but we couldn't because the coal company controlled the judges and the courts."

Corporations owned resources, production, commerce, trade, prices, jobs, politicians, judges, and the law. Over the next half century, as a United States Congressional committee concluded in 1941, "The principal instrument of the concentration of economic power and wealth has been the corporate charter with unlimited power . . ."

Today, many U.S. corporations are transnational. No matter how piratical or where they roam, the corrupted charter remains the legal basis of their existence.

TAKING BACK THE CHARTERS, TAKING BACK THE LAW

We are out of the habit of contesting the legitimacy of the corporation, or challenging concocted legal doctrines, or denying courts the final say over our economic lives.

For most of this century, citizens skirmished with corporations to stop doing harm, but failed to question the legitimacy of the harm-doers. We do not use the charter and the chartering process to stop corporate harm, or to define the corporation on our terms.

What passes for political debate today is not about control, sovereignty, or the economic democracy that many American revolutionaries thought they were fighting to secure.

Too many organizing campaigns accept the corporation's rules, and wrangle on corporate turf. We lobby Congress for limited laws. We have no faith in regulatory agencies, but turn to them for relief. We plead with corporations to be socially responsible, then show them how to increase profits by being a bit less harmful.

How much more strength, time, and hope will we invest in such dead ends?

Today, corporate charters can be gotten easily by filling out a few forms and by paying modest fees.

Legislatures delegate authority to public officeholders to rubber-stamp the administration of charters and the chartering process. The secretary of state and the attorney general are the officials most often involved. Sometimes they are elected; sometimes they are appointed.

In all states, legislatures continue to have the historic and the legal obligation to grant, to amend, and to revoke corporate charters. They are responsible for overseeing corporate activities. But it has been a long time since many legislatures have done what they are supposed to do.

In Illinois, the law reads:

> 12.50 Grounds for judicial dissolution. A Circuit Court may dissolve a corporation:
>
> (a) in an action by the Attorney General, if it is established that:
>
> 1. the corporation obtained its certificates of incorporation through fraud; or
>
> 2. the corporation has continued to exceed or abuse the authority conferred upon it by law, or has continued to violate the law. . . .
>
> 3. in an action by a shareholder, if it is established that . . . the directors or those in control of the corporation have acted, or are acting, or will act in a manner that is illegal, oppressive or fraudulent; . . . or if it is established that dissolution is reasonably necessary because the business of the corporation can no longer be conducted to the general advantage of its shareholders.

After entering an order of dissolution, "the court shall direct the winding up and liquidation of the corporation's business and affairs."

In Delaware, Section 284 of the corporation law says that chancery courts can revoke the charter of any corporation for "abuse or misuse of its powers, privileges or franchises."

New York requires dissolution when a corporation abuses its powers, or acts "contrary to the public policy" of the citizenry. The law calls for a jury trial in charter revocation cases.

CORPORATE DISSOLUTION

The Model Business Corporation Act, first written in 1931 by the committee on corporate laws of the American Bar Association, and revised twice since, is the basis for chartering laws in more than half the states and the District of Columbia. Although strongly protecting corporate property, this model law gives courts full power to liquidate the assets of a corporation if they are "misapplied or wasted."

It requires the Secretary of State "from time to time" to list the names of all corporations that have violated their charters along with the facts behind the violations. Decrees of involuntary dissolution can be issued by the Secretary of State and by courts.

Corporations chartered in other states are called foreign corporations. Corporations chartered in other nations are called alien corporations. Legislatures allow foreign or alien corporations to go into business in their states through this same chartering process. Either may establish factories or do business after obtaining a state's certificate of authority.

In Illinois, foreign corporations are "subject to the same duties, restrictions, penalties and liabilities now or hereafter imposed upon a domestic corporation of like character."

THINKING STRATEGICALLY

When we limit our thinking only to existing labor law, or only to existing environmental law, or only to the courts, or only to elections—or when we abide by corporate agendas—we abandon our Constitutional claim on the corporate charter and the chartering process.

When we forsake our Constitutional claim, we ignore historic tools we can use to define and to control the corporation. We pass up strategies that can inspire citizens to act. We fail to demand what we know is right.

We must name and stop what harms us. John H. Hunt, a member of the Equal Rights Party, wrote this resolution in 1837: "Whenever a people find themselves suffering under a weight of evils, destructive not only to their happiness, but to their dignity and their virtues; when these evils go on increasing year after year, with accelerating rapidity, and threaten soon to reach that point at which peaceable endurance ceases to be possible; it becomes their solemn duty coolly to search out the causes of their suffering—to state those causes with plainness—and to apply a sufficient and a speedy remedy."

His resolution was passed unanimously by cheering mechanics, farmers, and working people during a mass rally in a New York City park.

Around the nation, citizens are no less willing—and are quite well prepared—to educate, to organize, and to agitate.

Citizens who have been to folk schools or labor colleges understand that by learning together and teaching ourselves corporate history, we can hone the skills of citizen sovereignty and power.

We can read our state constitutions. Libraries containing our states' constitutional histories, corporate histories, and corporate case law can provide details about what earlier citizens demanded of corporations, what precedents they established, and which of their legal and organizing methods we can use to our advantage.

We can demand to see the charters of every corporation. We need to know what each charter prohibits, especially if it is an old charter. Armed with our states' rich legal precedents, and with our evidence of corporate misuse or abuse, we can amend or revoke charters and certificates ofauthority.

When corporations violate our Constitutional guarantees, we can take them to court ourselves. Corporate officers can be forced to give us depositions under oath, just as elected officials who spurned the Constitution were forced to do by the civil rights movement—often in courtrooms packed with angry citizens.

New Yorkers used to get sufficient and speedy remedy through injunctions against corporations. We can revive this tradition. Surrounded by citizens and their peers, judges can be encouraged to enjoin corporate officials from doing further harm, or from stripping the corporation's assets, or from moving the company away.

Stockholders have authority to seek injunctions and file dissolution suits if they fear managers are acting illegally, oppressively, fraudulently, or are misusing or wasting corporate assets.

REDEFINING THE CORPORATION

As in the first half of the nineteenth century, would-be or ongoing incorporators must be made to ask us for the privilege of a charter. We can set our own criteria: workers must own a significant or majority share of the company; the workforce must have democratic decision-making authority; charters must be renewed annually; corporate officers must prove all corporate harm has ceased. For starters.

Who defines the corporation controls the corporation. We cannot command the modern corporation with laws that require a few days' notice before the corporation leaves town, or with laws that allow the corporation to spew so many toxic parts per million. If we expect to define the corporation using the charters and putting legislators on our civic leash, we must also challenge prevailing judicial doctrines. We cannot let courts stand in the way of our stopping corporate harm.

Legal doctrines are not inevitable or divine. When the liberty and property rights of citizens are at stake, as former Supreme Court Justice Louis D. Brandeis said, "the right of property and the liberty of the individual must be remoulded to meet the changing needs of society."

The corporation is an artificial creation, and must not enjoy the protections of the Bill of Rights.

Corporate owners and officers must be liable for harms they cause. No corporation should exist forever. Both business judgment and managerial prerogative must meet the same end as the colonial trading companies' delusion of divine authority.

Our sovereign right to decide what is produced, to own and to organize our work, and to respect the earth is as American as a self-governing people's right to vote.

In our democracy, we can shape the nation's economic life any way we want.

33

Determinants of Health

A PROGRESSIVE POLITICAL PLATFORM

MILTON TERRIS

A major development in recent years is the increasing recognition that health promotion—the achievement of healthful living conditions—is one of the most important determinants of health. This was understood 150 years ago by William P. Alison in Scotland (1), Louis René Villermé in France (2, 3), and Rudolf Virchow in Germany. (4)

The term "health promotion" was used almost 50 years ago by Henry E. Sigerist, the great medical historian, when he defined the four major tasks of medicine as: (a) the promotion of health, (b) the prevention of illness, (c) the restoration of the sick, and (d) rehabilitation. He stated that "health is promoted by providing a decent standard of living, good labor conditions, education, physical culture, means of rest and recreation," and called for the coordinated efforts of statesmen, labor, industry, educators, and physicians to this end. (5) This call was repeated 40 years later in the Ottawa Charter for Health Promotion, adopted at the first International Conference on Health Promotion in November 1986, which stated that "the fundamental conditions and resources for health are peace, shelter, education, food, income, a stable ecosystem, sustainable resources, social justice, and equity. Improvement in health requires a secure foundation in these basic prerequisites." (6)

Unfortunately, this powerful call for action is still not heeded; the importance of health promotion through the improvement of living conditions may be honored in theory but is neglected in practice. Governments continue to identify medical care as the most important aspect of health policy, and focus their attention and concern primarily on treatment services.

However, equity in medical care has clearly failed to assure equity in health. In England and Wales, the inequality in mortality of social classes has not only failed to narrow since the establishment of the National Health Service in 1948, but has actually widened. The Standardized Mortality Ratio (SMR) of the two highest classes (professional and managerial) was 91 in 1951 and 80 in 1971. For the two lowest social classes (semi-skilled and unskilled workers), it was 110 in 1951 and 121 in 1971. The difference in SMRs more than doubled in 20 years, rising from 19 in 1951 to 41 in 1971. (7)

Perhaps one of the most dramatic examples of the importance of living conditions as a determinant of health status is provided by Canada, a country with a universal medical care system and the highest expectation of life in the Americas. In the late 1970s, the difference in life expectancy between people in the lowest and highest income fifths of the population was 4.4 years; for disability-free life expectancy, the difference was 11 years. Poor people in Canada have, on the average, only 55 years of healthy life, that is, life free from disability, as compared with 66 years of healthy life for rich Canadians. (8)

These data underline the fact that medical care is the least significant of the basic triad of public health. The most important determinants of health status are preventive services on the one hand, and living standards on the other.

A PROGRESSIVE POLITICAL PLATFORM FOR HEALTH

The Ottawa Charter outlines the fundamental social conditions for health. It also provides a general strategy consisting of three interlocking components:

a. Intersectoral action to achieve healthy public policy as well as public health policy.
b. Affirmation of the active role of the public in using health knowledge to make choices conducive to health and to increase control over their own health and over their environments.
c. Community action by people at the local level. Strengthening public participation and public direction of health matters is at the heart of the health promotion strategy.

The Ottawa Charter does not, however, present specific recommendations for national health programs. In 1940, Henry Sigerist outlined the essentials of such programs, and I shall use his categories as the basis for attempting to develop a progressive political program for health. He stated:

> The health program of every country can be summarized in a few points. Such a program must provide:

1. Free education to all the people, including health education.
2. The best possible working and living conditions.
3. The best possible means of rest and recreation.
4. A system of health institutions and medical personnel, available to all, responsible for the people's health, ready and able to advise and help them in the maintenance of health and in its restoration when prevention has broken down.
5. Centers of medical research and training.(9)

Free Education to All the People

It is significant that Henry Sigerist made free education to all the people the first item in a national health program. Inadequate education, resulting in both formal and functional illiteracy, is a serious obstacle to learning the use of preventive measures such as personal hygiene, immunization, and lifestyle changes. In Canada, for example, the prevalence of smoking among women declined by only 7 percent between 1977 and 1981, but among women with a postsecondary certificate or diploma it declined by 25 percent, and among women with a university degree it declined by 41 percent. (10) In the United States, the prevalence of cigarette smoking declined from 1974 to 1987 by only 7 percent in persons with less than 12 years of education, 13 percent in those with 12 years, 24 percent in those with 13-15 years, and 39 percent in those with 16 or more years of education. In 1987, the prevalence of cigarette smoking was 41 percent in the first group, 32 percent in the second, 27 percent in the third, and only 17 percent in the last, most highly educated group. (11)

Unfortunately, the educational system in the United States is in deep crisis as the result of serious cuts in state funding. In a letter in the September 24, 1993 issue of *Science*, titled "The End of Public Higher Education?" Brewster C. Denny of the University of Washington stated:

> Is the end near? It may be. Two indices—tuition and percentage of state spending on higher education—tell the grim tale. Tuitions, once free at some of the best places and almost nominal at most others, have been rising rapidly. State legislative support is in a tailspin. Nationally, higher education's share of the states' budgets has been dropping steadily, now averaging around 10 percent from more than twice that just a few years ago.

All of public education is in crisis, not only higher education. There is a profound equity crisis, based on the fact that schools are financed largely through local property taxes. In Texas, for example, in the early 1990s, spending on the 5 percent of students in the richest districts averaged $11,801 per pupil, while spending on the poorest 5 percent was $3,190. School financing lawsuits are pending in many states. (12) In Rhode Island, for example, "the state's three poorest school systems—Pawtucket, West Warwick, and Woonsocket—are suing the state, alleging officials discriminate against poorer districts by financing education in a way that favors wealthy communities." (13)

The federal government's response to this crisis has continued to be minimal: as Stephen O'Connor has pointed out, "The problem with both the Bush and the Clinton proposals is their shameless tokenism." (14) The Clinton administration has increased the funding for Chapter 1, the main federal program helping poor children reach grade level in reading, writing, and math, from $6.3 billion for fiscal year 1994 to $7 billion in 1995. New York State would receive an increase of $74.5 million. (15) When one considers that the school budget for New York City alone is $7.5 billion, the total budgetary increase for that city would amount to much less than 1 percent.

RECOMMENDATION 1: That the United States adopt the policy of *free public education for all the people*, including all levels of education; that it correct the current methods of financing public education to guarantee equity in the annual spending per pupil in all school districts of the nation, whether rich or poor, urban or rural; and that it achieve such equity, as well as major improvement in the quality of public education, through a large-scale program of federal grants-in-aid to the states, based on state plans that guarantee equity in spending per pupil in all school districts and that meet national standards of educational quality.

The Best Possible Working and Living Conditions

The great impact of working and living conditions on health is indicated by recent studies by the National Center for Health Statistics of the U.S. Public Health Service. In 1986, the age-adjusted death rates (deaths per 1,000) by income class for white and black men and women age 25-64 were:

	Men		Women	
Income	White	Black	White	Black
Under $9,000	16.0	19.5	6.5	7.6
$9,000-14,999	10.2	10.8	3.4	4.5
$15,000-18,999	5.7	9.8	3.3	3.7
$19,000-24,999	4.6	4.7	3.0	2.8
$25,000 or more	2.4	3.6	1.6	2.3

It should be noted that very large differences by income class occur in each of the four groups studied, and that these differences are much greater than the differences by race. Indeed, the authors report that "the differences in overall mortality according to race were eliminated after adjustment for income, marital status, and household size."

Furthermore, it was found that not only do "poor or poorly educated persons have higher death rates than wealthier or better educated persons," but that "these differences increased from 1960 through 1986." The authors state that:

> The presence of widening differences in mortality rates according to income and education level should come as no surprise, given the broad social changes in this country since 1960. Increasing inequalities in income, education, and housing and a falling standard of living for a large segment of the U.S. population have been reported. Access to health care is a problem for a growing number of Americans. Although Medicaid may have improved access and health outcomes for a portion of the population, it appears to have been insufficient to equalize the chances for survival among the poorest and least educated. The social distribution of behavior that presents health risks may be important in explaining the widening gap. It may be that people of higher socioeconomic status have adopted healthy lifestyles more rapidly. (16)

The studies on morbidity in the United States give similar results on the impact of income class. In 1989, the lowest income fifth of the population, with a family income of less than $15,000 a year, in comparison with the highest fifth, with a family income of $50,000 or more, had more than twice as many persons limited in activity due to chronic conditions, more than twice as many days of restricted activity, more than twice as many days in bed due to illness, injury, or impairment, and four times as many persons reported to be in fair or poor health. (17)

Clearly, health is a function of income, and the most important precondition for achieving a healthy population is greater equity in income and the elimination of poverty. Unfortunately, poverty is on the rise in our country. *The New York Times* stated on October 5, 1993, that "the number of poor people in the United States increased last year by 1.2 million, to 36.9 million, increasing three times as fast as the overall population, the Census Bureau reported today." Unemployment in 1992 was 7.4 percent, up from 6.7 percent in 1991. The number of people receiving food stamps climbed steadily, to 26.6 million in December 1992, from 24.9 million in December 1991. And the number of people on welfare, in the program for Aid to Families with Dependent Children, rose to 14 million in December 1992, from 13.4 million in December 1991.

The Census Bureau found that 25 percent of children under 6, and 21.9 percent of those under 18, were poor last year. Nearly half of black children under 18 were poor. The poverty rate was 11.6 percent for whites,

12.5 percent for Asian-Americans, 29.3 percent for Hispanics, and 33.3 percent for blacks.

Among poor people 16 years old and over, 40 percent worked in 1992, and 9.2 percent had full-time jobs throughout the year.

The Census Bureau also reported that the condition of the most affluent families improved substantially in the last 25 years. The most affluent one-fifth of all families had incomes averaging 8.4 times the poverty level in 1992, as against six times the poverty level in 1967. By contrast, the least affluent one-fifth of all families had incomes averaging 91 percent of the poverty level in 1992, down from 97 percent in 1967. (18)

How can poverty be eliminated? How can greater equity in income be achieved?

Full Employment

The most important goal for the American people must become "JOBS FOR ALL." Private industry has proved totally incapable of reaching this objective. It needs full-scale assistance through government action as follows:

RECOMMENDATION 2: That the federal, state, and local governments provide funds for a massive housing program that will not only eliminate homelessness but will provide decent, affordable housing for everyone in the United States. This will not only resurrect the construction industry, but will provide large numbers of construction workers with increased purchasing power that will help other American industries to recover from depression. Such a program must include full protection of labor's right to organize, effective affirmative action to employ blacks and other minorities as well as women, and close monitoring of expenditures and quality.

RECOMMENDATION 3: That a similar large-scale program be instituted for the renewal and expansion of the nation's neglected and decaying public infrastructure: public buildings, schools, hospitals, health centers, water supply and sewage disposal systems, roads, bridges, railways, subways, and other necessary public facilities.

RECOMMENDATION 4: That it be made national policy to move from the current norm of the eight-hour day to a six-hour day with full protection of wage and salary levels, in all branches of industry, and that the federal, state, and local governments take the initiative by instituting the six-hour day for all government employees and for those employed in private industry where the funding for their wages and salaries comes from governmental grants and contracts.

RECOMMENDATION 5: That obstacles to employment of women be removed by the development of a large-scale program of child care centers of good quality, financed by federal, state, and local governments; by national legislation mandating paid maternity leave for all women, which has been long established throughout Europe (19, 20); by enforcement of the principle of equal pay for equal work by men and women; and by affirmative action to overcome all other forms of gender discrimination.

RECOMMENDATION 6: That education and training programs be developed in all communities to enlarge the knowledge and skills of unemployed persons and to provide counseling and assistance in obtaining employment; such programs should be funded with federal financial support and meet national educational standards.

Elimination of Poverty

Full employment would clearly eliminate much but not all poverty. Additional measures are necessary.

RECOMMENDATION 7: Raise the current totally inadequate minimum wage so that full-time workers who receive that wage have incomes above the poverty level.

RECOMMENDATION 8: Revise the Social Security program to guarantee a level of benefits for low-income retirees that is above the poverty level.

RECOMMENDATION 9: Persons unable to work because of physical or mental impairments should receive government subsidy adequate to keep them from falling below the poverty level.

Unfortunately, there has been a marked change in the pattern of employment in the postwar period. U.S. corporations have actively pursued a policy of deindustrializing the United States by shifting factories to the Third World where labor is cheap; they are now beginning to accelerate this process by massive shifts to Mexico through the North American Free Trade Agreement (NAFTA). Manufacturing jobs—jobs with higher wages and more organized workers—have fallen since World War II from more than one-third of U.S. nonfarm jobs to less than one-fifth today. (21) Workers in the service industries that replaced them receive lower wages; many work for the minimum wage or slightly more, not enough to stay above the poverty level. The Census Bureau finding that 40 percent of poor people age 16 and over worked in 1992, and that 9.2 percent had full-time jobs throughout the year, is testimony to this harsh reality. (18)

RECOMMENDATION 10: Initiatives to strengthen the development of Third World countries and their achievement of higher living standards are essential to full employment and higher living standards in the United States. One such initiative is the Pan American Health Organization's (PAHO's) environmental health plan (22):

> The PAHO emergency plan is designed as a three-year effort to cope with the immediate consequences of the arrival of cholera in the Americas, at a total cost of about U.S. $610 million, to be funded 60 percent from national budgets with the balance from external donors. Over the longer term, PAHO calls for the full repair of existing water, sanitation, and health infrastructures and their extension to the more than 30 percent of the population currently without adequate services. The cost of the long-term plan is estimated at U.S. $220 billion over 10 years. Funding for this effort would come from several sources:
>
> Individual countries would direct investment of 1.5 percent to 2 percent of gross national product into investment in health, water quality, and sanitation. This would produce U.S. $140 billion to U.S. $160 billion.
>
> Approximately 20 percent of multilateral and bilateral lending would be directed to the health and environment sectors. Allocations at this level would provide U.S. $35 billion to U.S. $40 billion.
>
> Swaps of debt for investment in health, water quality, and environmental sanitation and grants from aid donors would provide approximately U.S. $20 billion.

The U.S. government should give full support to this program, and should support similar plans for Africa, Asia, and the Near East. It should also support United Nations action to fund markedly increased public expenditures for housing, education, and other developmental needs. These measures are based on the recognition that the continuing worldwide economic depression—in which the peoples of the world simply do not have the purchasing power to buy the goods produced by the industries—cannot be overcome unless the purchasing power of the population is greatly increased.

FUNDING A PROGRESSIVE PLATFORM FOR HEALTH

Space considerations do not permit completion of this presentation of a progressive political platform for health. Only two items, education and living conditions, have been covered. Working conditions, the environment, rest and recreation, health services for prevention and treatment, and health research and training have not been discussed. Many important issues, such as discrimination and segregation, women's rights, labor law reform, and electoral reform, have not been covered. These are areas and issues that must be included in a progressive political platform for health. At this point, however, it is essential to propose recommendations for funding the large-scale programs that have been proposed.

Progressive Taxation

Inequities in income in the United States have been deliberately widened by federal, state, and local tax policy. The maximum federal income tax rates stayed at 91 percent during the '50s, fell to 70 percent during the '60s and '70s, and were reduced to 28 percent in the '80s. (23) This sharp decline in the progressive character of the income tax was accompanied by a sharp increase in the regressive social security tax rates, from 3 percent in 1955 to 15 percent in 1990. (24) State and local taxes rely heavily on regressive taxes: sales taxes, excise taxes, and property taxes. A study by Citizens for Tax Justice estimates that for married couples with two children, the richest 1 percent (average 1991 family income $875,000) pay 7.6 percent of their income in state and local taxes. The middle 20 percent (earning $39,100) pay 10 percent. The poorest 20 percent (earning $12,700) pay 13.8 percent. (25)

In 1989, the maximum rate, in the 86 countries with an income tax, averaged 47 percent, compared with only 28 percent in the United States. Denmark, the Netherlands, and Sweden levied income taxes of 72 percent on the top tax bracket. Japan and Taiwan taxed the maximum income bracket at 50 percent, West Germany at 56 percent, and South Korea at 63 percent. (23) Since the U.S. maximum rate was 91 percent in the '50s, and 70 percent in the '60s and '70s, it would be reasonable to propose that we emulate such advanced industrial nations as Denmark, the Netherlands, and Sweden and return to the pre-Reagan maximum rate of 70 percent. This would not only help narrow the disparity in incomes, but would make it possible to carry out many of the massive programs needed to implement a progressive political platform. The Clinton administration's increase to a maximum of 39.6 percent for incomes of more than $250,000 is a welcome improvement, but it remains far short of what is needed.

Furthermore, state and local taxes need to be drastically revised. Currently, only 26 percent of their tax revenues come from income taxes, compared with 72 percent for the federal government. (25) Regressive taxes—sales, excise, and property taxes—account for 74 percent of local and state tax revenues. The situation needs to be reversed in order to achieve a more equitable tax policy and to further narrow the incredibly wide income disparities in our population.

RECOMMENDATION 11: That the federal maximum income tax rate be set at 70 percent, and that the share of local and state tax revenues raised by progressive, graduated income taxes be increased to 75 percent, in order to provide equity in taxation policy, help narrow the very wide income disparities, and provide the funds needed to implement a progressive political program.

Reduce the Military Budget

An editorial titled "Where's the Bottom?" in the September 3, 1993, issue of *The New York Times* states:

> With its only superpower enemy having retired from the global battlefield, does a fiscally strapped United States still need to spend more than $1.2 trillion over the next five years for defense? It's hard to see why. But that's the number on the bottom line of the Pentagon's new "bottom up review," unveiled Wednesday by Secretary of Defense Aspin.
>
> By cutting perhaps $120 billion more than the Bush Administration had planned over the five-year period and scrapping some of the forces and weapons aimed solely at the now-vanished Soviet threat, the review makes a modest start at adjusting defense budgets to new international realities. But the start is too modest to do justice to competing fiscal needs.

Robert O. Borosage, founder of the Campaign for New Priorities, put this issue more sharply:

> America is safer from external attack than at any time in this century. With the United States spending more on its military than the next 10 nations combined—all of which are our friends, or want to be—significant savings should be possible, even as we ante up our share of peacekeeping and humanitarian missions. So how much does the bottom-up review cut from the outdated base force? 50 percent? 30 percent? No, it pares the forces by a mere 10 percent, and plans to spend $1.3 trillion over the next five years. The defense budget will stay at or near cold war levels, declining to about $260 billion a year by the end of the century. (26)

RECOMMENDATION 12: That the military budget be cut by 50 percent now, with further cuts to follow, and that the funds so released be used exclusively to provide jobs, housing, and health for all Americans.

It is time to implement the Ottawa Charter's statement that "The fundamental conditions and resources for health are peace, shelter, education, food, income, a stable ecosystem, sustainable resources, social justice, and equity." Let us move forward together to achieve these goals.

ACKNOWLEDGMENTS

This chapter was originally presented to the Council on Living Standards, National Association for Public Health Policy, and the Health Administration Section, APHA, at the annual meeting of the American Public Health Association, San Francisco, October 25,1993. It is reprinted from the Spring 1994 issue of the *Journal of Public Health Policy*.

NOTES

1. Brotherston, J.H.F. *Observations on the Early Public Health Movement in Scotland.* London: H.K. Lewis & Co., 1952, pp. 58, 83.
2. Ackerknecht, E.H. "Hygiene in France, 1815-1848," *Bulletin of Historical Medicine* 22 (1948):117-155.
3. Coleman, W. *Death Is a Social Disease: Public Health and Political Economy in Early Industrial France*. Madison, WI: University of Wisconsin Press, 1982.
4. Sigerist, H.E. *Medicine and Human Welfare*. New Haven: Yale University Press, 1941, p. 93.
5. Sigerist, H.E. *The University at the Crossroads*. New York: Henry Schuman, 1946, pp. 127-128.
6. *Ottawa Charter for Health Promotion*. Ottawa: Canadian Public Health Association, 1986.
7. Wilkinson, R.G., ed. *Class and Health: Research and Longitudinal Data*. London: Tavistock Publications, 1986, p. 14.
8. Wilkins, R., and Adams, O.B. "Health Expectancy in Canada, Late 1970s: Demographic, Regional and Social Dimensions," *American Journal of Public Health* 73 (1983):1073-1080.
9. Sigerist, 1941, p. 104.
10. Terris, M. "Newer Perspectives on the Health of Canadians: Beyond the Lalonde Report," *Journal of Public Health Policy* 5 (1984):327-337.
11. *Health United States 1989*. U.S. Department of Health and Human Services, Public Health Service, 1990, p. 166.
12. *The New York Times*, May 30, 1993, p. E3.
13. *Burlington Free Press*, June 5, 1993, p. 6B.
14. O'Connor, S. "Death in the Everyday Schoolroom," *The Nation*, May 24, 1993, p. 704.
15. *The New York Times*, September 15, 1993, p. B15.
16. Pappas, G., Queen, S., Hadden, W., and Fisher, G. "The Increasing Disparity in Mortality Between Socioeconomic Groups in the United States, 1960 and 1986," *New England Journal of Medicine* 29 (1993):103-109.

17. *Educational Differences in Health Status and Health Care*. Vital and Health Statistics, Series 10, No. 179, Publication No. (PHS) 91-1507. Hyattsville, MD: U.S. Department of Health and Human Services, 1991.

18. *The New York Times*, October 5, 1993, p. A20.

19. Wagner, M.G. "Infant Mortality in Europe: Implications for the United States," *Journal of Public Health Policy* 9 (1988):473-484.

20. Williams, B.C. "Social Approaches to Lowering Infant Mortality: Lessons from the European Experience," *Journal of Public Health Policy* 14 (1993):19-26.

21. Miller, J. "Silent Depression," *Dollars & Sense*, April 1992.

22. Guerra de Macedo, C. "Overview of the Epidemic and the Regional Plan," in *Confronting Cholera: The Development of a Hemispheric Response to the Epidemic*, Conference Proceedings. Coral Gables, FL: University of Miami, p. 6.

23. Miller, J., and Goodno, J. "Much Ado About Nothing," *Dollars & Sense*, December 1990.

24. American Institute of Certified Public Accountants. *CPA Client Bulletin*, March 1993, p. 4.

25. "Soaking the Poor: State and Local Taxes are the Unfairest of All," *Dollars & Sense*, July/August 1991.

26. Borosage, R. "Disinvesting in America," *The Nation*, October 4, 1993, p. 346.

34

Women Changing the Times

AN ITALIAN PROPOSAL TO ADDRESS THE GOALS AND ORGANIZATION OF WORK

MARGARET M. QUINN
EVA BUIATTI

Occupational stress is a major concern in developed nations. In Italy, as in the United States, working women are faced with the double job duties of work inside and outside of the home, and many women see these dual responsibilities as major contributors to occupational stress. Lack of recognition for work that provides social support creates stress that is eroding the quality of life for both men and women. Solutions for stress generally focus on specific symptoms but do not address the broader organization of our lives and work that may cause or exacerbate symptoms. This chapter presents a women's response to occupational stress. The issues raised here challenge the definition and ultimate goals of work and force a reconsideration of the way our daily lives are organized.

This chapter is based on a proposed law entitled "Women Changing the Times: a law to make the time for work, the schedules of the community, and the rhythm of life more human." It was authored by a national committee of women in the Italian women's movement and was endorsed by women from a wide range of political parties. This mechanism was chosen explicitly for the opportunity it provided for a broad public discussion of the issues. The preamble to the bill acknowledges that a law cannot change social behavior completely, but the intent is to provide a vehicle for discussion and the development of tangible solutions.

We hope that the summary of the proposal presented here will encourage international dialogue about how to organize our work and personal lives in a more humane way—in a way that allows us time for work and for providing care for others and for ourselves. Women in Italy have described the movement of women into the labor market as creating a crisis of lack of time for the provision of care for themselves and others. In the United States, women employed inside and outside of the home have begun to feel like they are on a treadmill. Only somewhat jokingly has this period been dubbed the "long march." Many of us feel our lives are crowded but not full, busy but not satisfying.

Women traditionally have provided care, sustaining the "personal" or "private" world of ourselves and our families. This work can be satisfying as well as difficult, but it always has been undervalued compared with "productive," wage-earning work traditionally assigned to the "public" world of men. As women move to work outside the home, this distinction between private and public lives seems too rigid. Our time feels fractured by the boundaries of the two worlds and is spread too thin in each. Many women always have had to work inside and outside of the home and experience this stressful conflict. However, the problem is intensifying because the number of working women as well as the number of female heads of household has increased rapidly over recent decades. In addition, men generally are not assuming equal responsibilities in the personal world. The men who do assume caregiving roles receive little societal encouragement or rewards, and they, too, experience the conflict.

BACKGROUND

The women who authored the proposed law come from many social backgrounds, work experiences, and political affiliations. Their primary identity is with the women's movement. Some of the women also have been working to address women's issues within the context of progressive political parties. However, as happens in the United States, there is not a complete identity between women's issues and the broader political agendas, and so there is an ongoing struggle between the two.

This is a period of fundamental political reevaluation, especially for progressive European political parties. Many Italians are redefining the primary purpose of left politics to be that of maximizing human liberation, to find ways in which people may live most humanely through democracy, equal opportunity, justice, and peace. They hold that left politics should not be tied to a particular economic system such as public ownership of property, but rather should support whatever economic system can best provide human liberation in a particular place and time.

This does not mean that the economic aspects of a society are unimportant; human needs are very closely tied to economic systems and economic control; however, the *primary* goal should be humanistic. It has been argued that the human, nonmaterial aspects of our lives would fall into place if material needs were satisfied. This idea led to the development of economic structures for production while the quality-of-life issues remain implicit—and unrealized. It is time to shift the focus from the economic to the humanistic by redefining the goals for human development and making them explicit. From this perspective, many social movements can contribute to a broader, coherent political program. The women's, students', minority, labor, and environmental movements are all seen as closely linked because they represent different aspects of the same struggle for human liberation. They are not viewed as completely separate movements.

A progressive political program must address the quality of our lives as well as our material needs. The Italian women's committee emphasizes that the provision of care, traditionally the work of women, can bring richness and deep satisfaction to our lives. Everyone must be able to find his or her own balance of work outside the home, providing care to others and to ourselves. This balance may be different depending on the individual; what is important is that the opportunity exists for each person to develop in whatever manner s/he chooses.

European debates on the future of cities and culture in developed nations are emphasizing that we must work toward improving the quality of life, not simply toward quantitative growth. Many agree that the period of economic expansion of human societies, at least in Europe, has ended. Therefore, it is time to rethink the organization of our communities, of work, and even of our own lives—not only in physical terms but also in the quality of relations among people. And it is here that *time*, how and when it is distributed, becomes the key element. As we approach the limits to growth, production no longer needs to be the foremost consideration, but rather the needs, material and nonmaterial, of citizens in a community must get top priority. And who is this citizen? In Modena, a city in northern Italy with a female mayor, the model for the "average citizen," the person who most represents the needs and complexity of modern life, is the working woman. If a community can be reorganized with the needs of the working woman as its basis, then it is likely that the needs of most other members of the community will be met. Rather than being invisible, women should be a focus because women do more than one job and because their time often belongs to other members of the community. In Modena, some of the proposals presented in this law, such as coordinating the time schedules of some stores and public services, are currently in practice.

SUMMARY TRANSLATION OF PROPOSAL

Preamble

Everyone needs time to study, to work, and to care for others and for themselves. The work of providing care is an activity that enriches rather than penalizes a society. The time spent providing care must be considered "social time," for the benefit of society as a whole, and not just something we do in our personal lives. We think that a law will be useful to give a voice and force to the idea of reorganizing time to allow for caring relationships. We do not believe that modes of thought and behavior in private lives will change by decree, but we think that a law enacted by popular initiative can be a vehicle for ideas; it provides a mechanism for men and women to confront and to crystallize their thinking about the cultural issues. If it is successful, perhaps the law can encourage individual and social change. There is, however, another important reason why we have chosen to put this question into the legislative arena. This is the time for women to become a full and autonomous political force; not to renounce their needs, but to participate with their valid experience. The choice to involve the government is one to claim a place in the mainstream of democratic action.

To become the boss of our own time, to value all of the phases of our lives, and to extend the work of caregiving to both men and women is to give substance to the idea of a democracy of daily life. This everyday democracy can only be achieved by rethinking many behaviors and parts of our lives that are now considered natural or inevitable, such as the organization of time into public (masculine) and private (feminine) parts; even the organization of the cycle of our lives. Achieving a democracy of daily life also would mean redefining our relationship to work and transferring resources and power from single or small groups of individuals to all who participate in the work.

We are starved for time. Nurses work around the clock even on holidays; university faculty and researchers teach and work nights and weekends to publish; factory workers and managers and employees of private businesses do shift work; store employees get out of work when everything else is closed. And then women add thousands of other jobs for the home, children, elderly, other family members, and friends. We are employed in many hours of domestic work and the care of others including having a major role in the education of our children.

We need more day care centers with the characteristics and schedules that suit our needs; full-time programs; places and organized activities for our children to meet after school; social centers for the elderly; more vacation time and domestic assistance. But even with all of these social services, part of the job of providing care to others still remains and cannot be covered through social institutions and services. And there often are times when taking care of others *gives us joy and satisfaction*. However, there is no time to do it: not because we don't know how to organize our time well, but because our time depends on others, on work, the hours that stores and services are open, on school hours, on traffic and means of transportation. Only rarely can we really decide how to use our time. We are calling for a way to reconcile all of the different types of work that we do, and we are asking that all of the jobs, including "women's work," be shared by men.

Introduction to the Bill: The Cycle of Our Lives

Our entire life is conditional on a model that does not take into account the fact that we are women. Is it really natural that while we are young we are involved in school, the university, or other formative activities and then—if we are fortunate enough to find work—we work all day, all week for 11 months for 25 to 40 years until we retire? We are beginning to think not; to realize that this model, in reality, was thought of and made by men for men, and it doesn't suit us well. In the past, and still today, many women abandon work when they have children and then look for it again (often in vain) when their children are grown. Or else many women today do not leave their work or their cultural commitments but become acrobats trying to do it all. Many others begin to get established in their careers or public life or acquire seniority at work and then must choose whether or not to have children or to postpone maternity to a biologically advanced age with lower fertility and more risks for the pregnancy and the newborn. These examples demonstrate how the model of social and workplace organization truly violates the biological clock of women.

Is it any wonder then that some women refuse to live like men or to take on the great effort of a double job and choose to work in the home even if this choice costs in terms of reduced income and personal autonomy? One can count the rhythm of life in a way that is more consonant with all the needs of humans because all that goes well for women can improve everyone's life, including men's lives: to study, to work, to have time for oneself, to love and care for others, to enrich our life experiences and knowledge, to do sports, to travel, to participate in cultural life, and to be involved socially and politically. One life with many dimensions and not just a life of work. We know that to accomplish this proposal would take a cultural revolution. Businesses would have to think about an organization of work that is not modeled exclusively on the needs of productivity and profit or on the idea that employees should be available totally to the company without obligations other than to work and to earn. This means a new relationship of people to their work.

The state must adjust the allocation of resources to provide care in a way that recognizes not only the time needed for caregiving but also the autonomous rights of children, the elderly, and women by constructing the necessary social services for them. Therefore we propose a policy called the "new cycle of life" that combines periods of work during the school years and provides time for personal and professional development and study, parental and family leaves, during our working lifetime. We want to make it possible for men and women to take temporary leaves from their work to study, to play with a child, to give companionship to an elderly or sick person, to gain new professional qualifications, or simply to reflect on their own lives. This should be possible while maintaining a position at work, without greatly decreased salary, without losing health and other benefits, and without compromising one's career.

Excerpts from the Bill

We propose that every working person have the right to parental leave: a period of leave from work up to a maximum of 12 months that can be taken together or in portions until the child is 11 years old. In families with a handicapped child or with only one parent, the duration of the leave is extended to 24 months.

To address family emergencies such as a child in adolescent crisis, an elderly or seriously sick person, a death, all of the situations that require emotional care, we propose a leave for family reasons. This is the right to be absent from work for a period of not more than 30 days for every two years of work.

Whether the leave is for parental or family reasons, there should be some form of recognition that the time taken to provide care is socially useful—as productive as commercial work. For this, we think that one should have a salary during the leaves with one part paid by the state and another by the employers.

During the leaves, workers should receive a minimal, guaranteed salary equal to 50 percent of the national average salary for their occupation. In addition, if a worker wants, he or she should be permitted to have up to 100 percent of the normal salary during the year just prior to the leave, drawing on money that will be received upon leaving work. (A type of severance pay. In Italy, a worker gets a significant sum of money, in addition to his or her monthly pension, upon leaving work. The amount of severance pay is based on the amount earned during the working lifetime.)

If the time for providing care has a value for all of society, then the care given by men and women who do not have wage-earning work, such as the unemployed, students, housewives, or those who are self-employed, such as artists, trades people, shopkeepers, farmers, or professionals, should be recognized also. The first group has the time to provide care, but doesn't have a corresponding salary. The second may be able to decide to suspend work temporarily, but also will not have a salary. We propose that all members of our community, including immigrants, who do not have full-time jobs, also have the same rights to a guaranteed minimum income distributed by the state for parental leave.

We know very well that a law will not be enough to redistribute the work of caregiving between the sexes: how many men will ask for parental or family leave? As we have already said, this will take a cultural revolution. A decisive role in the new socialization could be developed by the schools. In addition, other traditionally masculine institutions could be employed. For example, the military service or alternative service could be used to provide care. Young men in the military service could satisfy part of their duty with a certain number of months (three) spent in caregiving activities such as infant and child care, domestic assistance, care of the elderly, assistance in preventive and medical services, and in recognized community organizations.

Above all, a different cycle of life means being able to use time for oneself without having to wait until retirement. There are dead-end jobs such as in the fast-food service sector, and then there are jobs—such as a pre-school, kindergarten, or secondary school teaching—that, although they may be gratifying, nevertheless become exhausting after many years. We propose that all employees, after working at least seven years, be eligible for a one-year sabbatical. Workers should be able to request the leave without specifying the reasons and the job positions would be held for them. The sabbatical would not be a concession of the employer, but a right of the employee. However, the employee would have to replace the time taken by delaying retirement age and working an additional year for every year of personal leave. The idea is that we could work longer when we are older in order to have time for ourselves when we are younger. The restitution of time in later years is important to make it clear that the sabbatical is time for oneself, and it is the responsibility of the person who takes it. The time cannot be viewed as a gift from the state or the employer, nor can the time be used to assume another salaried job. During the sabbatical, the person would receive part of the pension and severance pay settlement, and it would be repaid with interest.

Work Time

Legislators, employers, and union officials think only of three times: the times for work, rest, and "free time." For decades, the objective of workers has been to have eight hours of work, eight hours of rest, and eight hours of free time per day. However, the time for caregiving has never been explicitly included in the daily schedule and for this, women have never had free time—it is full of other work. For this, we propose a workweek limited by law to 35 hours. The daily work schedule should be flexible, and the hours should be established in a contract between the employer and the employee that acknowledges not only the demands of production but also the private life of the employee. We want to avoid having a discrepancy between the "real" hours required for the job and those set by law. It is essential that work time over the legal limit be viewed as truly extraordinary—not the norm, but the exception. Therefore overtime work must be voluntary and must not exceed two hours per day and eight hours per week. In each case, the employee should have the right to recover the extra time. In both small and large workplaces, public and private, each employee should have the right to four paid weeks of vacation with at least two of those weeks chosen by the employee.

There are certain jobs, such as public utilities, hospital work, transportation, and so forth, that must be conducted 24 hours a day. However, work should be confined to the day shift whenever possible. This especially refers to stores. Those workers who cannot work the night shifts because they have to provide care should have the right to refuse. Those who work the night shift or Sundays and holidays should be given 20 percent extra time off for every eight hours worked. Those who do very demanding work should have some compensation in time, rather than money, as well as the possibility of an early retirement or pension. Compensation in time instead of money is emphasized to make it explicit that extenuating work conditions take time from the provision of care and that the time for care is valuable. The purpose of this aspect of the law is to provide minimum, guaranteed rights. This should not preclude unions from negotiating even better conditions.

All of these proposals liberate time, and they also create new positions for those seeking employment. We have chosen solutions that limit the time at work for all workers. There are those who are proponents of other solutions, such as part-time work and job sharing. To us, these do not seem like good solutions because they are voluntary and because they are not imposed evenly throughout the workplace. It is also difficult to provide adequate benefits and union coverage. When permitted at all, these forms of work are used in an uneven fashion throughout workplaces and, overall, it is women who use them. It has become a way to reconcile "women's work" in the home with wage-earning work. This perpetuates the idea of women as not really serious about their jobs. It also prevents or limits the redistribution of work between the sexes and the social recognition of the value of time spent caregiving. In this way, it may be viewed as a means for not having to create so many social services outside the home. It is for these reasons that, in our opinion, the best way to address time in the workplace is to reduce the work schedules for all, men and women.

Time in the Community

The schedules of services and store hours seem made as if to play a nasty trick on us! For many of us, when we have to leave work, the banks, post offices, doctors' offices, state administrative services, and often even the stores are closed. As women continue to enter the workplace in even greater numbers, there is no one at home during the day to attend to all the services necessary to sustain a family. Upon reflection, we think that the schedule of the city is set on a model developed by men, of a city made for the "producers," a city in which women and their work are invisible, erased. For this, we propose that the city council be entrusted with developing a plan to coordinate the schedules of the city. This should be established in accord with the employers, employees, and with the users of the services. Instead of paying more taxes to plan these improved services, perhaps members of a community could donate time to the town government.

And finally, there is the time that is stolen from us: the time having to deal with various aspects of public administration. We call for the community to make the bureaucracy simpler.

COMMENTS ON THE PROPOSAL

Engaging in international exchange can teach us a great deal not only about a new culture but also about our own when we look at ourselves and our problems through the eyes of others. The Italian bill raises issues and new solutions that contribute to our understanding of stress in this country.

In the United States, debates on women's roles in society often have focused on the need for recognizing the different perspectives of women and on the lack of equal opportunity for women in the workplace. The Italian proposal acknowledges the importance of these two issues but poses the problem in a way that goes beyond them (see Table 34.1). It argues that equal opportunities in the workplace are only a partial solution, because the very definition of work is limited to those activities whose pri-

TABLE 34.1. Summary of the Italian Proposal

Time throughout the cycle of our lives

A "New Cycle of Life" policy combining periods of work during the school years and time for personal and professional development and study, parental, and family leaves during our working lifetime. This includes:

- Parental leave—available to all including the self-employed, unemployed, and immigrants
- Leave for family reasons
- Guaranteed income during leaves
- One-year sabbatical every seven years for all working people
- Three months of military service satisfied by caregiving activities

Worktime

- 35-hour workweek
- Flexible work hours
- Voluntary overtime
- 20 percent extra time off for every eight hours worked on a night shift, Sunday, or holiday
- Work confined to day shift when possible
- For very demanding work: compensation in time off and option for early retirement
- Minimum of four paid vacation weeks

Time in the community

- City planning of the schedules for services and businesses based on needs of employers, employees, and users/customers
- Reduction of state bureaucracy

mary objective is economic growth. Defining work so narrowly is eroding the quality of life for both men and women, creating stressful conflicts at work and at home. Women find a large part of their lives negated, while men receive little social recognition for participating in the nonmaterial aspects of life such as childrearing and other kinds of social support.

It is striking that working women in Italy have articulated the stress of dual job roles and the lack of time for others and for ourselves in much the same way as it has been described in the United States. Somewhat different, however, is their description of the problem as a crisis of care. That is not exactly the way it is framed in the United States, even though the Italian term "cura" is derived from the English use of the word "care." In the United States, the difficulty of trying to do the tasks required for work in and out of the home are discussed, but the "caring" part, the emotional investment, is often left out. Perhaps U.S. women are afraid to talk about the caring component because, in a world defined by men, it is not as socially valid as the "work" part. The caring part is womanly, weak, soft, not serious. And so U.S. women seek validation in a world defined by men by emphasizing how stress interferes with production or by quantifying particular aspects of the physical strain of working inside and outside of the home. Another very big part of the work, the emotional investment that goes into caregiving, usually is excluded from discussion. The emotional involvement can be very satisfying, but it is very demanding. It truly is a part of the "work," and it is an important factor in sustaining relationships and providing care.

A DIFFERENT DEFINITION

Perhaps one reason that Italian women define the problem differently is that there have been some important differences in the development of the women's movements in Italy and the United States. A large part of the American struggle has been to make men and women truly equal. The U.S. Constitution uses the phrase "blind to sex" as if there literally should be no difference between a man and a woman. In Italy, this is a somewhat strange idea, and the women's movement has focused not so much on showing that there is no difference between men and women but that the perspectives and capabilities of each are equally valid; that society should incorporate the experiences of each equally and recognize that there are many ways to do the same job. While U.S. women say that they do not want to become just like men, Italian women believe this on a very fundamental level. This difference in perspectives has perhaps allowed Italian women to call for recognition and social institutionalization of women's work, in this case the physical and emotional work of providing care, in a way that has not been as clear-cut for U.S. women.

First, society must recognize that caregiving is as socially productive as commercial work. Can we teach emotional engagement and the adequate provision of care as socially valuable? The changes must begin in the home and continue in school and other social institutions including workplaces. Perhaps it would be useful to extend the sabbatical idea to children and particularly to adolescents in school. Every few years, for periods of a month or two, children could participate in some form of caregiving, either to younger children or to the elderly, or to teach a skill that they have to others of the same age. Caregiving and social contribution should be taught to all children along with the traditional school subjects. For the early adult years, perhaps the model successfully employed by some universities that have combined semesters of work and study for rounded professional development and exploration is most appropriate.

INVOLVE INSTITUTIONS IN RESOCIALIZING MEN

The idea of involving traditionally male institutions in caregiving as a means to resocialize men is intriguing. There is a major difference between Italy and the United States regarding the military, because Italian service is mandatory for all men for one year, and so even young men who might choose to do caregiving are inducted. Alternatively, the United States gets a military population that voluntarily accepts war activities and thus may be less adaptable to caregiving. Perhaps there are other male institutions in the United States, such as fraternities, where caregiving could be structured into their regular activities. Universities could require that fraternities provide a certain number of hours of social service to the university or town community in exchange for representation on campus.

There should be legal and social structures to limit the amount of "real" time spent on the job. In some jobs in the United States, work time may be legally limited, but intense pressures to keep working remain. In white-collar jobs, those who "only" work 40 hours a week are considered not serious about work. In many blue- and pink-collar jobs, overtime is not completely voluntary. To achieve such a change, as the preamble notes, requires a restructuring of social and cultural expectations and a reexamination of our relationship to work.

The proposed law emphasizes compensation for overtime or difficult working conditions in time rather than money. This is an unpopular idea among many working people because they want more money. Many unions have explicit contract agreements that permit extended work shifts for extra pay. Such policies are popular with young male workers but older male workers often question it. Older workers are more likely to have experienced the toll of extenuated work in safety problems, physical and mental health problems, and family stress. The needs of older working men and working women may be more similar in regard to time at work.

A SENSE OF COMMON EXPERIENCE

Human beings have a need for nonmaterial enrichment of daily lives and relationships. This may include personal, emotional, and even—a subject almost taboo in U.S. progressive politics—spiritual development. It also could include sharing a sense of the common human experience through community. Recognition of the need for a nonmaterial component can mean improvement and enjoyment of the fullness of life. It is important to acknowledge, however, that personal fulfillment can be achieved in many ways for different people. The Italian proposal is not suggesting that women have the prescription for the way life should be; rather, that the primary focus on materialism does not allow time for men and women to find the combination of pursuits that fosters their own development or supports the development of those they care for. Neither labor nor the left have adequately addressed the need for a human component in everyday life. The importance of relieving people from economic need cannot be underestimated. However, the calls from progressive movements in the developed world for better quality of life, a deeper sense of community, improved relationships among peoples, validation of the importance of providing care and sustaining our emotional lives, and a more harmonious existence with nature are signs that it is time for the left to take into consideration the need that many people have for a nonmaterial aspect of life and to offer a way for people to find the time to develop it. The United States needs a way to integrate what the progressive movements are developing into a cohesive program. The point is not to co-opt them but to use the results of their work for social and political policy.

SUSTAINABLE DEVELOPMENT, IMPROVED QUALITY OF LIFE

Another important issue leading to the development of this referendum is the idea that human cultural development should no longer be based on the assumption of continued growth and expansion. Developed societies consume world resources and pollute excessively, particularly in comparison with developing nations. The emphasis should be on how we can reduce the impact of developed society on the world as a whole, and this creates a need to think about reorganizing our work and communities for sustainable development rather than increasing production. Issues related to quality rather than quantity should be addressed in our public and private lives and in trying to find satisfying balances between the two for all people. Since women traditionally have dealt with issues of the quality of life, perhaps we can make a particular contribution toward moving progressive politics forward during this period of fundamental reevaluation.

The women who authored the proposed law would like to start an international dialogue through this volume and the *New Solutions* journal. We invite comments about the specific proposals or general perspectives presented and particularly suggestions for new solutions and how they could be implemented.

EDITORS' NOTE

In the spring of 1991, people throughout Italy were asked to sign a petition endorsing this proposal. More than 300,000 signatures were obtained, enough to send the bill to Parliament. With the new Italian coalition of the Left in power, perhaps this proposal will finally receive serious attention.

ACKNOWLEDGMENTS

The authors wish to thank the many women of "Il Giardino dei Ciliegi" (named for Anton Chekhov's play "The Cherry Orchard"), an Italian national women's organization based in Florence, which contributed to the proposed law and made this presentation possible. We also appreciate the helpful comments of David Kriebel, Guido Dolara, Vieri Dolara, Lucia Miligi, Paolo Vineis, John Wooding, Tish Davis, Elise Morse, Joshua Cohen, Ellen Eisen, Nancy Fox, Joan Parker, Deborah Patten, Francesca Cappelletto, and Enzo Merler.

Index

Contributors

Frank T. Adams is an author who lives in Asheville, North Carolina.

Dean Baker, a physician, works at the University of California at Irvine.

Jim Baldauf has worked as special projects director for Texans United, a 70,000-member citizens environmental organization.

Geoffrey C. Beckwith formerly served three terms as a member of the Massachusetts House of Representatives where he was the legislative author of the Massachusetts Toxics Use Reduction Act. He is now the executive director of the Massachusetts Municipal Association.

Millie Buchanan is a project officer in the toxics program at the Jessie Smith Noyes Foundation in New York City. She formerly was director of the Clean Water Fund of North Carolina.

Eva Buiatti is director of epidemiology for the Center for the Study and Prevention of Cancer, which is operated by the National Health Service in Florence, Italy.

Janet Cahill is professor of psychology at Rowan University in Glassboro, New Jersey.

Lin Kaatz Chary is a doctoral student at the University of Illinois who is researching the effects of PCB exposure on electrical capacitor workers.

Caron Chess is director of the Center for Environmental Communication at Rutgers University in New Jersey.

David C. Christiani is professor of occupational medicine at Harvard School of Public Health and head of Harvard's occupational health program.

Richard Clapp, an epidemiologist, is a faculty member at Boston University's School of Public Health and a consultant to the John Snow Institute in Boston.

Mark R. Cullen, a physician, directs the Occupational/Environmental Medicine Program at Yale University's School of Medicine in New Haven, Connecticut.

Peter Dillard, an occupational physician, has worked in both Massachusetts and Ohio.

John R. Froines is professor of toxicology at the School of Public Health at the University of California in Los Angeles. He also directs the UCLA Center for Occupational and Environmental Health.

Louis Head is development coordinator for the South West Organizing Project in Albuquerque, New Mexico.

Kenneth Geiser is an associate professor in work environment at the University of Massachusetts Lowell. He is director of the Massachusetts Toxics Use Reduction Institute and a codirector of the Lowell Center for Sustainable Production.

Robert Ginsburg, a toxicologist, is research director for the Midwest Labor Research Center.

Richard L. Grossman is codirector of the Program on Corporations, Law, and Democracy and is based in Provincetown, Massachusetts.

Michael Leon-Guerrero is a codirector of the South West Organizing Project in Albuquerque, New Mexico.

Margrit K. Hugentobler is a research scientist at the Federal Technical University in Zurich, Switzerland.

Barbara A. Israel is a professor and the chair of the Department of Health Behavior and Health Education at the School of Public Health and the University of Michigan.

Renee Goldsmith Kasinsky is professor of criminal justice at the University of Massachusetts Lowell.

Karl T. Kelsey is an associate professor of occupational medicine and radiobiology at the Harvard School of Public Health.

Susan Klitzman is the assistant commissioner of environmental risk assessment and communication for the New York City Department of Health.

Sarah Kuhn is an assistant professor in the Department of Regional Economic and Social Development at the University of Massachusetts Lowell.

Paul Landisbergis is an assistant professor of epidemiology at the Hypertension Center of Cornell University Medical College in New York.

Philip J. Landrigan is a physician, professor, and the chair of the Department of Community Medicine and director of the Division of Environmental and Occupational Medicine at the Mount Sinai School of Medicine of the City University of New York.

Michael Lax, a physician, is medical director of the Central New York Occupational Health Clinical Center in Syracuse.

Charles Levenstein is an economist and is a professor of work environment policy at the University of Massachusetts Lowell. He is the editor of the Journal *New Solutions*.

Sanford J. Lewis is the director of the Good Neighbor Project for Sustainable Industries, the chair of the Network Against Corporate Secrecy, and an instructor in environmental law at Tufts University's graduate department of urban and environmental policy.

Michael Merrill is the director of education and training for the New Jersey branch of the AFL-CIO, based in Trenton.

Amy Mock is a health and safety trainer for the American Federation of State, County, and Municipal Employees and is based in Columbus, Ohio.

Elise Pechter Morse is a certified industrial hygienist in the occupational health surveillance program at the Massachusetts Department of Health.

Charles Noble is professor of political science at California State University at Long Beach.

Mary H. O'Brien works in Eugene, Oregon, for Hell's Canyon Preservation Council and is the chair of the toxics board for the city of Eugene, which is implementing a new right-to-know law.

Joan N. Parker is the director of safety in the office of the Massachusetts Attorney General. She is a certified industrial hygienist.

Richard M. Pfeffer is an attorney in the solicitor's office of the U.S. Department of Labor in Washington.

Laura Punnett is an ergonomist and epidemiologist who is an associate professor in work environment at the University of Massachusetts Lowell. She also is the director of the Lorin E. Kerr Ergonomics Institute for the Prevention of Occupational Injuries at the university.

Margaret M. Quinn is an assistant professor of industrial hygiene in the Department of Work Environment at the University of Massachusetts Lowell.

Peter L. Schnall is an adjunct associate professor of medicine at the University of California at Irvine and the director of the Center for Social Epidemiology.

Susan J. Schurman is the executive director of the George Meany Center for Labor Studies in Silver Springs, Maryland.

Gerry Scoppettuolo formerly worked at the North Carolina Occupational Safety and Health Project in Durham, North Carolina, and was a member of United Food and Commercial Workers Local 1557 in Nashville, Tennessee.

Ellen K. Silbergeld is professor of epidemiology at the Medical School of the University of Maryland and has long been associated with the Environmental Defense Fund.

Barbara Silverstein is an epidemiologist who specializes in ergonomics. She is research director at the SHARP (Safety and Health Assessment and Research for Prevention) program at the state of Washington's Department of Labor and Industries.

C. Mark Smith formerly was a fellow in the Department of Cancer Biology in the Laboratory of Radiobiology at Harvard School of Public Health and now works in the Office of Research and Standards at the Massachusetts Department of Environmental Protection.

Joel Swartz is a health risk analyst with the Toxicology Branch of the California Department of Health Services.

Eileen Senn Tarlau is an industrial hygienist who has long worked for labor unions and government agencies and has authored many worker education materials in occupational health.

Milton Terris is the editor of the *Journal of Public Health Policy*, based in South Burlington, Vermont.

Dominick J. Tuminaro is an attorney who formerly served as counsel to the chair of the New York State Assembly Labor Committee and who works at the Brooklyn, New York law firm Pasternack Popish Reiff.

Daniel Wartenberg is an associate professor at the Robert Wood Johnson Medical School.

James L. Weeks is an industrial hygienist who is an associate research professor in the Department of Environmental and Occupational Health at George Washington University's School of Public Health.

David Weil is an assistant professor of economics at the School of Management at Boston University. He is also a faculty member at the Harvard Trade Union Program.

John Wooding, a political scientist, is the chair of the Department of Regional Economic and Social Development and an adjunct professor of work environment policy at the University of Massachusetts Lowell. He specializes in international and regulatory policy.